Approaches of Computational Biophysics and Chemistry in Molecular Biology

Approaches of Computational Biophysics and Chemistry in

Molecular Biology

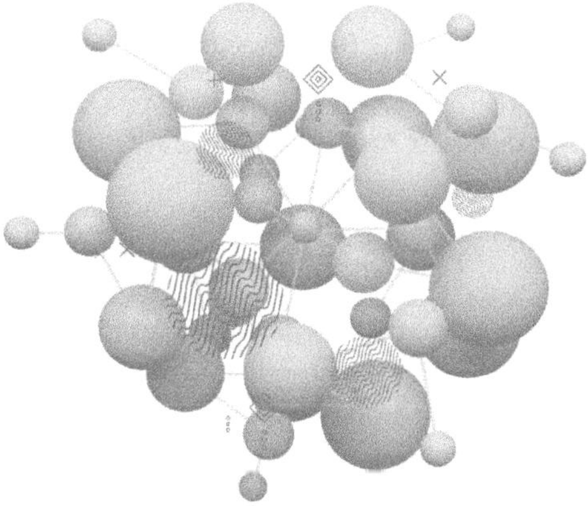

Edited by

Emil Alexov
Clemson University, USA

World Scientific

NEW JERSEY • LONDON • SINGAPORE • BEIJING • SHANGHAI • HONG KONG • TAIPEI • CHENNAI

Published by

World Scientific Publishing Co. Pte. Ltd.
5 Toh Tuck Link, Singapore 596224
USA office: 27 Warren Street, Suite 401-402, Hackensack, NJ 07601
UK office: 57 Shelton Street, Covent Garden, London WC2H 9HE

Library of Congress Control Number: 2024060643

British Library Cataloguing-in-Publication Data
A catalogue record for this book is available from the British Library.

APPROACHES OF COMPUTATIONAL BIOPHYSICS AND CHEMISTRY IN MOLECULAR BIOLOGY

ISBN 978-981-98-0818-2 (hardcover)
ISBN 978-981-98-0819-9 (ebook for institutions)
ISBN 978-981-98-0820-5 (ebook for individuals)

For any available supplementary material, please visit
https://www.worldscientific.com/worldscibooks/10.1142/14182#t=suppl

Desk Editor: Brittney Phitojo

Typeset by Stallion Press
Email: enquiries@stallionpress.com

https://doi.org/10.1142/9789819808199_fmatter

Preface

This book provides a broad range of computational biophysics and chemistry methods and their applications to study various phenomena in molecular biology. The emphasis is on recent advances to make computational modeling more accurate, achieve longer timescales, and enable simulation of biological macromolecules and their complexes of extreme sizes. The methods discussed range from molecular dynamics (MD) simulations, Poisson-Boltzmann MD surface area, to quantum mechanical methods. Particularly, advances in MD simulations are discussed in terms of the development of polarizable force fields, where polarizabilities are explicitly incorporated into the potential energy function. Alternatives to first principle-based methods include machine learning (ML) and artificial intelligence (AI) approaches, which often achieve better performance in terms of speed and modeling accuracy of the predictions. Both first principle-based approaches and ML/AI are used to predict the pKa of ionizable groups in biological macromolecules such as proteins, DNAs, and RNAs. This book outlines numerous applications of these methods to study biological phenomena at a molecular level of detail. An important aspect of the book is the application of computational biophysics and chemistry to address questions relevant to human health.

This book is unique as it offers a comprehensive coverage of nearly all existing methods in computational chemistry, rather than focusing on just one, making it attractive to a broad audience. Furthermore, the applications are also taken from a diverse set of cases, and thus illustrating that computational biophysics and chemistry can be successfully applied to study diverse biological phenomena. Finally, the applications of computational biophysics and chemistry methods to advance our understanding of molecular effects that cause human diseases are directly linked to personalized medicine.

This book is organized into five parts: (1) Polarizable Force Fields for Biomolecular Modeling; (2) Novel Methods in Computational Biophysics and Chemistry and their Applications to Biological Problems; (3) Computational Biophysics and Chemistry Methods to Predict pKa of Ionizable Groups in Proteins, RNAs, DNAs, and Small Molecules; (4) Artificial Intelligence in Biophysics and Chemistry, and (5) Computational Biophysics and Chemistry and Diseases.

1. Polarizable Force Fields for Biomolecular Modeling

Molecular modeling and simulation are important techniques to study the structure-dynamics-function relationship of biomolecules. However, their explanatory and predictive power critically depend on the accuracy of underlying computational models, typically molecular mechanics force fields. Recently, advances toward more accurate force field models have been achieved through the development of polarizable force fields, in which polarizabilities are explicitly incorporated into the potential energy function forms. This first part presents the latest developments and applications of polarizable force fields in biomolecular modeling and simulations.

Classical molecular dynamics force field models represent bonded interactions via harmonic function and ignore cross-coupling effects. As outlined in "**Chapter 1: High-Order *Ab Initio* Valence Force Field with Chemical Pattern-Based Parameter Assignment**", this limitation can lead to significant errors. To address this deficiency, bonded interaction parameters were developed for a wide range of chemistry based or quantum mechanics applications. The method was tested against 270 atomic types, and it was demonstrated that quantum mechanics produces more accurate vibrational frequency than classical harmonic models. Additionally, it is indicated that the model can be applied to any MM3(MM4)-based molecular mechanics force field.

A particular example of a widely used polarizable force field is the CHARMM Drude polarizable force field. "**Chapter 2: Testing and Optimizing the Drude Polarizable Force Field for Blocked Amino Acids Based on High-Level Quantum-Mechanical Energy Surfaces**" investigates its performance by evaluating the reorganization energies for all low lying energy minima occurring in all 15 neutral blocked amino acids on a quantum-mechanical (QM) energy surface at the MP2/cc-pVDZ level. The results showed that a reparameterization of five bond angles significantly improved the agreement with the QM energy surface. Furthermore, it was highlighted that the corrected Drude force field exhibits almost the same average reorganization energies relative to the MP2 energy surface as the AM1 and PM3 semi-empirical methods.

Electrostatic interactions are essential components of bonded energies, and their correct modeling is crucial for the accuracy of molecular dynamics simulations. However, most force fields (FFs) use fixed partial atomic charges to account for electrostatic interactions, which results in a lack of electronic polarization response, representing an intrinsic limitation. "**Chapter 3: Accurate Modeling of RNA Hairpins Through the Explicit Treatment of Electronic Polarizability with the Classical Drude Oscillator Force Field**" discusses this deficiency and reports the application of an all-atom polarizable (Drude) and non-polarizable (CHARMM) nucleic acid FFs in RNA hairpin systems to investigate the impact of polarization on structural properties, dipole moment distributions, and cation interactions. Results show that the presence of polarizability in the FF significantly improves the stabilization of RNA hairpin structure. The chapter highlights the importance of explicitly treating electronic polarizability in RNA molecular simulations, including in non-canonical regions.

2. Novel Methods in Computational Biophysics and Chemistry and their Applications to Biological Problems

Due to rapid developments and improvements in computational techniques, computational biophysics and chemistry have been increasingly recognized as significant approaches to understand biological phenomena, especially at the molecular level. Recently, computational tools and software have been widely used in many biology related fields, such as drug design, protein structure prediction, protein-protein complex structure prediction, and simulations of biomolecules. Therefore, this section will focus on the advances in computational chemistry and its applications.

One frequently used method to investigate problems in molecular biology is studying the binding of a ligand to a receptor. There are many approaches to model this process, including first principle-based methods such as docking to homology-based methods that use templates to predict the binding mode. However, predicting ligand-receptor binding is difficult without a template, especially when the binding site and its entry pathway are unknown. This prompted the development of a first principle-based approach that combines molecular dynamics (MD) simulations with steered forces calculated using DelPhiForce software, as discussed in "**Chapter 4: Ab-*initio* Binding of Barnase–Barstar with DelPhiForce Steered Molecular Dynamics (DFMD) Approach**". An essential component of the DFMD approach is enhancing the role of long-range electrostatic interactions to provide a guiding force for the monomers toward their correct binding orientation and position. The motivation for computing the electrostatic force via DelPhiForce instead of the MD protocol stems from the observation that the calculated force by MD often is insufficient to dock the partners. On the other hand, it has been demonstrated that the DFMD approach can successfully dock the barstar to the barnase even when their initial positions and orientations are completely different from the correct ones. Thus, electrostatic forces provide orientational guidance along with a pulling force, bringing the ligand into close proximity to the receptor.

Further evaluation of the prominent force fields was done to test their accuracy in modeling intrinsically disordered proteins (IDPs), as illustrated in "**Chapter 5: The Accuracy of Force Fields on the Simulation of Intrinsically Disordered Proteins: A Benchmark Test on the Human p53 Tumor Suppressor**". For this purpose, MD simulations were carried out to generate structural ensembles of the activation domain 1 (AD1) in p53 tumor suppressor using widely used force fields such as AMBER99SB-ILDN, CHARMM27, and CHARMM36m with different water models. The results showed that CHARMM36m generates more extended conformations than other force fields, while CHARMM27 prefers to sample the α-helical structure. Moreover, the chemical shifts obtained with CHARMM36m are the closest to experimental measurements. This indicates that the CHARMM36m force field performs the best in characterizing the structural properties of p53 AD1. Water models are also critical to describe the structural ensemble of IDPs. In this simulation, the TIP4P water model was shown to generate more extended conformations and produce more local helical conformations than the TIP3P model.

In contrast, when IDP binds to its partner, it adopts a well-defined 3D structure, which can be modeled via MD simulations. This was demonstrated by modeling the N-terminal transactivation domain (TAD) of p53, which is intrinsically disordered in isolation, but adopts a particular 3D structure when bound to the transcriptional adaptor zinc-binding 2 (Taz2) domain of p300. This is further discussed in "**Chapter 6: Changes in Structure and Flexibility of p53 TAD2 Upon Binding to p300 Taz2**". The MD simulation results showed very good agreement when compared with experimental data. The study also demonstrated the ability of MD protocol to change of the conformational ensemble from unbound to bound states.

3. Computational Biophysics and Chemistry Methods to Predict pKa of Ionizable Groups in Proteins, RNAs, DNAs, and Small Molecules

This part explores computational chemistry methods to predict the pKa of ionizable groups in proteins, RNAs, DNAs and small molecules. All computational chemistry methods including quantum mechanics (QM), molecular dynamics (MD), Monte Carlo (MC), Poisson-Boltzmann (PB) approaches, semi-empirical and machine learning (ML) methods are considered. This section provides a comprehensive overview of the state of the art in computational modeling of pKa values. The importance of predicting the pKa of ionizable groups is long-established due to the crucial role pKa plays in pH-dependent processes. The following chapters outline the contributions that address this topic, as well as computational studies utilizing existing pKa predictors to address important biological and medical problems, including the pKa effects of disease-causing mutations.

One of the most popular approaches to compute the pKa of ionizable groups is to utilize implicit solvent models via the Poisson-Boltzmann equation (PBE). Numerous methods were developed, each of them implementing different descriptions of the solute and solvent phases, the putative boundary between solute-solvent, and the treatment of the energies (typically only electrostatic energy). One example is the treecode-accelerated boundary integral (TABI) solver as discussed in, "**Chapter 7: Computing Protein pKas Using the TABI Poisson–Boltzmann Solver**". The pKa computing procedure is enclosed in a convenient Python wrapper which is publicly available for download. Predicted results were compared with experimental pKa values for several proteins, and good accuracy was reported.

In terms of levels of detail, the Multiconformation Continuum Electrostatics (MCCE) is the leading algorithm in the field of pKa predictions, as explored in "**Chapter 8: Characterizing the Water Wire in the Gramicidin Channel Found by Monte Carlo Sampling Using Continuum Electrostatics and in Molecular Dynamics Trajectories with Conventional or Polarizable Force Fields**". The MCCE allows limited number of water molecules to be treaded explicitly along with implicit water modeling of the rest of the water phase. This unique MCCE feature is used for the explicit treatment of water molecules in the Gramicidin A (gA) channel. The resulting distributions of water molecules and key hydration features are then compared against those obtained with explicit solvent MD simulations with the non-polarizable CHARMM36 and polarizable Drude force fields. CHARMM36 leads to an aligned water wire in the channel characterized by a large absolute net water dipole moment, while the MCCE and Drude analysis led to a small net dipole moment as the water molecules change orientation within the channel.

The application of another popular pKa predictor, the DelPhiPKa, to model large macromolecular complex formation is reported in **"Chapter 9: pH-Dependent Interactions of Apolipophorin-III with a Lipid Disk"**. This chapter computationally investigates the conformational changes of ApoLp-III and the pH-dependence of the binding free energy of ApoLp-III association with a lipid disk. The pH-dependence of the binding energy for ApoLp-III with the lipid disk is predicted to be significant, with the pH-optimum at pH=8.0. The calculations suggest that there are no direct interactions between the lipid head groups and titratable residues of ApoLp-III. In the physiological pH range from 6.0 to 9.0, the binding free energy of ApoLp-III with the lipid disk decreases significantly with respect to its optimal value at pH 8.0 (at pH=6.0, it is 1.02 kcal/mol and at pH=9.0 it is 0.23 kcal/mol less favorable than at the optimal pH=8.0), indicating that the pH is an important regulator of ApoLp-III lipid disk association.

4. Artificial Intelligence in Biophysics and Chemistry

Machine learning (ML) and artificial intelligence (AI) methods offer powerful alternatives to first principle-based, approaches, often achieving better performance. In addition, ML and AI are fast, making them suitable for genome-scale studies. This section outlines recent progress in ML and AI applications within computational biophysics and chemistry, and on their role in modeling various biological phenomena.

At the cellular level and in tissue development, an important aspect is predicting cell trajectory inference. However, due to the high noise and heterogeneity in single-cell data, it is challenging to infer cell trajectory in complex biological processes. To address this challenge, a new trajectory inference method, called the 'Metric Learning Bhattacharyya Kernel Feature Decomposition' is recently developed.

This approach is discussed in "**Chapter 10: MLBKFD: Probabilistic Model Methods to Infer Pseudo Trajectories from Single-cell Data**". The method was evaluated on four typical datasets, as well as seven recent human fetal lung datasets. It was shown to exhibit nearly twice the speed of other leading algorithms while maintaining high precision, particularly when dealing with large datasets.

Another important component of cell growth is gene expression, which often is heterogenetic, resulting in significant phenotypic consequences. One of the important sources of this heterogeneity is transcriptional bursting. "**Chapter 11: Inferring Transcriptional Bursting Kinetics Using Gene Expression Model with Memory and Crosstalk from scRNA-seq Data**" explores the usage of an approximate Bayesian computation algorithm combined with a gene expression model that considered crosstalk and memory, to infer the kinetics of transcriptional bursting. It was shown that genes are jointly regulated by several competitive pathways when activated by external signals, and there is molecular memory in the process of gene state switching. Furthermore, it was shown that some previously proposed models can be simplified as special cases of the gene expression model and inference algorithm proposed here under certain circumstances.

At the protein level, accurately predicting functionality is essential. For example, enzymes function as catalysts in biological reactions such as digestion, metabolism, DNA replication, and other physiological processes. "**Chapter 12: Investigating Enzyme Biochemistry by Deep Learning: A Computational Tool for a New Era**" provides an overview of various deep learning algorithms and recent advancements in their application to enzyme science. This chapter summarizes the ML algorithms for function prediction, enzyme kinetic parameters prediction, enzyme-substrate identification, condition optimization, thermophilic property prediction, enzyme catalytic site prediction, and enzyme design. It is estimated that each protein interacts with many partners resulting in protein–protein interaction (PPI) networks.

5. Computational Biophysics and Chemistry and Diseases

The rapid development of genetic methodologies generates large amounts of data that require the support of computational and statistical methods for accurate and efficient processing. Computational chemistry has grown into a field that proactively contributes to many biomedicine related fields and numerous aspects of drug discovery, including target identification and lead generation. This final part highlights achievements in computational chemistry and its applications in disease diagnosis, treatment, prevention, and the discovery of the molecular mechanisms of diseases.

Computational biophysics and chemistry have been applied to reveal the effect of mutations, primarily missense mutations, on the wild type function of the corresponding proteins and protein/DNA/RNA complexes. Understanding that there is a clear relation between thermodynamics parameters as folding and binding free energies and pathogenicity, researchers assess the effect of disease-causing mutations, by probing the effect of these mutations on protein stability and binding. "**Chapter 13: Computational Analysis of Hereditary Spastic Paraplegia Mutations in the Kinesin Motor Domains of KIF1A and KIF5A**" investigates the effect of pathogenic mutations causing hereditary spastic paraplegias (HSPs) disorders, which is a genetically heterogeneous collection of neurodegenerative disorders categorized by progressive lower-limb spasticity and frailty. These pathogenic mutations were found in the kinesin superfamily proteins (KIFs), which are microtubule-dependent molecular motors involved in intracellular transport. Kinesins directionally transport membrane vesicles, protein complexes, and mRNAs along neurites, thus playing important roles in neuronal development and function. It was shown that most disease-causing mutations change the folding free energy of kinesin motor domain and the binding free energy of kinesin–tubulin complex.

"**Chapter 14: In-Silico Analysis to Identify the Role of MEN1 Missense Mutations in Breast Cancer**" focuses on menin protein, coded by the multiple endocrine neoplasia type 1 (MEN1) gene located on chromosome 11q13, and missense mutations causing breast cancer. The analysis indicated that some of the variants found in breast cancer tissue show similar pattern of destabilizing the menin protein and its interactions as the pathogenic variants associated with neoplasia. Taking into account the results of an in-house in-silico consensus predictor, the authors classified missense mutations in the menin protein found in breast cancer tissue as either pathogenic or benign, suggesting their potential as an indicator for early detection of elevated breast cancer risk.

"**Chapter 15: Computational and Structural Studies of MeCP2 and Associated Mutants**" summarizes the general strategy of elucidating changes in thermodynamic characteristics, folding and binding properties, and pathogenicity risk, with a focus on the Rett syndrome. Rett syndrome is a rare genetic disorder exclusively seen in girls. Approximately 95% of RTT cases is caused by mutations in the MeCP2 gene which codes for Methyl-CpG-binding protein 2 (MeCP2). This chapter outlines Rett Syndrome, MeCP2 protein structure and function, mutation types and frequencies, and phenotype–genotype relationships. Comparison is made for the wild-type and mutant MeCP2 protein structure and dynamics, as well as its binding to DNA.

Professor Emil Alexov
Department of Physics and Astronomy, Clemson University

https://doi.org/10.1142/9789819808199_fmatter

About the Editor

Emil Alexov received his PhD degree in Physics from Sofia University, Bulgaria, and was assistant professor there, staff scientist in Bulgarian Academy of Sciences, visiting scientist at RIKEN, Japan, postdoctoral researcher at City College of New York, and senior scientist at Columbia University, USA. He is known for his expertise in computational biophysics and bioinformatics. Dr. Alexov has made significant contributions to the field, particularly in the areas of molecular modeling and simulations of biological macromolecules. His research often focuses on understanding the molecular mechanisms underlying various biological processes, including protein-protein interactions, ion channels, and the effects of mutations on protein stability and function, and their association with diseases. His research is supported by grants from NIH and NSF. He is a recipient of numerous awards including Outstanding Graduate Student Mentor, Faculty Achievement in the Science, Faculty Scholar at the School of Health Research, and Alumni Award for Outstanding Achievement in Research, Clemson, USA. Dr. Alexov is the Editor-in-Chief of '*Journal of Computational Biophysics and Chemistry*', and Editor for '*International Journal of Molecular Sciences*', '*Computational and Mathematical Methods in Medicine*', and '*Frontiers Molecular Biosciences*'.

https://doi.org/10.1142/9789819808199_fmatter

Contents

Part I
Polarizable Force Fields for Biomolecular Modeling

https://doi.org/10.1142/9789819808199_0001

Chapter 1

High-Order *Ab Initio* Valence Force Field with Chemical Pattern-Based Parameter Assignment*

Xudong Yang, Chengwen Liu and Pengyu Ren†

Department of Biomedical Engineering,
The University of Texas at Austin,
Austin, TX 78712, USA
†*pren@utexas.edu*

Bonded (or valence) interactions, which directly determine the local structures of the molecules, are fundamental parts of molecular mechanics force fields (FFs). Most popular classical FFs adopt the simple harmonic models for bond stretching and angle bending and ignore cross-coupling effects among the valence terms. This may lead to less accurate vibrational properties and configurations in molecular dynamics (MD) simulations. AMOEBA models utilize an MM3(MM4)-style bonded interaction model, in which the vibrational anharmonicity, the coupling effects among different energy terms, and the out-of-plane bending for *sp*2-hybridized atoms are considered. In this work, we report the development of bonded interaction parameters for a wide range of chemistry based on quantum mechanics (QM). About 270 atomic types defined by SMARTS strings were used to model the valence interactions. Our results indicate that the resulting valence parameters produce accurate vibrational frequencies (RMSD from QM is less than $\sim 36.6\,\text{cm}^{-1}$) over a large set of molecules with diverse functional groups (445 molecules). By contrast, the harmonic models usually give an RMS error greater than $60\,\text{cm}^{-1}$. Meanwhile, this model accurately reflects the potential energy surface of the out-of-plane bending.

*To cite this article, please refer to its earlier version published in *J. Comput. Biophys. Chem.* **2022**, *21* (4), 431–447, DOI: 10.1142/S2737416521420047
†Corresponding author.

Our model can generally be applied to the AMOEBA family and any MM3 (MM4)-based molecular mechanics FFs.

Keywords: AMOEBA potential; force field; bonded interactions; normal mode frequencies.

1. Introduction

Classical molecular dynamics (MD) simulations are widely applied by researchers to investigate various biomolecular systems. One of the essential elements determining the reliability of MD simulations is the underlying classical potentials, or force fields (FFs),[1–3] which describe the intra- and intermolecular interactions using empirical functional forms. Abundant studies focus on improving modeling the intermolecular (nonbonded) interactions as they are responsible for thermodynamic properties in protein-ligand binding,[4–7] protein folding,[8] enzymatic catalysis,[9] and other critical biomolecular processes. Nonetheless, bonded interactions are equally important as they determine the local structural ensemble sampled during the MD simulations.[10]

Most popular FFs, such as AMBER,[11–14] OPLS,[15–17] and CHARMM,[18–20] adopt simple harmonic functions to describe bond stretching and angle bending without considering the anharmonic nature of bonds and angles. AMBER omits cross-coupling effects between bond and angle terms, and it uses the Fourier style to describe out-of-plane effects. OPLS uses a similar functional form to AMBER. CHARMM employs a unique improper potential to represent the out-of-plane bending and a Urey–Bradley potential to account for the cross-coupling. Those methods are computationally efficient, but their accuracy is limited in complex molecular structures, especially in vibrational analysis. By contrast, AMOEBA-based models[21–23] use the sophisticated MM3(MM4)-style[24,25] functional forms to describe these bonded interactions, in which the anharmonicity, cross-coupling effects, and out-of-plane bending (opbend) have been included. First, higher-order polynomial terms were added to the bond stretching and angle bending terms to capture the anharmonicity while avoiding abnormal effects on strained bonds or angles due to the negative contributions from odd polynomial terms. Second, the cross-coupling terms were proven to be essential for capturing the coupling effects and improving the vibrational frequencies (VF). For example, it is evidenced that the combination of a high-order polynomial model and cross-terms can greatly enhance the description of equilibrium structure and VF of formate ion.[26,27] Specifically, the average absolute deviation of VF ($|\Delta\nu|_{\text{ave}}$) plunges from 155.6 cm^{-1} by the simple harmonic model without any cross-terms to 7.2 cm^{-1} by a quartic model with third-order cross-terms. Similar improvements have been found on systems including amides, peptides, and others.[28,29] Finally, the opbend[30] is a necessary and important energy term to maintain planar structure around trigonal centers as a significant pyramidalization can form without seriously distorting any of the three angles of bonds involved.

MM3/MM4 parameters were derived almost three decades ago. Quantum mechanics (QM) calculations have significantly improved over the years, with energy, structures, VF closely matching experimental measurements. In addition, a more comprehensive range of coverage in chemical space is needed, and better ways for parameter assignment and typing (e.g., chemical pattern based) are readily available. This work aims to provide an up-to-date database of valence parameters for common organic molecules using atomic types determined by SMARTS strings. These parameters will be used for AMOEBA[21] and AMOEBA+[22] FFs but are also compatible with other classical FFs that use MM3/MM4-style valence terms. We have developed an automated Python program that implements the modified Seminario method[31] to generate bond and angle parameters from *ab initio* Hessian matrices. In addition, a reversed searching with the ranking tree was implemented inside this program to assign valence parameters for new molecules based on SMARTS patterns. Moreover, the anharmonicity, coupling effects, and opbend have been analyzed on a large set of diverse molecules. The database of parameters will allow us to parameterize and simulate new biological and synthetic molecules, such as ligands and drug compounds by using AMOEBA and AMOEBA + FFs.

2. Methodologies

2.1. *Theoretical background*

The bonded interactions of AMOEBA[21] and AMOEBA+[22,23] include five components: bond stretching, angle bending, opbend, torsion, and cross-coupling terms for bond/angle/torsion, as shown in Eq. (1).

$$E_{\text{bonded}} = E_{\text{stretch}} + E_{\text{bend}} + E_{\text{torsion}} + E_{\text{opbend}} + E_{\text{cross}}. \quad (1)$$

This study focuses on bond stretching, angle bending, stretch–bend coupling, and opbend terms, which are the standard contributing elements to the total potential energy shared by almost all the molecules. The torsional energy involving four atoms in connection is usually coupled with other interactions such as electrostatics, van der Waals and polarization. The torsion term and related cross-coupling terms are treated as a residual correction to match the *ab initio* QM reference energy after all the other terms of AMOEBA(+), including nonbonded terms are determined.

As mentioned above, the AMOEBA valence terms have been mostly inherited from MM3/MM4 model.[24,25] The advantage of using the higher-order polynomials is that the energy variation within a wider range of geometry changes can be accurately captured. Bond stretching term adopts the quartic form (Eq. (2)), which can generate reliable energy for bond variation in the range of ± 0.3 Å from the equilibrium structure. Also, the sextet bending model is utilized, which can reliably reproduce *ab initio* energy for the angle variation in the range of $\pm 70^\circ$ and beyond.[32] The opbend term is used to keep the planar structures for *sp*2 atoms at and around the trigonal centers such as the amide group. Without the out-of-plane or improper torsion term used by some FFs, large pyramidalization may occur during the MD simulations without seriously distorting the bond lengths and angles. In AMOEBA, the opbend term (Eq. (4)) shares a similar functional form to the angle bending term. The χ is defined as the bending angle between the plane and the central atom (Fig. 1), which means each out-of-plane structure has three components contributed by each peripheral atom.

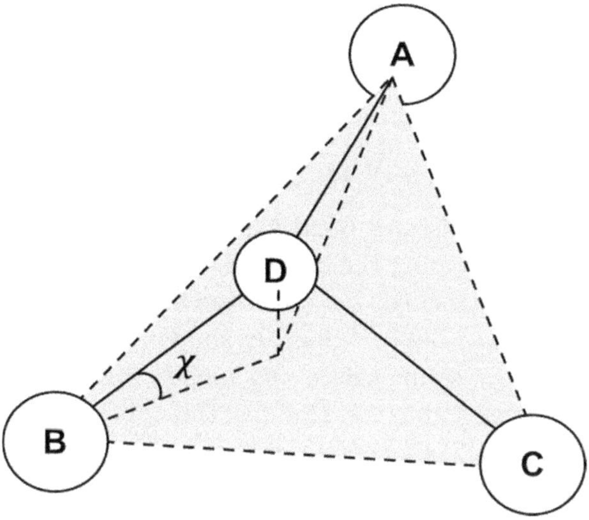

Fig. 1. An illustration of a trigonal center D pyramidalizing from plane ABC. The out-of-plane angle χ has been defined as illustrated.

$$E_{\text{stretch}} = k_b \Delta r^2 \left[1 - 2.55\Delta r + \frac{7}{12} \times 2.55^2 \Delta r^2\right], \tag{2}$$

$$E_{\text{bend}} = k_a \Delta\theta^2 [1 - 0.014\Delta\theta + 5.6(10)^{-5}\Delta\theta^2 - 7.0(10)^{-7}\Delta\theta^3 + 2.2(10)^{-8}\Delta\theta^4], \tag{3}$$

$$E_{\text{oop}} = k_{\text{opbend}} \Delta\chi^2 [1 - 0.014\Delta\chi + 5.6(10)^{-5}\Delta\chi^2 - 7.0(10)^{-7}\Delta\chi^3 + 2.2(10)^{-8}\Delta\chi^4]. \tag{4}$$

Here, $\Delta r = r - r_0$ is the difference between the real bond length and the reference r_0, also, we have $\Delta\theta = \theta - \theta_0$ and $\Delta\chi = \chi - \chi_0$. The cross-terms (Eq. (5)) describe the coupling effects among bond stretching, angle bending, and rotational torsion. In this study, we focus on the stretch–bend cross-term only. Torsion and related cross-terms are included in AMOEBA functional forms and will be considered later. These torsion-related cross-terms are needed to describe the hyperconjugation effects for conjugated molecules.[33] The stretch–bend cross-term is expressed in Eq. (6) for any angle A–B–C, two parameters $k_{b,AB}$ and $k_{b,BC}$ denote the coupling of the bond AB,

BC, respectively, with angle ABC.

$$E_{\text{cross}} = E_{\text{str-bnd}} + E_{\text{str-str}} + E_{\text{bnd-bnd}} + E_{\text{str-tors}} + E_{\text{bnd-tors}} + E_{\text{tors-tors}}, \quad (5)$$

$$E_{\text{str-bnd}} = [k_{ba,AB}\Delta r_{AB} + k_{ba,BC}\Delta r_{BC}]\Delta\theta_{ABC}. \quad (6)$$

In this work, the distinct bond types are defined by the atomic types of two atoms forming the bond (A–B). Also, the different angle types have been classified by the atomic types of three atoms forming the angle (A–B–C).[23] Slightly simple types are used for the stretch–bend energy term, where any angles with the same central atom use the same parameters. Besides, the opbend terms have been classified by the type of central *sp*2 atom with the type of one peripheral atom.

2.2. *Initial bond/angle parameters from QM Hessian matrix*

In this work, we adopt the modified Seminario method[31,34] to generate our initial bond stretching and angle bending parameters directly from the QM Hessian matrix. The modified Seminario method is a convenient and efficient way to derive accurate stretching and bending parameters with all other contributions ruled out (as opposed to fitting).

The Seminario method was initially designed for the generation of OPLS-AA[35] style parameters and written in MATLAB. We have re-written the code in Python based on the atom types we defined here for organic molecules.[23] The OPLS-AA FF uses the classical harmonic model for stretching and bending, which means the modified Seminario cannot be directly applied to the parameterization based on our model. The resulting parameters from this method serve as excellent initial values for further optimization against experimental data, which is significantly improved over the previous simple estimates used by Tinker[36] or Poltype[37] tools.

The mathematical details and principles about the modified Seminario are described here. The Hessian matrix is written based on the Cartesian coordinates of each atom, which is further divided into many small 3×3 matrices belonging to different bonds, such as H_{AB} for the bond AB in Eq. (7). Stretching parameters can be obtained straightforwardly by using Eq. (7), where $\lambda_i^{AB}(i = 1, 2, 3)$ are the eigenvalues of the small Hessian for the bond AB and $\mathbf{v}_i^{AB}(i = 1, 2, 3)$ are the corresponding eigenvectors, $\mathbf{u}_{AB}$ is the bond vector (all the vectors mentioned here and in the next paragraph are normalized).

$$H_{AB} = \begin{bmatrix} \frac{\partial^2 E}{\partial x_A \partial x_B} & \frac{\partial^2 E}{\partial x_A \partial y_B} & \frac{\partial^2 E}{\partial x_A \partial z_B} \\ \frac{\partial^2 E}{\partial y_A \partial x_B} & \frac{\partial^2 E}{\partial y_A \partial y_B} & \frac{\partial^2 E}{\partial y_A \partial z_B} \\ \frac{\partial^2 E}{\partial z_A \partial x_B} & \frac{\partial^2 E}{\partial y_A \partial z_B} & \frac{\partial^2 E}{\partial z_A \partial z_B} \end{bmatrix},$$

$$k_{AB} = \sum_{i=1}^{3} \lambda_i^{AB} |\mathbf{u}_{AB} \cdot \mathbf{v}_i^{AB}|. \tag{7}$$

As to deriving angle bending parameters, additional work is involved. Inside the equations (8) below, we need to first derive the intermediate force constants in the direction perpendicular to the bonds forming the angle (vectors $\mathbf{u}_{PA}$ and $\mathbf{u}_{PC}$ with the orientation pointing to the interior in Fig. 2(a)). Then, by considering the relation of those normal vectors and their contributions to the angle bending, Eqs. (9) and (10) are used to calculate the final bending force constants. More mathematical and analytical details about this method can be found elsewhere.[31] The N is the number of angles in the FF with a central atom B and involves the bond AB. Take oxalic acid as an example in Fig. 2(b); atom A is the peripheral atom for two coupled angles: ABC and ABE. To obtain the force constant for angle $ABCk_{\theta(ABC)}$, we can see the vector $\mathbf{u}_{PA,1}$ perpendicular to bond AB points to ABC and $\mathbf{u}_{PA,2}$ perpendicular to bond AB points to ABE. However, this method is not appropriate to deal with angles involving *sp*-hybridized atoms, because it cannot detect the orientation. *sp* atoms are not common, so some more rough estimates are used here for initial values in the subsequent fitting to QM VF.

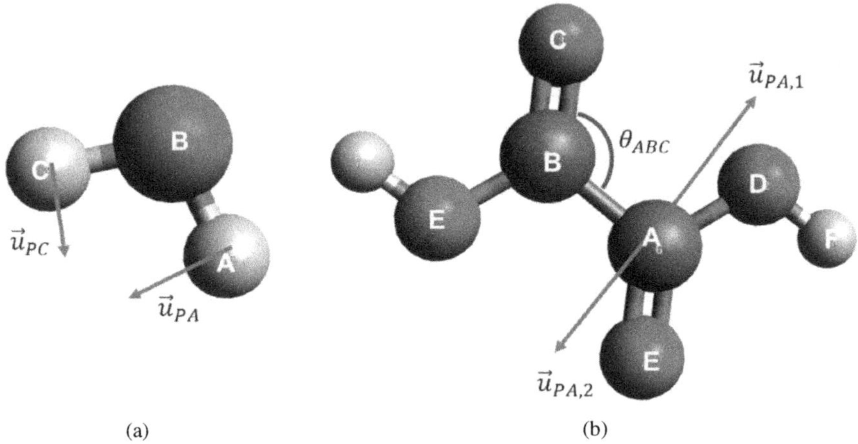

Fig. 2. (Color online) An illustration of the vectors used in the Seminario method. (a) The two vectors $\mathbf{u}_{PA}$ and $\mathbf{u}_{PC}$ contributing to the force constant of angle ABC in water molecule; (b) the two vectors pointing to two coupling angles with A as a side atom in oxalic acid.

$$k_{PA} = \sum_{i=1}^{3} \lambda_i^{AB} |\mathbf{u}_{PA} \cdot \mathbf{v}_i^{AB}|, \quad k_{PC} = \sum_{i=1}^{3} \lambda_i^{CB} |\mathbf{u}_{PC} \cdot \mathbf{v}_i^{CB}|, \tag{8}$$

$$s_{PA} = \begin{cases} 1 + \dfrac{\sum_{j=1}^{N} |\mathbf{u}_{PA,1} \cdot \mathbf{u}_{PA,j}|^2 - 1}{N-1} & (N > 1) \\ 1 & (N = 1) \end{cases}, \tag{9}$$

$$\frac{1}{k_{\theta(ABC)}} = \frac{s_{PA}}{R_{AB}^2 k_{PA}} + \frac{s_{PC}}{R_{CB}^2 k_{PC}}. \tag{10}$$

It is convenient to derive the initial bond/angle parameters using the modified Seminario method from the QM Hessian matrix of a given molecule. However, since this model is appropriate mainly for the harmonic model, the parameters need to be further fitted to the QM VF. The whole procedure has been illustrated in Fig. 5 and implemented in an automated program which will be described in Sec. 2.4.

2.3. *Computational details*

2.3.1. *Selection of training set molecules*

The training set consists of 445 molecules, with the distribution of the number of heavy atoms in each molecule shown in Fig. 3. The whole range is 1–18 heavy atoms with most molecules having 5–9 heavy atoms. All the molecules have been classified into categories that span from alkane to bio-fragments, as illustrated in Fig. 4 and Table 1. In total, they are grouped into seven categories: (i) alkanes and aliphatic rings; (ii) molecules comprised of only single bonds; (iii) molecules with double bonds or triple bonds; (vi) benzene derivatives; (v) heterocyclics; (vi) bio-fragments; (vii) combinations of multiple aromatic cycles. The detailed list of molecules with their SMARTS and specific categories is in Table S1. The whole procedure of parametrization has been illustrated by the workflow in Fig. 5.

2.3.2. *Initial bond/angle parameters from Hessian matrix*

The frequencies of the normal modes have been calculated by using Gaussian09[38] with MP2/6-31g** (6-311g** for iodine atom) after the

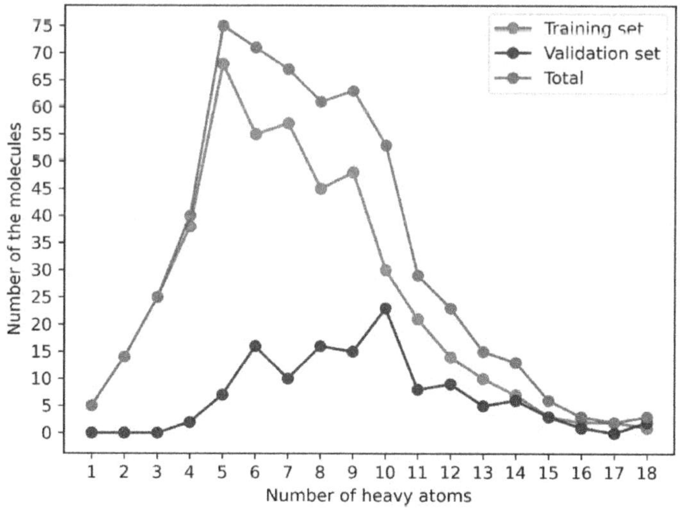

Fig. 3. (Color online) Distribution of the number of heavy atoms in different molecules. (Red: training set; blue: validation set; green: total.)

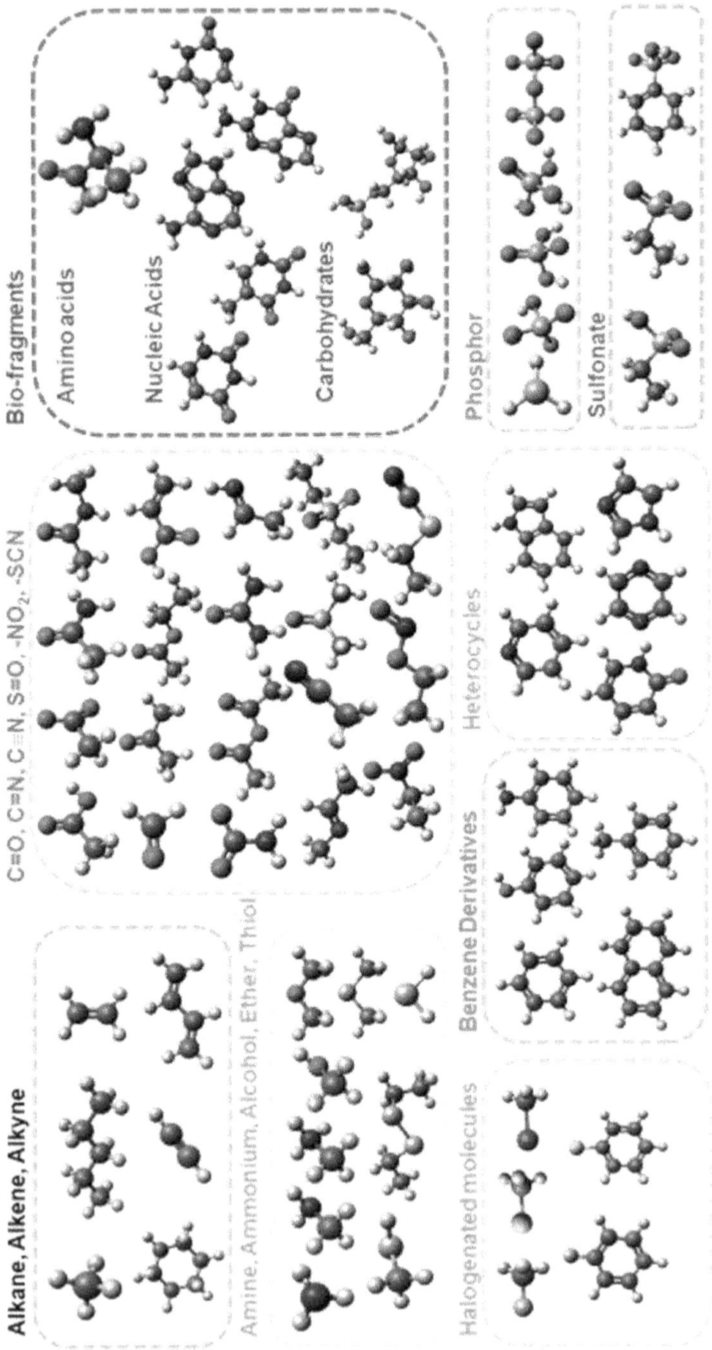

Fig. 4. (Color online) Representatives of the organic molecules covered in this study.

Table 1. A general illustration of the functional groups and distinct chemical structures covered in this study.

Group	Chemical structure	Group	Chemical structure
Alkane	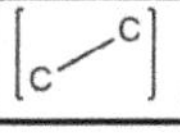 (chain & cyclic)	Alkene	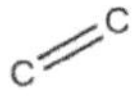
Alkyne	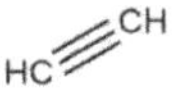	Conjugated molecules	
Amine	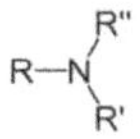 (all R = H, C)	Ammonium	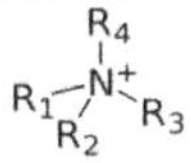 (all R = H, C)
Alcohol	 (R = C)	Ether	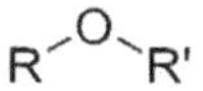 (all R = C)
Di-oxyl	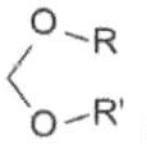 (all R = C)	Thiol	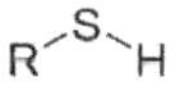 (R = C)
Sulfide	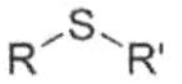 (all R = C)	Disulfide	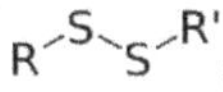 (all R = C)
Halide	R-X (R = C; X=F, Cl, Br, I)	Aldehyde	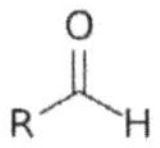 (R=C, H)
Ketone	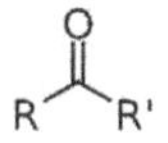 (R=C)	Carboxylic acid / Carboxylate	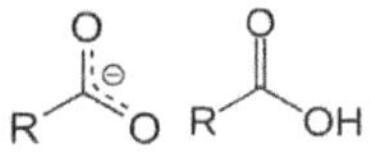 (R=C, H)
Carboxylate ester	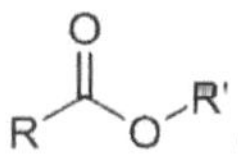 (R=C, H; R'=C)	Carbamic acid / Carbamate	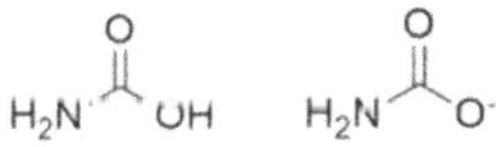
Amide	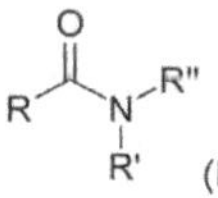 (R, R', R''=C, H)	Carbonate / Bicarbonate	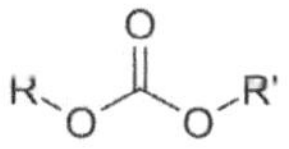 (R=C, H; R'=C)
Anhydride	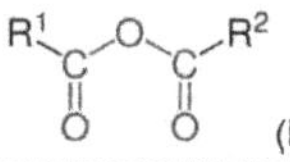 (R^1, R^2=C)	Acyl halide	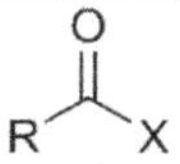 (R=C; X=F, Cl, Br)
Oxalic acid Oxalate	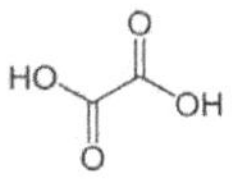 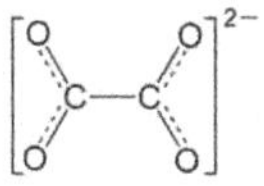	α-keto acetate / β-keto acetate	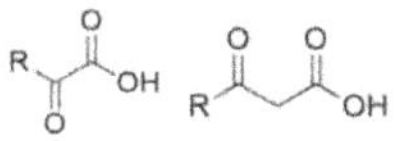 (R=C)

Table 1. (*Continued*)

Malonic acid / Malonate	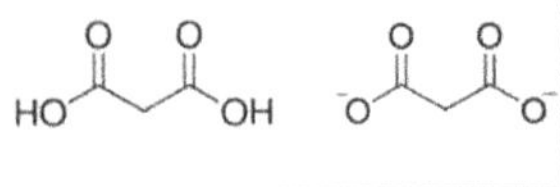	Acrylic acid/ Acrylate	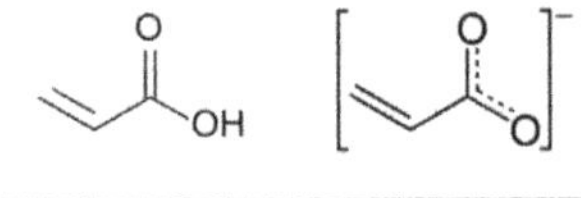
Imide	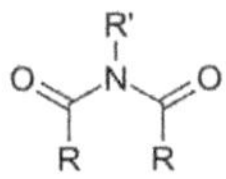 (R, R'=C)	Imine (Aldimine/ Ketimine)	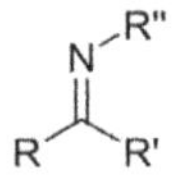 (R, R', R" remain 1 C)
Azide	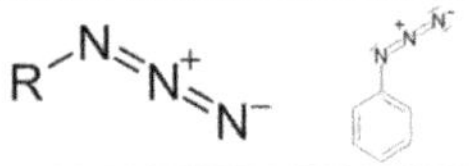	Azo-compound	
Nitro-compound	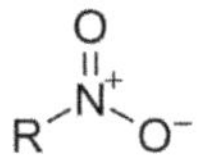 (R=C)	Guanidine	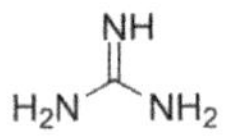
Amino acid	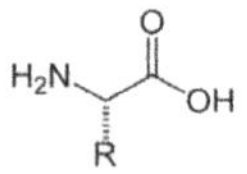	Oxime	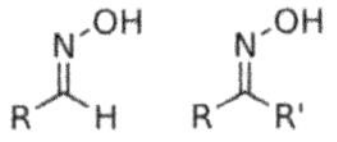 aldoxime ketoxime (R=C, H; R' =C)
Nitrite	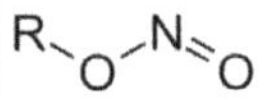 (R=C)	Nitrile	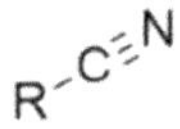 (R=C)
Sulfoxide	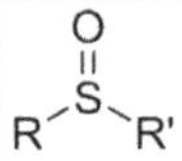 (R, R'=C)	Sulfone	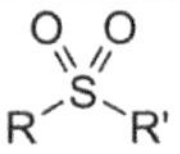 (R, R'=C)
Sulfonic acid / Sulfonate	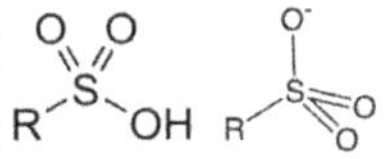 (R=C)	Sulfonate ester	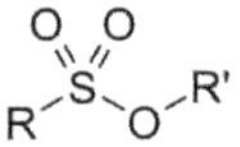 (R, R'=C)
Thiocyanate	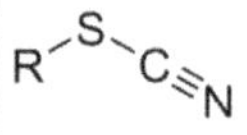 (R=C)	Phosphine	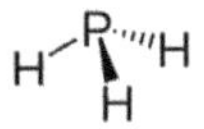
Phosphoric acid / Phosphate	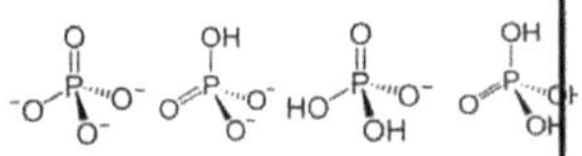	Pyridine / Pyrazine	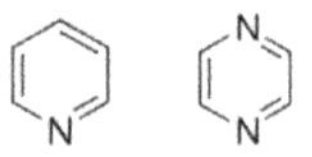
Benzene Derivatives		Conjugated lactam	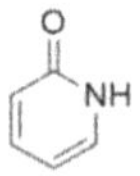
		Indole	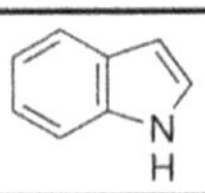

Table 1. (*Continued*)

	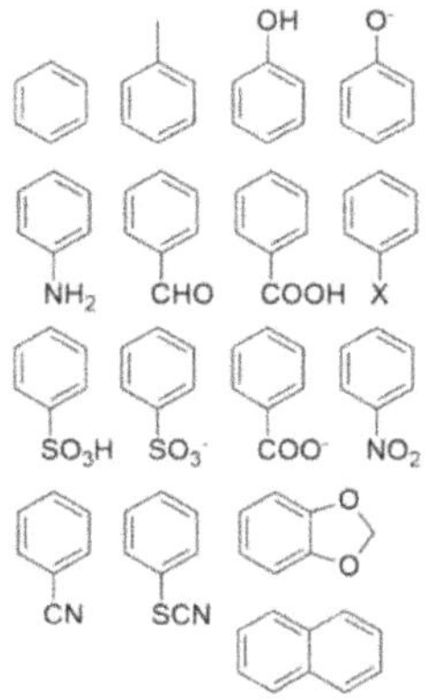	Imidazole / Imidazolium	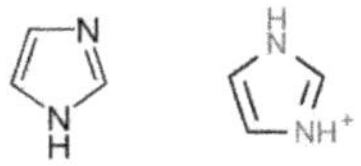
NA Base	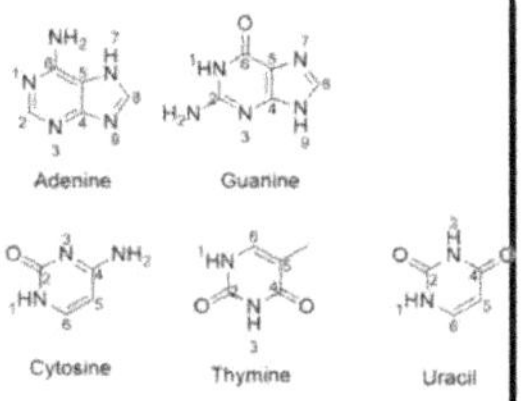	Aliphatic cycles and derivatives	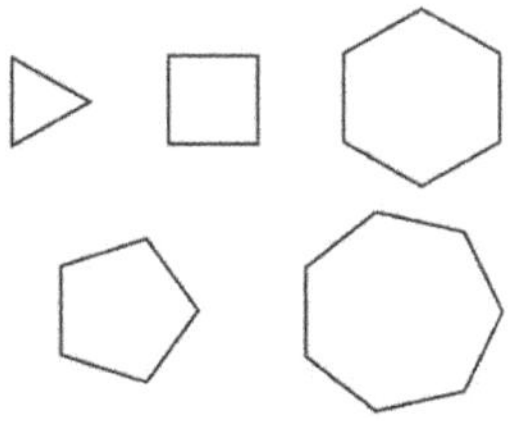
Pyran	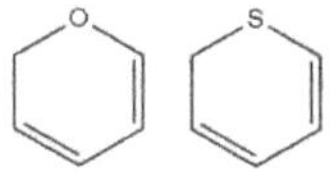	Benzoquinone	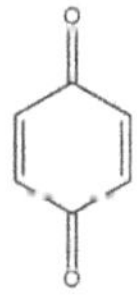
Thioacetate	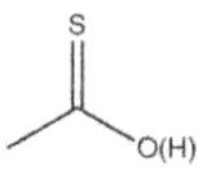	Tetrazole	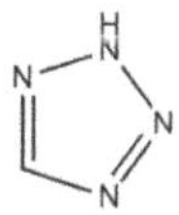
Peracetic acid/ Peroxoate	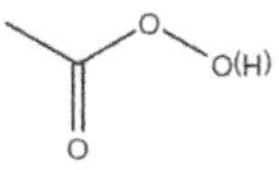	Pyrrolidinone	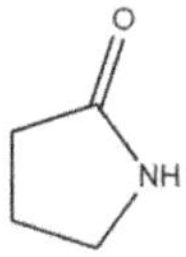

Table 1. (*Continued*)

Pyrrole	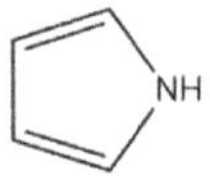	Oxazole	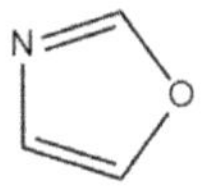
Thiazole	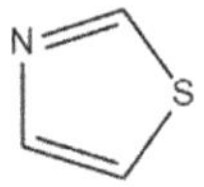	Thiophene	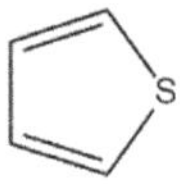
Connected structure of double heterocyclics, benzenes or aliphatic cycles	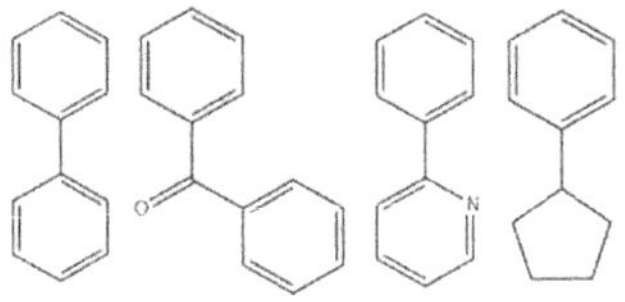		

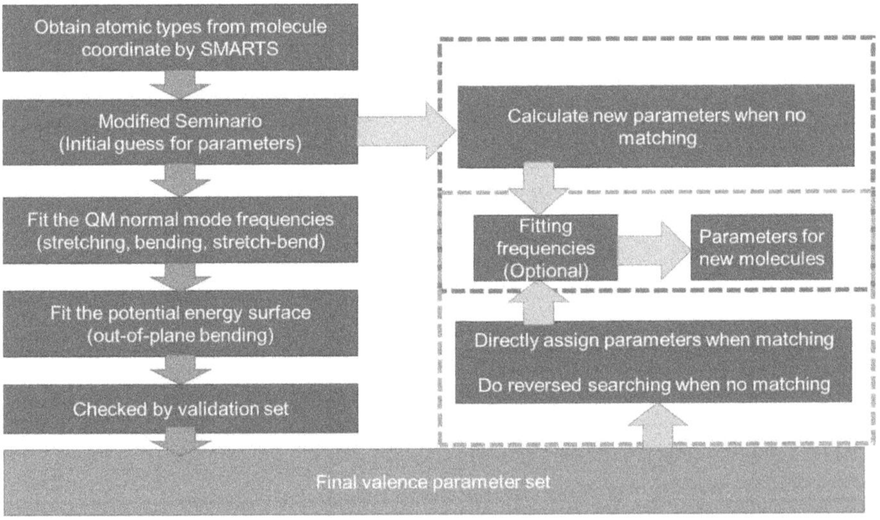

Fig. 5. (Color online) The procedure to generate the final valence parameter set (green arrows) and the methods to obtain and assign parameters for any new molecule (yellow arrows). Assignment by modified Seminario is shown inside the red dash box, and assignment by reversed searching is shown inside the blue dash box.

tight optimization with the same method and basis set. The bond/angle equilibrium values at MP2/6-31g** are close to the experiment for the organic molecules.[39] In this work, the reference bond lengths and angles were extracted from the final optimized structures and averaged based on the pre-defined bond and angle types. Then, the initial force constant parameters for bonds and angles were derived from the Hessian matrix by using the modified Seminario method.[31] These parameters were further refined in the next step (Sec. 2.3.3), targeting the QM frequencies.

2.3.3. *Bond/angle parameters refinement using QM frequency*

As suggested previously,[40,41] the frequencies calculated by MP2/6-31g** (6-311g** for iodine) are scaled by a factor of 0.943 for better consistency with the experimental values. The vibrational analysis program (*vibrate.x*) in Tinker was used to generate normal mode frequencies for our models. Since we only focus on the interactions related to bonds and angles here and the frequencies of high range (usually higher than $1000\,\mathrm{cm}^{-1}$) are mainly contributed by those terms while the low range depends mainly on torsional terms, the reference data lower than $1000\,\mathrm{cm}^{-1}$ are not used in the refinement. The least-square method from scipy[42] is used to optimize the parameters for bond, angle, and stretch–bend cross-terms, with the cost, root-mean-square-error (RMSE in VF), as given by Eq. (11).

$$\mathrm{RMSE(frequency)} = \sqrt{\frac{1}{n}\sum_{i}(\nu^{MM} - \nu^{QM,\mathrm{scaled}})^2}. \qquad (11)$$

2.3.4. *Opbend parameters derivation from potential energy surface fitting*

The potential energy surfaces of the training set molecules were utilized to derive the opbend parameters (further details in Table S2). We focus on those molecules containing the *sp*2 atoms that form trigonal planar structures. The central *sp*2 atom was pulled out of the plane by 0.05 Å, 0.1 Å, and 0.15 Å (opbend-small subset hereafter), and 0.2 Å, 0.25 Å, and 0.3 Å (opbend-large subset hereafter), respectively, to create deformed configurations and corresponding energies. We obtained 4318 configurations for the training set, and the single point energy was calculated using Psi4[43] at MP2/aug-cc-pvtz (aug-cc-pvtz-pp for iodine) level of theory.

The least-square method was used to fit the opbend parameters by optimizing the weighted RMSE of the energy difference (shown by Eq. (12)). To ensure that our parameters are not biased towards the high-energy (rare) structures far from equilibrium, we used a smaller weight (0.2) for configurations with energy 20 kcal/mol higher than that of equilibrium.

$$\text{wRMSE(energy)} = \sqrt{\begin{aligned}&\frac{1}{n}\sum_i[(E^{\text{MM,small}} - E^{\text{QM,small}})^2 \\ &+ 0.2 \times (E^{\text{MM,large}} - E^{\text{QM,large}})^2]\end{aligned}}. \quad (12)$$

2.3.5. *Validation*

After obtaining the parameters by fitting frequencies and potential energy surface, we chose 120 new molecules as the validation set (see Tables S3 and S4 for details). The distribution of the number of heavy atoms in molecules of the validation set has been illustrated in Fig. 3. Overall, the molecules in the validation set contain 4–14 heavy atoms. Those molecules are made of functional groups included in the training set molecules (Table 1). We mainly evaluated the bond/angle and coupling parameters by comparing the high-range frequencies of normal mode for these molecules. The opbend parameters were validated by comparing the energy differences between QM and our model for 829 configurations stemming from the 120 molecules in the validation set. This validation is intended to show the transferability of the current valence parameters.

2.4. *Automated generation and assignment of valence parameters for new molecules*

Although we have covered a range of chemistry (see Table 1) widely encountered in the practical simulations, there remains the possibility of failure to find the parameters for a certain molecule. For the molecules that the exact SMARTS patterns are missing in our database, we attempt to assign parameters based on the highest similarity to the existing chemical fragments.

Here, we provide an automated program to determine the best matching atomic type for any new molecule and assign both valence parameters and our previously published charge-flux parameters[23] to each atom. It will attempt to find an exact matching from our database at first. When the exact matching fails, this program has implemented two methods to estimate those missing parameters: (i) modified Seminario (in the red dash-line rectangle of Fig. 5); (ii) Reversed searching by ranking tree (in cyan dash-line rectangle of Fig. 5). In this work, we used the second method to generate the new parameters for the validation set.

The second method depends on the current parameter set to generate new parameters suitable for our model. We established a ranking tree where different types were placed in some positions in different levels determined by the complexity of the structure (Fig. 6). For example, given a bond *A*–*B* which has no direct equivalent in the

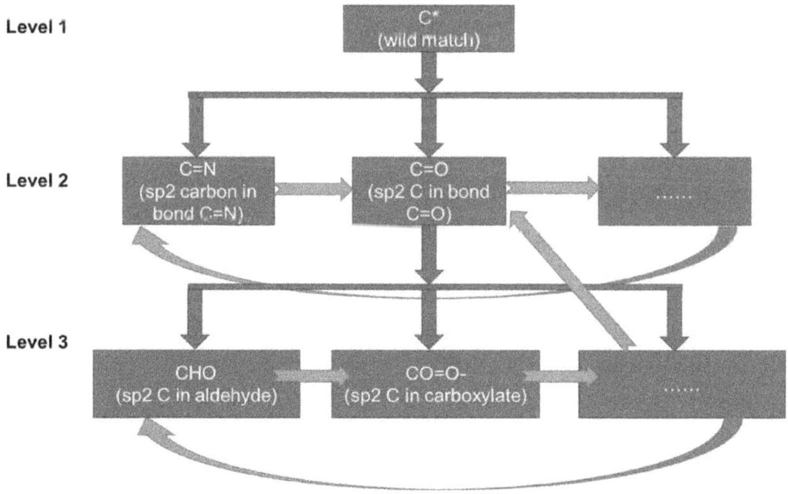

Fig. 6. (Color online) The basic structure of the ranking tree (simplified). The green arrows show the sequence of searching in the same level and reversely to the upper levels.

database, the program will check the position of types A and B in the ranking tree. It undergoes a reversed searching starting from the bottom (most complex), during which the types A and B are sequentially replaced by other types in the same level under the same root. If there is still no matching item, the searching will move to the upper level and continue this operation. The final parameter will be given by averaging the first five parameters found during the reversed searching.

For the common *sp*3 carbon, we have introduced canonical types (*nsp*3 C) based on the number of H atoms attached. They are designated as C30, C31, C32, C33, or C34 (see Table S5). Given a molecule, the detailed atom types of any *sp*3 C atoms will be identified by comparing their local environment (SMARTS) with the database. If there is no complete match, these canonical types will be used to generate parameters. For instance, if we have a molecule CH3–CH (CH3)–CH2–CHO and find no angle parameter of CH–CH2–CHO, then we will collect all the known angle parameters for the structure as *nsp*3 C–*nsp*3 C–CHO in the database to compute the average value and assign its value to the above angle. If still no appropriate parameter is found, we will continue reversed searching on the carbon atoms which are not *sp*3. If high accuracy is required under this method, we can use the frequency fitting to improve all the valence terms except the opbend terms.

3. Results and Discussion

3.1. *Overview of the atomic types and valence parameters*

Currently, we have determined the valence parameters based on about 270 atomic types. All the parameters were derived by both fitting to QM data or the ranking tree. To be specific, we generated 768 bond terms (of which 42 terms from the ranking tree, with each term including k_b and b_0), 1946 angle terms (of which 186 terms from the ranking tree, with each term involving k_a and θ_0), 109 stretch–bend terms (one term involves $k_{ba,AB}$ and $k_{ba,BC}$), and 565 opbend terms (of which 34 from the ranking tree, one term involves k_{opbend}).

The force constants have been determined for different structures by using our version of the modified Seminario program for bond and angle parameters. All the reference bond (b_0) and angle (θ_0) are averaged over the same atomic types. The current atomic types can keep the overall RMSD of the reference bond length less than 0.05 Å and angle degree less than 5° for our training set comparing to the QM optimized structures, which is acceptable in MD simulations.[23] As to cross-terms, our current setting only turns on the stretching–bending coupling effects. The stretch–bend parameters are only distinguished by the atomic type of the central atom, i.e., all the angle structures with the same central atom share the same stretch–bend parameters. Next, the classification of opbend terms only considers the atomic type of the central atom and the side atom, which possibly pop out of the triangle plane. We will analyze this in detail in the following sections.

3.2. *Analysis on parameters of the valence terms*

We examined the force constants (k_b) and equilibrium bond lengths (b_0) between different *sp*3 C and H. It is observed from Table 2 that the *sp*3 C with different numbers of the bonded hydrogens does not show a remarkable change in the bond length of C-H and corresponding force constant, especially the negligible difference among C(H3)–H, C(H2)–H and C(H1)–H. So, the bond type for terminal C and H can be equated with that for middle C and H. For most cases, the local functional groups have minimal effects on C–H because the RMSD of the lengths of all the matching bonds in the database is below 0.03 Å.

Next, we listed the possible pairs of the bonds between the canonical *sp*3 C in Table 2. The bond lengths between *sp*3 C and *sp*3 C with the different number of connected hydrogens are mostly within the range of 1.52–1.53 Å, but it is observed that less number of bonded hydrogens lead to longer bond lengths and smaller force constants. The C(H3)–C(H1) has the largest force constant, but it involves a terminal carbon with 3 H atoms. For those carbons of the interior, this trend is mostly correct, suggesting a strong coupling

Table 2. The comparison of equilibrium bond lengths and corresponding force constants among *sp*3 C and H.[a]

Bond[b]	Equilibrium length (database)[c]	Force constant (database)[c]
C[H4]–H	1.0897 (4) 0.0000	383.57
C[H3]–H	1.0940 (1297) 0.0297	345.98
C[H2]–H	1.0957 (448) 0.0028	340.53
C[H1]–H	1.0980 (52) 0.0034	338.72
C[H3]–C[H3]	1.5255 (1) 0.0000	266.49
C[H3]–C[H2]	1.5225 (153) 0.0065	203.31
C[H3]–C[H1]	1.5240 (82) 0.0052	280.35
C[H3]–C[H0]	1.5306 (16) 0.0044	210.15
C[H2]–C[H2]	1.5236 (30) 0.0046	205.76
C[H2]–C[H1]	1.5231 (9) 0.0112	214.73
C[H2]–C[H0]	1.5352 (3) 0.0094	190.84
C[H1]–C[H1]	1.5420 (1) 0.0000	172.75
C[H0]–C[H0]	1.5697 (1) 0.0000	133.12

Notes: [a]Statistics are reported from all the configurations in our database.
[b]Here, C[Hn] means the *sp*3 C with n hydrogens bonded.
[c]The numbers in the parenthesis represent the number of matching structures in the database, and the underlined number means the RMSD. (Unit: length =Å; force constant = kcal · mol^{-1}·Å^{-1}.)

between bonds and the chemical environment (other bonds and angles) inside the molecule.

As for other kinds of bonds and angles, the analysis is becoming complicated, and no clear trend is observed.

3.3. *Normal mode frequency*

The frequencies of the normal modes for all 445 molecules in the training set have been utilized for determining force constants after a tight structure optimization. The correlation between QM reference and our model is illustrated in Fig. 7(a). The RMSD of the frequency of the normal modes reaches 36.6 cm^{-1} for the whole training set; at the same time, the relative error percentage is around 1.7% (Table 3). By comparison, it was observed that OPLS combined with the modified Seminario method could generate frequencies with the

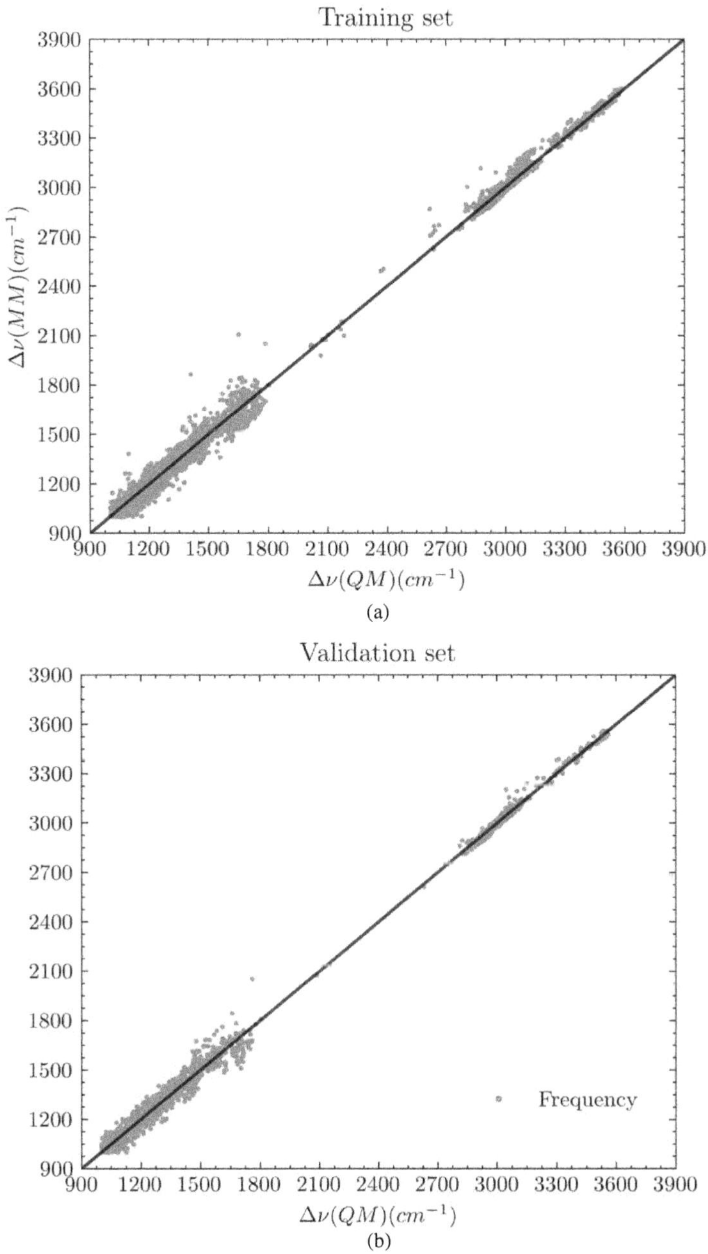

Fig. 7. (Color online) The correlation between the QM and MM normal mode frequencies (cm^{-1}) for the molecules in (a) the training set and (b) the validation set.

Table 3. Statistical comparison of QM scaled frequencies of normal modes and the values calculated by MM model.

	Training set[b]			Validation set[b]		
Statistics[a]	RMSE	MRE	Num.	RMSE	MRE	Num.
Group A	38.4279	1.81%	4846	37.9629	1.82%	1181
Group B	34.2120	1.72%	354	35.2745	1.59%	146
Benzene derivatives	28.5765	1.45%	1433	32.1555	1.48%	982
Heterocyclics	41.0277	1.91%	1122	33.3864	1.71%	541
Bio-fragments	39.0039	1.84%	831	38.6137	1.94%	81
Combination of multiple aromatic cycles	28.8252	1.41%	752	26.0445	1.30%	449
Total	36.6139	1.73%	9338	34.0284	1.62%	3410
	$R^2 = 0.9981$			$R^2 = 0.9983$		

Notes: [a]Group A includes those molecules with less than 10 heavy atoms (nonhydrogen) and Group B means those molecules with $\geq$ 10 heavy atoms. Both Group A and Group B contain those molecules which are not involved in other categories listed in the table.
[b]RMSE: cm^{-1}; MRE: mean relative error $(\frac{1}{N}\sum_{i=1}^{N}\frac{|\nu(QM)-\nu(MM)|}{\nu(QM)} \times 100\%)$. The number of respective unique data points is marked with Num. R^2: correlation coefficient.

relative errors of 7.4% (RMSD: 59.4 cm^{-1}) in general.[31] AMBER-type valence terms combined with a strategy of partial hessian fitting gives out RMSD $\sim$ 81.4 cm^{-1} from a much smaller database,[44] and GAFF would show frequency error mostly larger than 100 cm^{-1}.[1] Meanwhile, these previous works on vibration analysis used small databases limited in the number and the variety of molecules, which in turn demonstrates the reliability of the valence model and parameters derived in this work using a large and diverse set of molecules.

As seen in Fig. 7(a), most valence related vibration frequencies concentrate on two areas (1000–1800 cm^{-1} and 2800–3600 cm^{-1}). The valence terms can capture those modes in the range 2800–3600 cm^{-1}, which are contributed mainly by the fast stretching between heavy atoms and hydrogen such as C–H, N–H, or O–H. These motions occur at the terminal of the molecule, indicating limited coupling effects with the local angle bending. It is a common practice to restrain the H-related high frequency stretching during molecular simulation to allow larger time steps,[45,46] while we have found reversible reference system propagator algorithms (RESPA) is a more effective approach — a small time step (e.g., 0.5 fs) can be used for valence forces and a large time step (e.g., 2 fs) is applied to the other contributions. The situation becomes more complicated when it comes to the range of 1000–1800 cm^{-1}. These involve double bond stretching, angle bending, and the mixture of several other motions. Because these motions directly affect the properties of interest in MD, the use of cross-terms and high-order polynomial models (for anharmonicity) becomes important. The detailed frequency comparisons for several typical molecules have been displayed in Fig. S1, proving the reliability of our current model and parameters.

Some data points in Fig. 7(a) with relatively large deviations arise from carbamate and guanine derivatives (several molecules having such functional groups). In carbamic acid and carbamate ion, we found that the normal mode contributed by the stretching of C=O is mixed with certain local angles bending (O–C–O, N–C–O, and more) lower frequency. The QM gives 1752 cm^{-1} in carbamate ion and 1893 cm^{-1} in carbamic acid for this coupled motion, while our model gives the frequencies inside 2000–2100 cm^{-1}. For guanine, the deviation also happens on the C6=O stretching. It involves the angle bending N1–C6–C5 and both two ring structures show large deformation. Because we only utilized stretch–bend cross-terms in this work, there might be part of coupling effects that cannot be fully described. More coupling terms between different structures (such as those involving torsion) can be included to improve the performance for these complex situations further.

Here, we assessed how much the cross-terms and the opbend terms contribute to the overall performance in Table 4. We re-classify Groups (i)–(ii) into 2 groups: Group A and Group B by the number of heavy atoms in each molecule (Group A: < 10; Group B: ≥ 10). The cross-terms impact the overall performance of vibration analysis, but opbend terms show minimal influence since the previous works about the simple quadratic model for stretching and bending reported the best RMSD up to nearly 60 cm^{-1}. It reveals that anharmonicity plays the most significant role in generating accurate

Table 4. Statistical comparison of QM scaled frequency of normal modes and the values calculated by MM model with or without coupling terms and opbend terms in the training set.

	With coupling/ with opbend[b]		With coupling/ without opbend[b]		Without coupling/ with opbend[b]	
Group name[a]	RMSE	MRE	RMSE	MRE	RMSE	MRE
Group A	38.4279	1.81%	39.0842	1.83%	41.9415	1.95%
Group B	34.2120	1.72%	33.4050	1.69%	33.2874	1.73%
Benzene derivatives	28.5765	1.45%	28.5561	1.45%	39.3057	1.93%
Heterocyclics	41.0277	1.91%	39.8838	1.89%	48.8816	2.38%
Bio-fragments	39.0039	1.84%	38.7749	1.88%	46.6939	2.33%
Combination of multiple cycles	28.8252	1.41%	28.8945	1.42%	42.4678	2.11%
Total	36.6139	1.73%	36.7788	1.75%	42.6315	2.04%

Notes: [a]Group A means those molecules with less than 10 heavy atoms (nonhydrogen) and Group B means those molecules with ≥ 10 heavy atoms. Both Group A and Group B contain those molecules which are not involved in other categories listed in the table.

[b]RMSE: cm^{-1}; MRE: $\frac{1}{N}\sum_{i=1}^{N}\frac{|\nu(QM)-\nu(MM)|}{\nu(QM)} \times 100\%$ has been included in the parenthesis. R^2: correlation coefficient.

frequencies. Besides, the results are improved consistently for all the groups except Group B when stretch–bend cross-terms have been used in this work, indicating that cross-terms can capture more details in structural vibrations. The molecules in Group B mostly include long carbon chain or large carbon ring, which demand more coupling effects considered. Notably, the improvement for those molecules with largely conjugated structures is remarkable: the RMSD for benzene derivatives is decreased by 27.3%, 16.1% for heterocyclics, 16.5% for bio-fragments, and 31.6% for the combinations of multiple aromatic cycles. These molecules possess complex structures, so the motion of interior atoms can easily impact the surrounding bonds and angles. In general, the RMSD of frequencies for the training set goes down by 13.6% by adding cross-terms. Opbend terms have very little effect on the frequencies and made them slightly worse for certain groups. These are likely vibration modes involving the motion in which some *sp*2 atoms popping out of the plane and adding opbend helps keep the correct structure and energy while introducing noise to VF.

3.4. *Potential energy surface for opbend parameters*

Out-of-plane parameters are used to avoid the abnormal pyramidalization of the planar structures with *sp*2 atoms in the center. The results of the current polynomial model with the corresponding parameters of the training set for opbend fitting have been illustrated in Fig. 8. In Fig. 8. Most of the data points are well matched between the model and QM energy within a wide range up to ~90 kcal/mol, while the model collectively underestimates the QM reference values. This suggests that our model can reflect the shape of potential energy surface during pyramidalization with high accuracy, even when the deformation of the structure is very large. The energy RMSD for the training set is 0.8375 kcal/mol for the opbend-small set (maximum QM energy 22.9956 kcal/mol with average 2.3185 kcal/mol), 3.5226 kcal/mol for the opbend-large set (maximum QM energy 91.0101 kcal/mol with average 12.9017 kcal/mol) and 2.5582 kcal/mol for the whole set (Table 5). Although Allinger's model can improve the description of potential energy surface in the out-of-plane bending, the simple model cannot capture the effect fully. As a result, the model energy tends to be smaller than expected from overall observation. Nonetheless, this model separates the out-of-plane motion into three components from each side atoms, which makes the parameters more transferable and flexible in applying in many distinct planar structures.

3.5. *Validation*

To test the accuracy and transferability of the valence parameters derived in this work, we applied these parameters to all the 120 molecules in the validation set. All the parameters for the validation set molecules are automatically assigned or generated via the ranking tree method described in Methodologies. Frequency validation results are shown in Fig. 7(b) and Table 3. The RMSD in frequencies for each group of validation set is mostly consistent with

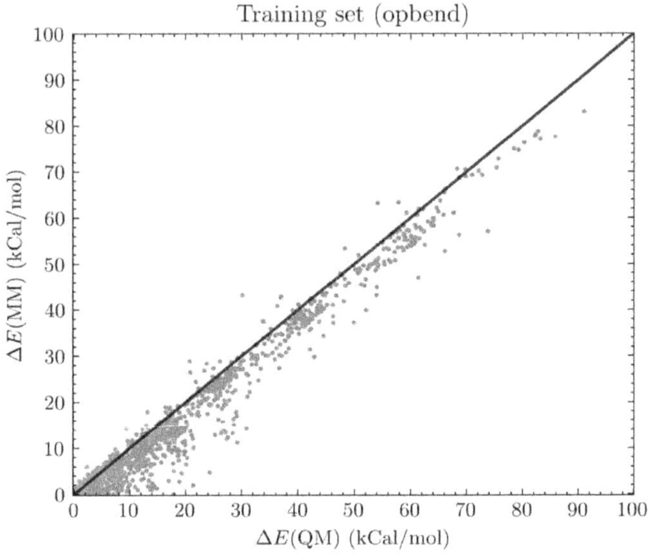

(a)

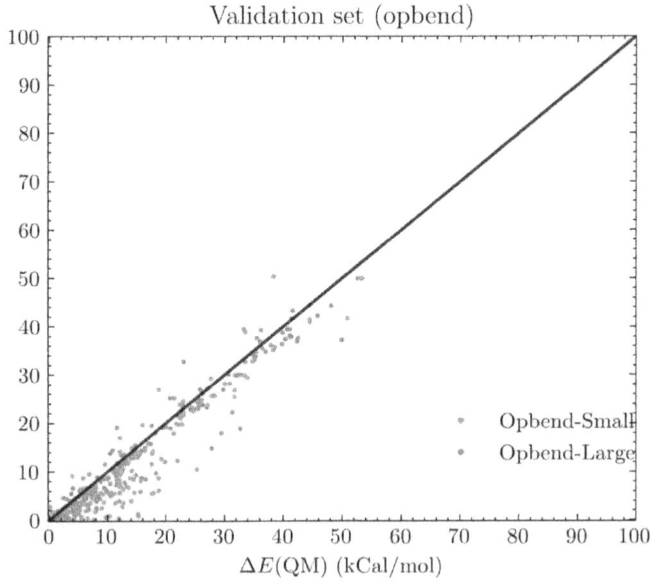

(b)

Fig. 8. (Color online) The correlation between the energy difference calculated by QM and MM for all the configurations in (a) the training set and (b) the validation set. Opbend-small refers to small deformation and opbend-large set refers to large deformation of the specified structures, as described in Sec. 2.3.4.

Table 5. Statistical comparison of QM energy differences and the values calculated by the MM model.

Group name[b]		Training set[c]	Validation set[d]
Opbend-small[a]	RMSE	0.8375	1.3453
	QM_Max	22.9956	17.6173
	QM_Ave	2.3185	3.4736
Opbend-large[a]	RMSE	3.5226	4.1265
	QM_Max	91.0101	53.3861
	QM_Ave	12.9017	15.4029
Total	RMSE	2.5582	2.8082
	R^2	0.9743	0.9471
	QM_Ave	7.6003	8.2367

Notes: [a]Opbend-small and opbend-large groups rely on the scale of deformation of the specified structures (see Sec. 2.2).
[b]QM_Max: maximum of the absolute QM ΔE; QM_Ave: average of the absolute QM ΔE (unit: kcal·mol^{-1}).
[c]R^2: correlation coefficient. The total number of configurations for the training set is 4318.
[d]R^2: correlation coefficient. The total number of configurations for the validation set is 829.

the results from the training set, around $\sim$34 cm^{-1}. In addition, we compared the equilibrium b_0 and θ_0 parameters assigned based on atom type with QM-optimized geometry for the 120 molecules. The overall RMSD is less than 0.05 Å for b_0 and less than 5° for θ_0. For the out-of-plane effects, the performance of the opbend parameters is satisfying as shown in Fig. 8(b) and Table 5. The RMSD between QM and MM energy for all the 829 configurations from 120 molecules is $\sim$2.8 kcal/mol. Overall the validation results demonstrated high accuracy and transferability of our parameters for a wide range of small organic molecules.

4. Conclusion

The classical harmonic equations have been used for decades to calculate the potential energy of bonded interactions by many FFs. While computationally attractive, it shows deficiencies in describing the potential energy surface and VF of molecules compared to quantum calculations. AMOEBA-family FFs apply the MM3(MM4)-style, anharmonic valence interaction model to improve the accuracy of structure and spectroscopic properties. The model employs high-order polynomial functionals in bond stretching and angle bending, the cross-terms that couple different valence motions, and opbend terms. The benefits of these high-order and coupling terms are evidenced in this work. They noticeably improve the accuracy of normal mode frequencies, indicating a strong anharmonicity in bonds and angles. The cross-terms capture more details of the intramolecular motions and significantly improve normal mode frequencies with consistently high accuracy, especially for large structures and conjugation molecules. In addition, our results indicate that opbend terms generate correct potential energy surface around the planar structure for those *sp*2 atoms.

The current parameter set can be applied to a wide range of distinct functional groups, and it is expandable to tackle additional chemistry in the future. The modified Seminario method has been combined with the current atom types of the AMOEBA(+) model, providing reliable initial force constant parameters of bond stretching and angle bending as reference values from QM Hessian. Refinement of bond, angle, and bond-angle coupling explicitly targeting the QM frequency data further increases the quality of these force constant parameters. The opbend parameters are derived by fitting to the potential energy surface of distorted structures composing *sp*2 atoms. It is interesting to examine how good our model can describe the overall geometry of simple compounds comparing to high-level QM or experimental structures. However, this will not be practical until all the other energy components are parameterized, as both the valence and nonbonded interactions are responsible for the accurate prediction of the molecular geometry. An automated program to assign the parameters has also been presented in this work, which is accessible to the public. This program as well as the parameters, will be further implemented in the canonical Tinker[36] and Poltype software packages.[37] The parameters, programs, and additional data are publicly available through https://github.com/prenlab at amoebaplus_data and amoebaplus_parameter repository.

Acknowledgments

This work was supported by the National Institutes of Health (Nos. R01GM106137 and R01GM114237), and National Science Foundation (No. CHE-1856173).

References

1. Dauber-Osguthorpe, P.; Hagler, A. T. Biomolecular Force Fields: Where Have We Been, Where Are We Now, Where Do We Need To Go and How Do We Get There? *J. Comput. Aided Mol. Des.* **2019**, *33* (2), 133–203.
2. Hagler, A. T. Force Field Development Phase II: Relaxation of Physics-Based Criteria... or Inclusion of More Rigorous Physics Into the Representation of Molecular Energetics. *J. Comput. Aided Mol. Des.* **2019**, *33* (2), 205–264.
3. Jing, Z.; Liu, C.; Cheng, S. Y.; Qi, R.; Walker, B. D.; Piquemal, J. P.; Ren, P. Polarizable Force Fields for Biomolecular Simulations: Recent Advances and Applications. *Ann. Rev. Biophys.* **2019**, *48*, 371–394.
4. Gilson, M. K.; Zhou, H. X. Calculation of Protein-Ligand Binding Affinities. *Ann. Rev. Biophys. Biomol. Struct.* **2007**, *36*, 21–42.
5. Gallicchio, E.; Levy, R. M. Recent Theoretical and Computational Advances for Modeling Protein-Ligand Binding Affinities. *Adv. Protein. Chem. Struct. Biol.* **2011**, *85*, 27–80.
6. Harger, M.; Lee, J. H.; Walker, B.; Taliaferro, J. M.; Edupuganti, R.; Dalby, K. N.; Ren, P. Computational Insights into the Binding of IN17 Inhibitors to MELK. *J. Mol. Model* **2019**, *25* (6), 151.
7. Kaoud, T. S.; Johnson, W. H.; Ebelt, N. D.; Piserchio, A.; Zamora-Olivares, D.; Van Ravenstein, S. X.; Pridgen, J. R.; Edupuganti, R.; Sammons, R.; Cano, M.; Warthaka, M.; Harger, M.; Tavares, C. D. J.; Park, J.; Radwan, M. F.; Ren, P.; Anslyn, E. V.; Tsai, K. Y.; Ghose, R.; Dalby, K. N. Modulating Multi-Functional ERK Complexes by Covalent Targeting of a Recruitment Site In Vivo. *Nat. Commun.* **2019**, *10* (1), 5232.
8. Neupane, K.; Manuel, A. P.; Woodside, M. T. Protein Folding Trajectories can be Described Quantitatively by One-Dimensional Diffusion Over Measured Energy Landscapes. *Nat. Phys.* **2016**, *12* (7), 700–703.
9. Lonsdale, R.; Ranaghan, K. E.; Mulholland, A. J. Computational Enzymology. *Chem. Commun.* **2010**, *46* (14), 2354–2372.
10. Qiu, Y.; Smith, D.; Boothroyd, S.; Jang, H.; Wagner, J.; Bannan, C. C.; Gokey, T.; Lim, V. T.; Stern, C.; Rizzi, A.; Lucas, X.; Tjanaka, B.; Shirts, M. R.; Gilson, M.; Chodera, J.; Bayly, C. I.; Mobley, D.; Wang, L.-P. Development and Benchmarking of Open Force Field v1. 0.0, the Parsley Small Molecule Force Field. *J. Chem. Theory Comput.* **2021**, *17* (10), 6262–6280.
11. Maier, J. A.; Martinez, C.; Kasavajhala, K.; Wickstrom, L.; Hauser, K. E.; Simmerling, C. ff14SB: Improving the Accuracy of Protein Side Chain and Backbone Parameters from ff99SB. *J. Chem. Theory Comput.* **2015**, *11* (8), 3696–3713.

12. Debiec, K. T.; Cerutti, D. S.; Baker, L. R.; Gronenborn, A. M.; Case, D. A.; Chong, L. T. Further along the Road Less Traveled: AMBER ff15ipq, an Original Protein Force Field Built on a Self-Consistent Physical Model. *J. Chem. Theory Comput.* **2016**, *12* (8), 3926–3947.
13. Wang, L.-P.; McKiernan, K. A.; Gomes, J.; Beauchamp, K. A.; Head-Gordon, T.; Rice, J. E.; Swope, W. C.; Martínez, T. J.; Pande, V. S. Building a More Predictive Protein Force Field: A Systematic and Reproducible Route to AMBER-FB15. *J. Phys. Chem. B* **2017**, *121* (16), 4023–4039.
14. Salomon-Ferrer, R.; Case, D. A.; Walker, R. C. An overview of the Amber Biomolecular Simulation Package. *WIREs Comput. Mol. Sci.* **2013**, *3* (2), 198–210.
15. Jorgensen, W. L.; Maxwell, D. S.; Tirado-Rives, J. Development and Testing of the OPLS All-Atom Force Field on Conformational Energetics and Properties of Organic Liquids. *J. Am. Chem. Soc.* **1996**, *118* (45), 11225–11236.
16. Kaminski, G. A.; Friesner, R. A.; Tirado-Rives, J.; Jorgensen, W. L. Evaluation and Reparametrization of the OPLS-AA Force Field for Proteins via Comparison with Accurate Quantum Chemical Calculations on Peptides. *J. Phys. Chem. B* **2001**, *105* (28), 6474–6487.
17. Jorgensen, W. L.; Tirado-Rives, J. The OPLS [Optimized Potentials for Liquid Simulations] Potential Functions for Proteins, Energy Minimizations for Crystals of Cyclic Peptides and Crambin. *J. Am. Chem. Soc.* **1988**, *110* (6), 1657–1666.
18. MacKerell, A. D.; Bashford, D.; Bellott, M.; Dunbrack, R. L.; Evanseck, J. D.; Field, M. J.; Fischer, S.; Gao, J.; Guo, H.; Ha, S.; Joseph-McCarthy, D.; Kuchnir, L.; Kuczera, K.; Lau, F. T. K.; Mattos, C.; Michnick, S.; Ngo, T.; Nguyen, D. T.; Prodhom, B.; Reiher, W. E.; Roux, B.; Schlenkrich, M.; Smith, J. C.; Stote, R.; Straub, J.; Watanabe, M.; Wiórkiewicz-Kuczera, J.; Yin, D.; Karplus, M. All-Atom Empirical Potential for Molecular Modeling and Dynamics Studies of Proteins. *J. Phys. Chem. B* **1998**, *102* (18), 3586–3616.
19. Vanommeslaeghe, K.; Hatcher, E.; Acharya, C.; Kundu, S.; Zhong, S.; Shim, J.; Darian, E.; Guvench, O.; Lopes, P.; Vorobyov, I.; Mackerell, Jr., A. D. CHARMM General Force Field: A Force Field for Drug-Like Molecules Compatible with the CHARMM All-Atom Additive Biological Force Fields. *J. Comput. Chem.* **2010**, *31* (4), 671–690.
20. Lopes, P. E.; Guvench, O.; MacKerell, Jr., A. D. Current Status of Protein Force Fields for Molecular Dynamics Simulations. *Methods Mol. Biol.* **2015**, *1215*, 47–71.
21. Ren, P.; Ponder, J. W. Polarizable Atomic Multipole Water Model for Molecular Mechanics Simulation. *J. Phys. Chem. B* **2003**, *107* (24), 5933–5947.
22. Liu, C.; Piquemal, J.-P.; Ren, P. AMOEBA+ Classical Potential for Modeling Molecular Interactions. *J. Chem. Theory Comput.* **2019**, *15* (7), 4122–4139.
23. Yang, X.; Liu, C.; Walker, B. D.; Ren, P. Accurate Description of Molecular Dipole Surface with Charge Flux Implemented for Molecular Mechanics. *J. Chem. Phys.* **2020**, *153* (6), 064103.
24. Allinger, N. L.; Yuh, Y. H.; Lii, J. H. Molecular Mechanics: The MM3 Force Field for Hydrocarbons. 1. *J. Am. Chem. Soc.* **1989**, *111* (23), 8551–8566.

25. Allinger, N. L.; Chen, K.; Lii, J.-H. An Improved Force Field (MM4) for Saturated Hydrocarbons. *J. Comput. Chem.* **1996**, *17* (5–6), 642–668.
26. Maple, J. R.; Dinur, U.; Hagler, A. T. Derivation of Force Fields for Molecular Mechanics and Dynamics from Ab Initio Energy Surfaces. *Proc. Natl. Acad. Sci. USA* **1988**, *85* (15), 5350–5354.
27. Lii, J. H.; Allinger, N. L. Molecular Mechanics: The MM3 Force Field for Hydrocarbons. 2. Vibrational Frequencies and Thermodynamics. *J. Am. Chem. Soc.* **1989**, *111* (23), 8566–8575.
28. Maple, J. R.; Hwang, M. J.; Jalkanen, K. J.; Stockfisch, T. P.; Hagler, A. T. Derivation of Class II Force Fields: V. Quantum Force Field for Amides, Peptides, and Related Compounds. *J. Comput. Chem.* **1998**, *19* (4), 430–458.
29. Palmo, K.; Mirkin, N. G.; Pietila, L. O.; Krimm, S. Spectroscopically Determined Force Fields for Macromolecules. 1. n-Alkane Chains. *Macromolecules* **1993**, *26* (25), 6831–6840.
30. Hagler, A. T. Quantum Derivative Fitting and Biomolecular Force Fields: Functional Form, Coupling Terms, Charge Flux, Nonbond Anharmonicity, and Individual Dihedral Potentials. *J. Chem. Theory Comput.* **2015**, *11* (12), 5555–5572.
31. Allen, A. E. A.; Payne, M. C.; Cole, D. J. Harmonic Force Constants for Molecular Mechanics Force Fields via Hessian Matrix Projection. *J. Chem. Theory Comput.* **2018**, *14* (1), 274–281.
32. Jensen, F. *Introduction to Computational Chemistry.*Wiley, 2017, p. 28.
33. Nevins, N.; Lii, J.-H.; Allinger, N. L. Molecular Mechanics (MM4) Calculations on Conjugated Hydrocarbons. *J. Comput. Chem.* **1996**, *17* (5–6), 695–729.
34. Seminario, J. M., Calculation of Intramolecular Force Fields From Second-Derivative Tensors. *Int. J. Quantum Chem.* **1996**, *60* (7), 1271–1277.
35. Robertson, M. J.; Tirado-Rives, J.; Jorgensen, W. L. Improved Peptide and Protein Torsional Energetics with the OPLS-AA Force Field. *J. Chem. Theory Comput.* **2015**, *11* (7), 3499–3509.
36. Rackers, J. A.; Wang, Z.; Lu, C.; Laury, M. L., Lagardère, L.; Schnieders, M. J.; Piquemal, J.-P.; Ren, P.; Ponder, J. W. Tinker 8: Software Tools for Molecular Design. *J. Chem. Theory Comput.* **2018**, *14* (10), 5273–5289.
37. Wu, J. C.; Chattree, G.; Ren, P. Automation of AMOEBA Polarizable Force Field Parameterization for Small Molecules. *Theor. Chem. Acc.* **2012**, *131* (3), 1138.
38. Frisch, M. J. T. G. W.; Schlegel, H. B.; Scuseria, G. E.; Robb, M. A. C. J. R.; Scalmani, G.; Barone, V.; Mennucci, B.; Petersson, G. A. N., H.; Caricato, M.; Li, X.; Hratchian, H.; P.; Izmaylov, A. F. B., J.; Zheng, G.; Sonnenberg, J. L.; Hada, M.; Ehara, M. T., K.; Fukuda, R.; Hasegawa, J.; Ishida, M.; Nakajima,; T.; Honda, Y. K., O.; Nakai, H.; Vreven, T.; Montgomery, J. A., Jr.; Peralta, J. E. O., F.; Bearpark, M.; Heyd, J. J.; Brothers, E.; Kudin,; K. N.; Staroverov, V. N. K., R.; Normand, J.; Raghavachari, K.; Rendell, A. B., J. C.; Iyengar, S. S.; Tomasi, J.; Cossi, M.; Rega, N.; Millam, J. M. K., M.; Knox, J. E.; Cross, J. B.; Bakken, V.; Adamo, C. J., J.; Gomperts, R.; Stratmann, R. E.; Yazyev, O.; Austin, A. J. C., R.; Pomelli, C.; Ochterski, J. W.; Martin, R. L.; Morokuma, K. Z., V. G.; Voth, G. A.; Salvador, P.; Dannenberg, J. J. D., S.; Daniels, A. D.; Farkas, Ö.;

Foresman, J. B. O., J. V.; Cioslowski, J.; Fox, D. J. Gaussian 09. *Gaussian*, Inc., Wallingford CT, **2009**.

39. Hehre, W. J. *A Guide to Molecular Mechanics and Quantum Chemical Calculations*, Vol. 2, Wavefunction, Irvine, CA, 2003.
40. Johnson, R. D. III. NIST Computational Chemistry Comparison and Benchmark Database, http://cccbdb.nist.gov/.
41. Irikura, K. K.; Johnson, R. D.; Kacker, R. N. Uncertainties in Scaling Factors for Ab Initio Vibrational Frequencies. *J. Phys. Chem. A* **2005**, *109* (37), 8430–8437.
42. Virtanen, P.; Gommers, R.; Oliphant, T. E.; Haberland, M.; Reddy, T.; Cournapeau, D.; Burovski, E.; Peterson, P.; Weckesser, W.; Bright, J.; van der Walt, S. J.; Brett, M.; Wilson, J.; Millman, K. J.; Mayorov, N.; Nelson, A. R. J.; Jones, E.; Kern, R.; Larson, E.; Carey, C. J.; Polat, İ.; Feng, Y.; Moore, E. W.; VanderPlas, J.; Laxalde, D.; Perktold, J.; Cimrman, R.; Henriksen, I.; Quintero, E. A.; Harris, C. R.; Archibald, A. M.; Ribeiro, A. H.; Pedregosa, F.; van Mulbregt, P.; Vijaykumar, A.; Bardelli, A. P.; Rothberg, A.; Hilboll, A.; Kloeckner, A.; Scopatz, A.; Lee, A.; Rokem, A.; Woods, C. N.; Fulton, C.; Masson, C.; Häggström, C.; Fitzgerald, C.; Nicholson, D. A.; Hagen, D. R.; Pasechnik, D. V.; Olivetti, E.; Martin, E.; Wieser, E.; Silva, F.; Lenders, F.; Wilhelm, F.; Young, G.; Price, G. A.; Ingold, G.-L.; Allen, G. E.; Lee, G. R.; Audren, H.; Probst, I.; Dietrich, J. P.; Silterra, J.; Webber, J. T.; Slavič, J.; Nothman, J.; Buchner, J.; Kulick, J.; Schönberger, J. L.; de Miranda Cardoso, J. V.; Reimer, J.; Harrington, J.; Rodríguez, J. L. C.; Nunez-Iglesias, J.; Kuczynski, J.; Tritz, K.; Thoma, M.; Newville, M.; Kümmerer, M.; Bolingbroke, M.; Tartre, M.; Pak, M.; Smith, N. J.; Nowaczyk, N.; Shebanov, N.; Pavlyk, O.; Brodtkorb, P. A.; Lee, P.; McGibbon, R. T.; Feldbauer, R.; Lewis, S.; Tygier, S.; Sievert, S.; Vigna, S.; Peterson, S.; More, S.; Pudlik, T.; Oshima, T.; Pingel, T. J.; Robitaille, T. P.; Spura, T.; Jones, T. R.; Cera, T.; Leslie, T.; Zito, T.; Krauss, T.; Upadhyay, U.; Halchenko, Y. O.; Vázquez-Baeza, Y.; SciPy, C. SciPy 1.0: Fundamental Algorithms for Scientific Computing in Python. *Nat. Methods* **2020**, *17* (3), 261–272.
43. Parrish, R. M.; Burns, L. A.; Smith, D. G. A.; Simmonett, A. C.; DePrince, A. E.; Hohenstein, E. G.; Bozkaya, U.; Sokolov, A. Y.; Di Remigio, R.; Richard, R. M.; Gonthier, J. F.; James, A. M.; McAlexander, H. R.; Kumar, A.; Saitow, M.; Wang, X.; Pritchard, B. P.; Verma, P.; Schaefer, H. F.; Patkowski, K.; King, R. A.; Valeev, E. F.; Evangelista, F. A.; Turney, J. M.; Crawford, T. D.; Sherrill, C. D. Psi4 1.1: An Open-Source Electronic Structure Program Emphasizing Automation, Advanced Libraries, and Interoperability. *J. Chem. Theory Comput.* **2017**, *13* (7), 3185–3197.
44. Wang, R.; Ozhgibesov, M.; Hirao, H. Partial Hessian Fitting for Determining Force Constant Parameters in Molecular Mechanics. *J. Comput. Chem.* **2016**, *37* (26), 2349–2359.
45. Ryckaert, J.-P.; Ciccotti, G.; Berendsen, H. J. C. Numerical Integration of the Cartesian Equations of Motion of a System with Constraints: Molecular Dynamics of n-Alkanes. *J. Comput. Phys.* **1977**, *23* (3), 327–341.
46. Ciccotti, G.; Ryckaert, J. P. Molecular Dynamics Simulation of Rigid Molecules. *Comput. Phys. Rep.* **1986**, *4* (6), 346–392.

https://doi.org/10.1142/9789819808199_0002

Chapter 2

Testing and Optimizing the Drude Polarizable Force Field for Blocked Amino Acids Based on High-Level Quantum-Mechanical Energy Surfaces*

Jinfeng Chen[†,‡,§] and Gerhard König[¶,||]

[†]*Key Laboratory of Structural Biology of Zhejiang Province, School of Life Sciences, Westlake University, Hangzhou, Zhejiang, P. R. China*

[‡]*Westlake Laboratory of Life Sciences and Biomedicine, Hangzhou, Zhejiang, P. R. China*

[§]*Institute of Biology, Westlake Institute for Advanced Study, Hangzhou, Zhejiang, P. R. China*

[¶]*Centre for Enzyme Innovation, School of Biological Sciences, Faculty of Science & Health, University of Portsmouth St. Michael's Building, Portsmouth PO1 2DT, UK*

[||]*gerhard.koenig@port.ac.uk*

The correct reproduction of conformational substates of amino acids was tested for the CHARMM Drude polarizable force field. This was achieved by evaluating the reorganization energies for all low lying energy minima occurring in all 15 neutral blocked amino acids on a quantum-mechanical (QM) energy surface at the MP2/cc-pVDZ level. The results indicate that the bonded parameters of the N-acetyl (ACE) and N-Methylamide (CT3) blocking groups lead to significant discrepancies. A reparametrization of five bond angles significantly improved the agreement with the QM energy surface. The corrected Drude force field exhibits almost the same average

*To cite this article, please refer to its earlier version published in *J. Comput. Biophys. Chem.* **2022**, *21* (4), 405–413, DOI: 10.1142/S2737416521420023
[||]Corresponding author.

reorganization energies relative to the MP2 energy surface as the AM1 and PM3 semi-empirical methods.

Keywords: Polarizable force field; validation; QM; amino acids.

1. Introduction

Proteins are the most important biomolecules both biologically and for industrial applications. The structure and dynamics of proteins can be understood in terms of two factors[1–9]: First, the interactions of the protein with itself, which correspond to the conformational equilibria in the gas phase. Those equilibria are governed by the potential energies of different conformational states. The second kind of driving forces for protein dynamics are the interactions of the protein with its environment, particularly through the hydrophobic effect that is caused by the surrounding water. Those solute–solvent interactions modulate the well depths and positions of the conformational energy minima according to their solvation free energy.[10,11] The solvation free energy is defined as the free energy of transfer from the gas phase to solution.[12–14] To simulate a protein, both the potential energies of the conformational substates in the gas phase and the solute–solvent interactions have to be correctly represented in the model.[15]

Polarizable force fields are particularly suited to gauge the effect of solvation.[16–23] In contrast to the fixed charge force fields, which employ the same charge distribution in the gas phase and in solution, the charge distributions of polarizable force fields can respond to the changes of the environment. The hydration free energies of simple solutes with the Drude polarizable force field[17,24] were previously compared to the ones from the fixed charge CHARMM36 force field,[25] exhibiting a 40% reduction of the root mean square error from the experimental reference results.[26] Quantum-mechanical (QM) and hybrid quantum-mechanics/molecular mechanics (QM/MM) methods are also particularly suited to determine solvation free energies if they are combined with a suitable representation of the solvent environment.[27–33] However, the computational expenses of simulations with QM methods are significantly higher than with polarizable force fields, which makes the use of force fields much more attractive for most practical applications.

This leaves the question in how far polarizable force fields are able to reproduce conformational equilibria in the gas phase. To this end, high-level quantum-mechanical calculations can be used as accurate reference results to test the performance of force fields. To evaluate the correct reproduction of the conformational substates of proteins, it is possible to resort to the YMPJ database.[34,35] This database contains the low-energy conformers of all amino acids at the MP2/cc-pVDZ[36,37] level of theory. It includes 11 low-energy structures for Ala, 12 for Asn, 21 for Gln, 9 for Gly, 28 for Leu, 25 for Ile, 30 for Phe, 5 for Pro, 26 for Ser, 17 for Thr, 26 for Trp, 17 for Tyr and 15 structures for Val. This makes in total 242 different structures of the energy minima of conformational substates. For these structures, the differences between the Drude polarizable force field and QM at the MP2/cc-pVDZ level of theory are evaluated.

The thermodynamically relevant measure for the agreement between two energy surfaces is the phase space overlap.[38] The phase space overlap is proportional to the required sampling on one energy surface to cover all relevant low-energy regions of the other energy surface, which also drives the convergence of free energy calculations.[39] For harmonic systems, like a single energy well, the phase space overlap is proportional to the reorganization energy ($\Delta U_{\mathrm{reorg}}$).[39] To determine the reorganization energy for an MM force field relative to a QM energy surface in one conformational substate, one can simply perform an MM energy minimization starting from the structure of the QM energy minimum and measure how much the potential energy drops as the MM energy minimum is reached (Fig. 1). Thus, the reorganization energy measures how much energy has to be invested in an MM simulation to visit the correct QM energy minimum. This principle was recently employed to evaluate the quality of several fixed charged force fields.[40]

Preliminary calculations of the reorganization energies for blocked serine using the Drude polarizable force field indicated high deviations between the MM and QM energy surfaces.[40] Those discrepancies could be addressed by automatically adapting the reference bond

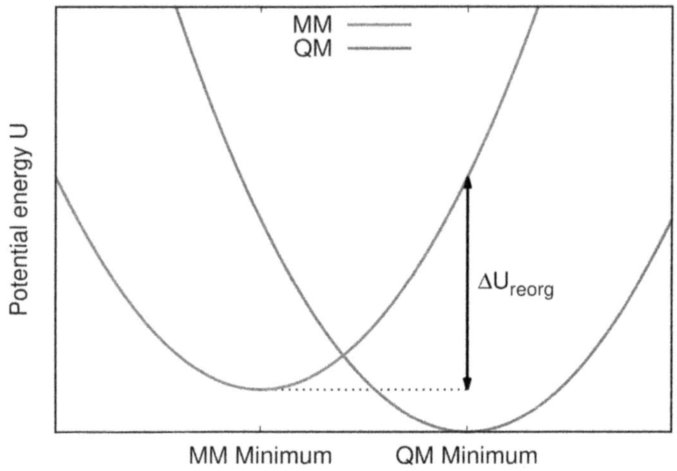

Fig. 1. The reorganization energy $\Delta U_{\mathrm{reorg}}$ is a measure for the energy that has to be invested to sample the QM energy minimum within the MM energy surface when starting from the MM energy minimum. The reaction coordinate on the x-axis and the potential energy on the y-axis are in arbitrary units.

lengths and reference bond angles of the force field, which indicates that at least one of those types of energy terms is responsible for the observed deviations from the QM energy surface. Here, the discrepancies are systematically assessed for all 15 neutral amino acids and then corrected by reparametrizing the degrees of freedom with the highest deviations between the QM and the MM optimal structures. The average reorganization energies for all 15 amino acids are then evaluated with the new parameters and compared to the outcome with the original parameters. Finally, the performance of the new parameters is compared with the reorganization energies from the AM1[41] and PM3[42] semi-empirical methods. Semi-empirical approaches represent the next higher level of theory with increased computational demands[43–46] and, therefore, serve as a useful reference for the relative merits of polarizable force fields.

2. Theory

The main criterion for the similarity of two energy surfaces is the phase space overlap.[38] Formally, the phase space overlap between an

MM energy surface and a QM energy surface can be determined by exhaustively sampling every point of both phase spaces, and evaluating the potential energies and probabilities of each conformation for both MM and QM. However, this procedure is extremely computationally expensive. If the phase space overlap only has to be evaluated for one particular conformational substate, it is possible to resort to linear response theory[47–55] and Marcus theory[56] to estimate the expected phase space overlap via the reorganization energy ($\Delta U_{\mathrm{reorg}}$).[39,40] As shown previously,[39] the expected variance of a free energy estimate between the MM and the QM energy surface with the Zwanzig equation[57] (also known as free energy perturbation or exponential formula) is then given by

$$\sigma^2_{\Delta G} \approx \frac{e^{\frac{2}{kT}\Delta U_{\mathrm{reorg}}} - 1}{n}, \tag{1}$$

where k is the Boltzmann constant, T is the temperature and n represents the number of independent potential energy difference samples.[39,58,59] If multiple low-lying energy mimima are present in the system, Eq. (1) has to be evaluated for all relevant conformational substates. If all conformational substates are equally important for the comparison, i.e., if one is interested in the overall similarity of the two phase spaces, then the relevant metric for the similarity is the average reorganization energy over all conformational substates ($|\Delta U_{\mathrm{reorg}}|$). To estimate the similarity between the MM and QM energy surfaces in proteins we resort to the overall performance for all amino acid types, which is given by the average over all $|\Delta U_{\mathrm{reorg}}|$ values over all considered amino acids (denoted as $||\Delta U_{\mathrm{reorg}}||_{aa}$).

3. Methods

3.1. *Determination of the reorganization energies*

All molecular mechanics calculations were carried out with the CHARMM simulation package,[60,61] using the Drude polarizable force field.[17,24] For the determination of the average reorganization energies $|\Delta U_{\mathrm{reorg}}|$, all energy minima of the N-acetyl-X-methylamide versions of the 15 neutral amino acids alanine (Ala), asparagine (Asn), cysteine (Cys), glutamine (Gln), glycine (Gly), isoleucine (Ile), leucine (Leu), methionine (Met), phenylalanine (Phe), proline (Pro), serine (Ser), threonine (Thr), tryptophan (Trp), tyrosine (Tyr) and valine (Val) were evaluated. The structures of the energy minima at the MP2/cc-pVDZ level of theory[36,37] were obtained from the work of Kesharwani *et al.*[34] The same protocol as in previous calculations[40] of ΔU_{reorg} was used for the Drude polarizable force field. The only exception is the addition of the amino acid Trp to the benchmark, which was omitted in the previous calculations. All protonizable amino acids were excluded from the benchmark to avoid the introduction of net charges in the system, which might bias the results. The gas phase energy minimizations used 100 steps of steepest descent, followed by up to 10,000 steps of Adopted Basis Newton–Raphson (ABNR) energy minimization. To verify that the reorganization energies are converged, the minimizations were repeated with the conjugate gradient method. This only leads to insignificant changes of the results (on average about 0.05%). The nonbonded interactions employed a switching function between 10 Å and 12 Å. For the MM calculations, the contributions of the individual energy terms to the reorganization energy were determined by evaluating the potential energy difference of each energy term between the MM and the QM optimized structure. The energy minimizations with the AM1[41] and PM3[42] semi-empirical methods were performed with the MNDO package, using the default settings.[62]

3.2. *Reparametrization of the bond angles*

As will be discussed in the next section, the deviations of $\|\Delta U_{\text{reorg}}\|_{aa}$ mainly arise from the bond angle terms of the common backbone. Therefore, the simple alanine dipeptide (ALAD) system was used to perform the reparametrization. The optimized geometry of ALAD at the molecular mechanics level was generated with 100 steps of steepest descent, followed by 10,000 steps of ABNR energy minimization using the lowest energy QM geometry at the MP2/cc-pVTZ level of theory as the initial structure. Thus, different basis sets are used for the reparametrization (cc-pVTZ) and the reorganization energies from the reference YMPJ database (cc-pVDZ), but this is necessary for consistency with the CHARMM parametrization strategy. To conduct the potential energy surface scan, the geometries of every QM scan were extracted with a fixed scanning angle, followed by the same MM optimization procedure as above. The vibrational frequency analysis was performed using MOLVIB module of CHARMM. All quantum-mechanical scan calculations were carried out using the Gaussian16 program.[63] The relaxed potential energy scan and the vibrational frequency analysis were performed at the MP2/cc-pVTZ level of theory.[36,37]

4. Results

4.1. *Reorganization energies with the original parameters*

The left side of Table 1 lists the total average of the average reorganization energies $||\Delta U_{\mathrm{reorg}}||_{aa}$ for the original parameters of the Drude polarizable force field for proteins,[17,24] considering all energy minima of the 15 neutral blocked amino acids. The new parameters are listed in Tables S1 and S2 of the SI. In preliminary calculations,[40] this force field exhibited high reorganization energies (above 100 kJ mol^{-1})

Table 1. Total average of the average reorganization energies for the 15 neutral blocked amino acids ($||\Delta U_{\mathrm{reorg}}||_{aa}$) with the original (old) and the new parameters for the Drude polarizable force field, as well as the average contributions from different MM energy terms.

Old parameters		New parameters	
$\Vert\Delta U_{\mathrm{reorg}}\Vert_{aa}$	31.5 kcal mol^{-1}	$\Vert\Delta U_{\mathrm{reorg}}\Vert_{aa}$	13.1 kcal mol^{-1}
Bond	5 %	Bond	33 %
Angle	66 %	Angle	7 %
Dihedral	4 %	Dihedral	20 %
Improper	5 %	Improper	1 %
CMAP	1 %	CMAP	2 %
vdW	15 %	vdW	31 %
Elec	4 %	Elec	8 %

for blocked serine relative to the energy surfaces of both MP2 and OM2[64,65] semi-empirical methods. The high reorganization energies relative to OM2 was interpreted as an indicator that semi-empirical calculations of ΔU_{reorg} can help to detect errors of the force field. A decomposition of the reorganization energies in regard to the individual force field terms revealed that the main contribution of this error arises from the bond angle terms. An automatic procedure[26] to adjust the bond length and bond angle parameters based on the quantum-mechanical reference structure with the lowest energy was able to reduce the average reorganization energy by 69%. This further indicates that the main discrepancies between the polarizable force field and the reference QM energy surface arise from the bonded terms.

To verify that this problem is not restricted to serine alone, the average reorganization energies and the individual contributions from the force field terms were determined. While a decomposition of the potential energy into contributions is only rigorous if there is no coupling between the energy terms,[66] this analysis can still help to guide the interpretation of the reorganization energies.[67,68] The results for the original parametrization are listed on the left side of Table 1. The total average reorganization energy of 31.5 kcal mol^{-1} (131.8 kJ mol^{-1}) shows that the discrepancies are high for all amino acids, which indicates that the main source of error lies in the common protein backbone or the N-acetyl (ACE) and N-Methylamide (CT3) blocking groups.

Table 1 also provides the average contributions of the individual force field energy terms, such as bond stretching (Bond), angle bending (Angle), dihedral angles (Dihedral), improper dihedrals (Improper), the cross term correction potential for the backbone dihedrals (CMAP), the van der Waals interactions (vdW) and electrostatic interactions (Elec). Strikingly, 66% of the observed discrepancies can be attributed to bond angle terms. Therefore, the reparametrization strategy aimed mainly at the bond angle terms of the backbone and the blocking groups. Because the reparametrization does not affect specific residues but the common backbone, the simple alanine dipeptide system (ALAD) is used for reference QM calculations at the MP2/cc-pVTZ level to derive the new parameters.

The different basis set is necessary for consistency with the CHARMM parameterization strategy.

4.2. *Reparametrization*

The main targets of the reparametrization were identified based on the differences of the bond angle values between the QM global energy minimum at the MP2/cc-pVTZ level and the polarizable force field optimal geometry of ALAD. The naming scheme for the atoms in ALAD can be found in Fig. 2. The interface between the protein backbone and the terminal blocking groups exhibited the largest discrepancies, particularly the angles formed by CAY, CY, OY and N. The optimal values of these angles at the MP2/cc-pVTZ level of theory and the Drude polarizable force field are listed in Table 2. Importantly, these discrepancies only arise when using blocking groups together with amino acids, so the present reparametrization does not affect the polypeptide chains in proteins.

The differences between the QM lowest energy structure at the MP2/cc-pVTZ level of theory and the original parameters of the Drude polarizable force field are 1.5°, 8.5°, 3.5°, 5° and 11° for the angles OY–CY–N, OY–CY–CAY, C–NT–CAT, N–CY–CAY and NT–CAY–HT1, respectively. Combining those discrepancies with the corresponding force constants of the bond angles leads to reorganization energies for the individual terms between 0.08 kcal mol^{-1} and 3.89 kcal mol^{-1}, as listed on the right side of Table 2. In particular, the angles N–CY–CAY and NT–CAT–HT1 exhibit high deviations from the QM energy minimum.

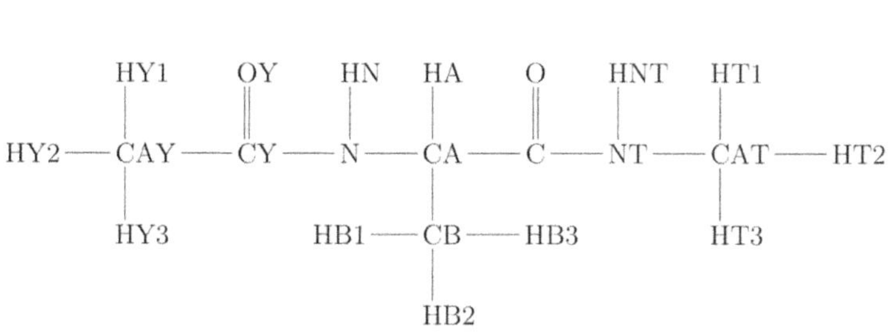

Fig. 2. Atom names of ALAD.

Table 2. Comparison of the optimal angle values for the lowest energy structures at the MP2/cc-pVTZ level of theory and with the original and new force field parameters.

	Optimal angle [°]			$\Delta U_{\mathrm{reorg}}$[kcal mol^{-1}]	
Angle	MP2	Old MM	New MM	$\Delta U^{\mathrm{old}}_{\mathrm{reorg}}$	$\Delta U^{\mathrm{new}}_{\mathrm{reorg}}$
OY-CY-N	122.5	124.0	121.5	0.08	0.03
OY-CY-CAY	122.5	114.0	121.5	3.53	0.08
C-NT-CAT	119.5	123.0	120.5	0.52	0.04
N-CY-CAY	115.0	120.0	116.5	2.38	0.18
NT-CAT-HT1	109.0	120.0	110.0	3.89	0.02

The new force field parameters are generated by fitting to a scan of the corresponding degrees of freedom. The corresponding results of the new parameters are shown in Table 2. The differences between the new reference angle values and the optimal geometry at the MP2 level of theory are all below 1.5°. The corresponding reorganization energies of the new parameters for those degrees of freedom range between 0.02 kcal mol^{-1} and 0.18 kcal mol^{-1}. The low reorganization energies signify that the phase space overlap between the new parameters and the QM energy surface of the MP2/cc-pVTZ level of theory is significantly higher. The underlying potential energy scans can be found in the Supporting Information.

4.3. *Comparison of original and new parameters*

The total average of the average reorganization energies over all amino acids and the contributions of the individual force field terms for the new parameters are listed on the right side of Table 1. By adjusting five bond angle parameters of the Drude polarizable force field, the average reorganization energy drops by 58% from 31.5 kcal mol^{-1} to 13.1 kcal mol^{-1}. The contribution from the bond angle energy terms to the total average of the reorganization energies is merely 7% with the new parameters. Instead, the dominant sources of discrepancies between the polarizable force field and the quantum mechanical energy surface are the bond lengths (33%), van der Waals parameters (31%) and dihedral angles (33%). Those contributions are similar to the ones that were observed for the CHARMM22[69] and CHARMM36[25] force fields in a previous publication.[40] However, due to the extra degrees of freedom from the Drude particles to incorporate the effects of polarization, the reorganization energies and the contributions are not directly comparable.

Notably, all deviations of the original parameters are related to the peptide bonds. As shown in Table 2, the largest deviations are observed for the angles of NT–CAT–HT1, OY–CY–CAY and N–CY–CAY, respectively. The atoms CY and CAT are bonded to the C_α atom (denoted as CA in the CHARMM force field) of the neighboring residues in proteins. The original parameters mostly reflect the situation in a protein environment. However, in this work, these atoms are bonded to the terminal blocking groups, rather than to other amino acids. The extrapolation of the parameters that were developed for a protein backbone to terminal blocking groups leads to the observed discrepancies. Because there are only two termini in proteins, but many more normal peptide bonds between amino acids, the original parameters are optimized for the latter type. Vice versa, the new parameters for the termini will not significantly affect protein dynamics. However, when studying small peptides, like the ones in the YMPJ database, the original parameters might lead to biased results, especially in long molecular dynamics simulations. Thus, special parameters for the terminal peptide bonds are necessary.

4.4. *Comparison with semi-empirical methods*

To provide a comparison of the new parameters of the Drude polarizable force field to other methods that include polarization effects, we resort to the semi-empirical quantum-mechanical methods AM1[41] and PM3.[42] Since the potential energies of the semi-empirical methods are not composed from individual energy terms like a force field, it is only possible to compare the average reorganization energies for each amino acid type, as listed in Table 3. The last column shows the average reorganization energies of the

Table 3. Average reorganization energies $|\Delta U_{\mathrm{reorg}}|$ for the 15 neutral blocked amino acids with the semi-empirical methods AM1 and PM3, as well as the original and new force field parameters (MM). The last row gives the total average of the reorganization energies over all amino acids. All values are in kcal mol^{-1}.

	AM1	PM3	Old MM	New MM
Ala	10.6	11.1	28.2	10.4
Asn	13.3	14.7	33.2	13.6
Cys	11.7	12.3	30.6	12.0
Gln	14.5	16.6	36.6	16.3
Gly	9.4	10.3	27.5	10.9
Ile	13.3	12.4	29.9	11.9
Leu	13.8	12.8	29.6	12.0
Met	14.3	13.3	31.0	12.2
Phe	12.4	12.6	31.7	13.1
Pro	16.1	14.2	29.3	15.5
Ser	12.4	11.6	30.1	12.1
Thr	13.1	13.4	33.9	15.3
Trp	15.0	15.5	33.4	15.4
Tyr	13.1	12.8	32.9	14.3
Val	12.4	11.6	30.5	12.2
$\|\|\Delta U_{\mathrm{reorg}}\|\|_{aa}$	**13.0**	**13.0**	**31.5**	**13.1**

Drude polarizable force field with the original (Old MM) and new parameters (New MM). The last row gives the total average of the reorganization energies over all amino acids. Surprisingly, although the average reorganization energies for the individual amino acids vary among the methods, the resulting total averages of the semi-empirical methods and the new Drude force field parameters are very similar with 13.0 kcal mol^{-1}, 13.0 kcal mol^{-1}, and 13.1 kcal mol^{-1}. This indicates that those three methods reproduce the QM energy surface at the MP2/cc-pVDZ level of theory equally well.

5. Conclusion

Polarizable force fields are known to be superior to conventional fixed charge force fields when it comes to determining solute–solvent interactions. However, to evaluate whether polarizable force fields are also able to correctly capture protein dynamics, it also needs to be shown that they correctly reproduce the conformational equilibria in the gas phase. Here, this is achieved by determining the phase space overlap of the Drude polarizable force field with QM-generated conformational substates from the YMPJ database.[34,35] The previously reported weaknesses of the bonded terms of N-acetyl-X-methylamide amino acids in the CHARMM polarizable force field[40] were corrected by adapting the bond angles at the interface between the protein backbone and the terminal blocking groups. While the original parameters exhibited a total average reorganization energy $||\Delta U_{\text{reorg}}||_{aa}$ of 31.5 kcal mol^{-1} relative to the QM reference structures at the MP2/cc-pVDZ level, the new parameters only exhibit a total average reorganization energy of 13.1 kcal mol^{-1} over all neutral amino acids. Thus, the agreement between the QM energy surface and the Drude polarizable force field has been greatly improved.

Some challenges remain in terms of the bond length parameters (which contribute 33% to the average reorganization energy), as well as the van der Waals parameters (31%) and dihedral angles (20%). However, similar discrepancies were observed previously for the CHARMM22 and CHARMM36 force fields,[40] which indicates that those deviations arise from the underlying parametrization strategy of the CHARMM force field. Most of the deviations of the bond length parameters are on the order of picometers,[40] but due to the stiffness of the bonded terms such small deviations can lead to sizeable potential energy differences. However, small deviations of the bond length can be easily addressed in multiscale free energy simulations by automatically updating the bond parameters to create a tailored force field for the QM target energy surface,[26,70,71] or by calculating corrections during post-processing.[72]

The average reorganization energies with respect to quantum-mechanical calculations serve as a useful indicator for the quality of the force field. To put the average reorganization energies of the Drude polarizable force field into perspective, the same test was also performed for the AM1 and PM3 semi-empirical methods. The semi-empirical methods represent the next higher level of theory in terms of rigour. In contrast to conventional fixed-charge force fields, quantum-mechanical semi-empirical methods are able to incorporate polarization effects, which makes them a suitable reference for polarizable force fields in terms of performance. The total average reorganization energies over all energy minima of all considered 15 amino acids are 13.0 kcal mol^{-1} for both the AM1 and PM3 methods. Thus, the accuracy of the new parameters of the Drude polarizable force field (13.1 kcal mol^{-1}) is comparable to the semi-empirical methods. Notably, the polarizable force field based on Drude particles exhibits lower computational demands and a better scalability than the semi-empirical methods. This makes the Drude polarizable force field an interesting option for biomolecular simulations, particularly for proteins.

Acknowledgments

GK gratefully acknowledges the financial support by the Research England for E3 funding. The authors also thank Jing Huang for his help with the reparametrization of the polarizable force field.

References

1. Kauzmann, W. Some Factors in the Interpretation of Protein Denaturation. *Adv. Protein Chem.* **1959**, *14*, 1.
2. Makhtadze, G. I.; Privalov, P. L. Contribution of Hydration to Protein Folding Thermodynamics the Enthalpy of Hydration. *J. Mol. Biol.* **1993**, *232*, 639.
3. Privalov, P. L.; Makhtadze, G. I. Contribution of Hydration to Protein Folding Thermodynamics the Entropy and Gibbs Energy of Hydration. *J. Mol. Biol.* **1993**, *232*, 660.
4. Dyson, H. J.; Wright, P. E.; Scheraga, H. A. The Role of Hydrophobic Interactions in Initiation and Propagation of Protein Folding. *Proc. Natl. Acad. Sci. U. S. A.* **2006**, *103*, 13057–13061.
5. Langhorst, U.; Backmann, J.; Loris, R.; Steyaert, J. Analysis of Water Mediated Protein–Protein Interactions within RNase T1. *Biochemistry* **2000**, *39*, 6586.
6. Kokubo, H.; Hu, C. Y.; Pettitt, B. M. Peptide Conformational Preferences in Osmolyte Solutions: Transfer Free Energies of Decaalanine. *J. Am. Chem. Soc.* **2011**, *133*, 1849–1858.
7. Tarek, M.; Tobias, D. Environmental Dependence of the Dynamics of Protein Hydration Water. *J. Am. Chem. Soc.* **1999**, *121*, 9740.

8. Makarov, V.; Pettitt, B.; Feig, M. Solvation and Hydration of Proteins and Nucleic Acids: A Theoretical View of Simulation and Experiment. *Acc. Chem. Res.* **2002**, *35*, 376–384.
9. Fulle, S.; Gohlke, H. Molecular Recognition of RNA: Challenges for Modelling Interactions and Plasticity. *J. Mol. Recognit.* **2010**, *23*, 220–231.
10. König, G.; Boresch, S. Hydration Free Energies of Amino Acids: Why Side Chain Analog Data Are Not Enough. *J. Phys. Chem. B* **2009**, *113*, 8967–8974.
11. König, G.; Bruckner, S.; Boresch, S. Absolute Hydration Free Energies of Blocked Amino Acids: Implications for Protein Solvation and Stability. *Biophys. J.* **2013**, *104*, 453–462.
12. Ben-Naim, A. Standard Thermodynamics of Transfer. uses and Misuses. *J. Phys. Chem.* **1978**, *82*, 792–803.
13. Ben-Naim, A.; Marcus, Y. J. Solvation Thermodynamics of Nonionic Solutes. *J. Chem. Phys.* **1984**, *81*, 2016–2027.
14. Wilhelm, E. Fugacities and All That in Solution Chemistry. *J. Solution Chem.* **2015**, *44*, 1004–1061.
15. van Gunsteren, W. F.; Daura, X.; Hansen, N.; Mark, A. E.; Oostenbrink, C.; Riniker, S.; Smith, L. J. Validation of Molecular Simulation: An Overview of Issues. *Angew. Chem. Int. Ed.* **2018**, *57*, 884–902.
16. Inakollu, V. S.; Geerke, D. P.; Rowley, C. N.; Yu, H. Polarisable Force Fields: What Do They Add in Biomolecular Simulations? *Curr. Opin. Struct. Biol.* **2020**, *61*, 182–190.
17. Lemkul, J. A.; Huang, J.; Roux, B.; MacKerell Jr, A. D. An Empirical Polarizable Force Field Based on the Classical Drude Oscillator Model: Development History and Recent Applications. *Chem. Rev.* **2016**, *116*, 4983–5013.
18. Bradshaw, R. T.; Essex, J. W. Evaluating Parametrization Protocols for Hydration Free Energy Calculations with the Amoeba Polarizable Force Field. *J. Chem. Theory Comput.* **2016**, *12*, 3871–3883.
19. Visscher, K. M.; Vosmeer, C. R.; Luirink, R. A.; Geerke, D. P. A Systematic Approach to Calibrate a Transferable Polarizable Force Field Parameter Set for Primary Alcohols. *J. Comput. Chem.* **2017**, *38*, 508–517.
20. Huang, J.; Simmonett, A. C.; Pickard, F. C.; MacKerell Jr, A. D.; Brooks, B. R. Mapping the Drude Polarizable Force Field onto a Multipole and Induced Dipole Model. *J. Chem. Phys.* **2017**, *147*, 161702.
21. Belsare, S.; Esser, A.; Marx, D.; Head-Gordon, T. Studying Solvation of Small Biomolecules Via Molecular Dynamics Using a Polarizable Force Field. *Biophys. J.* **2017**, *112*, 497a.
22. Bedrov, D.; Piquemal, J.-P.; Borodin, O.; MacKerell Jr, A. D.; Roux, B.; Schröder, C. Molecular Dynamics Simulations of Ionic Liquids and Electrolytes Using Polarizable Force Fields. *Chem. Rev.* **2019**, *119*, 7940–7995.
23. Lin, F.-Y.; Lopes, P. E.; Harder, E.; Roux, B.; MacKerell Jr, A. D. Polarizable Force Field for Molecular Ions Based on the Classical Drude Oscillator. *J. Chem. Inf. Model.* **2018**, *58*, 993–1004.
24. Lopes, P. E.; Huang, J.; Shim, J.; Luo, Y.; Li, H.; Roux, B.; MacKerell Jr, A. D. Polarizable Force Field for Peptides and Proteins Based on the Classical Drude Oscillator. *J. Chem. Theory Comput.* **2013**, *9*, 5430–5449.

25. Best, R. B.; Zhu, X.; Shim, J.; Lopes, P. E. M.; Mittal, J.; Feig, M.; MacKerell Jr, A. D. Optimization of the Additive CHARMM All-Atom Protein Force Field Targeting Improved Sampling of the Backbone phi, psi and Side-Chain chi(1) and chi(2) Dihedral Angles. *J. Chem. Theory Comput.* **2012**, *8*, 3257–3273.
26. König, G.; Pickard, F.; Huang, J.; Thiel, W.; MacKerell, A.; Brooks, B.; York, D. A Comparison of QM/MM Simulations with and without the Drude Oscillator Model Based on Hydration Free Energies of Simple Solutes. *Molecules* **2018**, *23*, 2695.
27. Pickard, F. C.; König, G.; Simmonett, A. C.; Shao, Y.; Brooks, B. R. An Efficient Protocol for Obtaining Accurate Hydration Free Energies Using Quantum Chemistry and Reweighting From Molecular Dynamics Simulations. *Bioorg. Med. Chem.* **2016**, *24*, 4988–4997.
28. König, G.; Pickard, F. C.; Mei, Y.; Brooks, B. R. Predicting Hydration Free Energies with a Hybrid QM/MM Approach: An Evaluation of Implicit and Explicit Solvation Models in SAMPL4. *J. Comput.-Aid. Mol. Design* **2014**, *28*, 245–257.
29. Huang, J.; Mei, Y.; König, G.; Simmonett, A. C.; Pickard IV, F. C.; Wu, Q.; Wang, L.-P.; MacKerell Jr, A. D.; Brooks, B. R.; Shao, Y. An Estimation of Hybrid Quantum Mechanical Molecular Mechanical Polarization Energies for Small Molecules Using Polarizable Force-field Approaches. *J. Chem. Theory Comput.* **2017**, *13*, 679–695.
30. Jia, X.; Wang, M.; Shao, Y.; König, G.; Brooks, B. R.; Zhang, J. Z. H.; Mei, Y. Calculations of Solvation Free Energy through Energy Reweighting from Molecular Mechanics to Quantum Mechanics. *J. Chem. Theory Comput.* **2016**, *12*, 499–511.
31. König, G.; Mei, Y.; Pickard, F. C.; Simmonett, A. C.; Miller, B. T.; Herbert, J. M.; Woodcock, H. L.; Brooks, B. R.; Shao, Y. Computation of Hydration Free Energies Using the Multiple Environment Single System Quantum Mechanical/Molecular Mechanical Method. *J. Chem. Theory Comput.* **2016**, *12*, 332–344.
32. Pickard, F. C.; König, G.; Tofoleanu, F.; Lee, J.; Simmonett, A. C.; Shao, Y.; Ponder, J. W.; Brooks, B. R. Blind Prediction of Distribution in the SAMPL5 Challenge with QM based Protomer and pK_a Corrections. *J. Comput.-Aided Mol. Des.* **2016**, *30*, 1087–1100.
33. König, G.; Pickard, F. C.; Huang, J.; Simmonett, A. C.; Tofoleanu, F.; Lee, J.; Dral, P. O.; Prasad, S.; Jones, M.; Shao, Y.; Thiel, W.; Brooks, B. R. Calculating Distribution Coefficients Based on Multi-scale Free Energy Simulations: An Evaluation of MM and QM/MM Explicit Solvent Simulations of Water-cyclohexane Transfer in the SAMPL5 Challenge. *J. Comput.-Aided Mol. Des.* **2016**, *30*, 989–1006.
34. Kesharwani, M. K.; Karton, A.; Martin, J. M. Benchmark ab initio Conformational Energies for the Proteinogenic Amino Acids Through Explicitly Correlated Methods. Assessment of Density Functional Methods. *J. Chem. Theory Comput.* **2015**, *12*, 444–454.

35. Yuan, Y.; Mills, M. J.; Popelier, P. L.; Jensen, F. Comprehensive Analysis of Energy Minima of the 20 Natural Amino Acids. *J. Phys. Chem. A* **2014**, *118*, 7876–7891.
36. Møller, C.; Plesset, M. S. Note on an Approximation Treatment for Many-Electron Systems. *Phys. Rev.* **1934**, *46*, 618–622.
37. Dunning, T. H., Jr. Gaussian basis sets for use in Correlated Molecular Calculations. I. The Atoms Boron through neon and Hydrogen. *J. Chem. Phys.* **1989**, *90*, 1007–1023.
38. Bennett, C. H. Efficient Estimation of Free Energy Differences from Monte Carlo Data. *J. Comput. Phys.* **1976**, *22*, 245–268.
39. König, G.; Brooks, B. R.; Thiel, W.; York, D. M. On the Convergence of Multi-scale Free Energy Simulations. *Mol. Simulat.* **2018**, *44*, 1062–1081.
40. König, G.; Riniker, S. On the Faithfulness of Molecular Mechanics Representations of Proteins Towards Quantum-mechanical Energy Surfaces. *Interface Focus* **2020**, *10*, 20190121.
41. Dewar, M.; Zoebisch, E.; Healy, E.; Stewart, J. A New General Purpose Quantum Mechanical Molecular Model. *J. Am. Chem. Soc.* **1985**, *107*, 3902–3909.
42. Stewart, J. Optimization of Parameters for Semiempirical Methods I. Method. *J. Comput. Chem.* **1989**, *10*, 209–220.
43. Thiel, W. Semiempirical Quantum–chemical Methods. *Wiley Interdiscip. Rev. Comput. Mol. Sci.* **2014**, *4*, 145–157.
44. Korth, M.; Thiel, W. Benchmarking Semiempirical Methods for Thermochemistry, Kinetics, and Noncovalent Interactions: OMX Methods are Almost as Accurate and Robust as DFT-GGA Methods for Organic Molecules. *J. Chem. Theory Comput.* **2011**, *7*, 2929–2936.
45. Dral, P. O.; Wu, X.; Spörkel, L.; Koslowski, A.; Thiel, W. Semiempirical Quantum-Chemical Orthogonalization-Corrected Methods: Benchmarks for Ground-State Properties. *J. Chem. Theory Comput.* **2016**, *12*, 1097–1120.
46. Margraf, J. T.; Dral, P. O. What is Semiempirical Molecular Orbital theory Approximating? *J. Mol. Model.* **2019**, *25*, 1–12.
47. Åqvist, J.; Medina, C.; Samuelsson, J. E. New Method for Predicting Binding-affinity in Computer-aided Drug Design. *Prot. Eng.* **1994**, *7*, 385.
48. Hansson, T.; Marelius, J.; Åqvist, J. Ligand Binding Affinity Prediction by Linear Interaction Energy Methods. *J. Comput. Aid. Mol. Des.* **1998**, *12*, 27.
49. Almlof, M.; Carlsson, J.; Åqvist, J. Improving the Accuracy of the Linear Interaction Energy Method for Solvation Free Energies. *J. Chem. Theory Comput.* **2007**, *3*, 2162–2175.
50. Stjernschantz, E.; Marelius, J.; Medina, C.; Jacobsson, M.; Vermeulen, N. P. E.; Oostenbrink, C. Are Automated Molecular Dynamics Simulations and Binding Free Energy Calculations Realistic Tools in Lead Optimization? An Evaluation of the Linear Interaction Energy (LIE) Method. *J. Chem. Inf. Model.* **2006**, *46*, 1972–1983.
51. Kolb, P.; Huang, D.; Dey, F.; Caflisch, A. Discovery of Kinase Inhibitors by High-throughput Docking and Scoring Based on a Transferable Linear Interaction Energy Model. *J. Med. Chem.* **2008**, *51*, 1179–1188.

52. Guitiérrez-de Terán, H.; Åqvist, J. *Computational Drug Discovery and Design*, Methods in Molecular Biology, Vol. 819, Humana Press, 2012, pp. 305–323.
53. Hummer, G.; Pratt, L. R.; García, A. E. Free Energy of Ionic Hydration. *J. Phys. Chem.* **1996**, *100*, 1206–1215.
54. Åqvist, J.; Hansson, T. On the Validity of Electrostatic Linear Response in Polar Solvents. *J. Phys. Chem.* **1996**, *100*, 9512–9521.
55. Heid, E.; Moser, W.; Schröder, C. On the Validity of Linear Response Approximations Regarding the Solvation Dynamics of Polyatomic Solutes. *Phys. Chem. Chem. Phys.* **2017**, *19*, 10940–10950.
56. Marcus, R. Electron-Transfer Reactions in Chemistry – Theory and Experiment (Nobel Lecture). *Ang. Chem. Int. Ed.* **1993**, *32*, 1111–1121.
57. Zwanzig, R. W. High-temperature Equation of State by a Perturbation Method. I. Nonpolar Gases. *J. Chem. Phys.* **1954**, *22*, 1420.
58. Zuckerman, D. M.; Woolf, T. B. Theory of a Systematic Computational Error in Free Energy Differences. *Phys. Rev. Lett.* **2002**, *89*, 180602.
59. Shirts, M. R.; Pande, V. S. Comparison of Efficiency and Bias of Free Energies Computed by Exponential Averaging, the Bennett Acceptance Ratio, and Thermodynamic Integration. *J. Chem. Phys.* **2005**, *122*, 144107–1–144107–16.
60. Brooks, B. R. *et al.* CHARMM: The Biomolecular Simulation Program. *J. Comput. Chem.* **2009**, *30*, 1545–1614.
61. Brooks, B. R.; Bruccoleri, R. E.; Olafson, B. D.; States, D. J.; Swaminathan, S.; Karplus, M. CHARMM - A Program for Macromolecular Energy, Minimization, and Dynamics Calculations. *J. Comput. Chem.* **1983**, *4*, 187.
62. Thiel, W. MNDO2005, version 7.1., Mülheim an der Ruhr, Max-Planck-Institut für Kohlenforschung, Germany, 2006.
63. Frisch, M. J. *et al.* Gaussian 16 Revision B.01. 2016; Gaussian Inc. Wallingford CT.
64. Weber, W.; Thiel, W. Orthogonalization Corrections for Semiempirical Methods. *Theor. Chem. Acc.* **2000**, *103*, 495–506.
65. Dral, P. O.; Wu, X.; Spörkel, L.; Koslowski, A.; Weber, W.; Steiger, R.; Scholten, M.; Thiel, W. Semiempirical Quantum-Chemical Orthogonalization-Corrected Methods: Theory, Implementation, and Parameters. *J. Chem. Theory Comput.* **2016**, *12*, 1082–1096.
66. Smith, P. E.; van Gunsteren, W. When are Free Energy Components Meaningful? *J. Phys. Chem.* **1994**, *98*, 13735–13740.
67. Boresch, S.; Archontis, G.; Karplus, M. Free Energy Simulations — the Meaning of Individual Contributions from a Component Analysis. *Proteins: Struct. Funct. Genet.* **1994**, *20*, 25–33.
68. Boresch, S.; Karplus, M. The Meaning of Component Analysis: Decomposition of the Free Energy in Terms of Specific Interactions. *J. Mol. Biol.* **1995**, *254*, 801–807.
69. MacKerell, A. D. *et al.* All-atom Empirical Potential for Molecular Modeling and Dynamics Studies of Proteins.*J. Phys. Chem. B* **1998**, *102*, 3586–3616.
70. Giese, T. J.; York, D. M. Development of a Robust Indirect Approach for MM → QM Free Energy Calculations that Combines Forcematched Reference Potential

and Bennett's Acceptance Ratio Methods. *J. Chem. Theory Comput.* **2019**, *15*, 5543–5562.

71. Hudson, P. S.; Boresch, S.; Rogers, D. M.; Woodcock, H. L. Accelerating QM/MM Free Energy Computations via Intramolecular Force Matching. *J. Chem. Theory Comput.* **2018**, *14*, 6327–6335.
72. König, G.; Brooks, B. R. Correcting for the Free Energy Costs of Bond or Angle Constraints in Molecular Dynamics Simulations. *Biochim. Biophys. Acta - Gen. Subj.* **2015**, *1850*, 932–943.

https://doi.org/10.1142/9789819808199_0003

Chapter 3

Accurate Modeling of RNA Hairpins Through the Explicit Treatment of Electronic Polarizability with the Classical Drude Oscillator Force Field*

Mert Y. Sengul and Alexander D. MacKerell Jr.†

Department of Pharmaceutical Sciences, School of Pharmacy, University of Maryland, Baltimore, Maryland 21201, USA

†*alex@outerbanks.umaryland.edu*

Molecular dynamics (MD) simulations play a crucial role in modeling biomolecular systems in which the electrostatic interactions are critical in dictating the structural and dynamical properties. Thus, the treatment of the electrostatic interactions defined in the underlying force field (FF) strongly affects the simulation accuracy. Most FFs use fixed partial atomic charges to include electrostatic interactions, and therefore lack the electronic polarization response, representing an intrinsic limitation. To address this limitation, polarizable FFs have been developed that treat atomic polarizabilities explicitly. Here we present the application of the all-atom polarizable (Drude) and non-polarizable (CHARMM) nucleic acid FFs in RNA hairpin systems to investigate the impact of polarization on structural properties, dipole moment distributions, and cation interactions. Results show that the presence of polarizability in the FF significantly improves the stabilization of RNA hairpin structure. As expected, the distributions of dipole moments show more fluctuations when simulated using the polarizable FF, with the variation in dipoles contributing to the stabilization of the structures of the loop regions of the RNAs. Contact map

*To cite this article, please refer to its earlier version published in *J. Comput. Biophys. Chem.* **2022**, *21* (4), 461–471, DOI: 10.1142/S2737416521420060

†Corresponding author.

analyses between the bases and cations show that the variation of the ion distribution around the entire hairpin is larger for the polarizable FF and the cations occupy the outer hydration shell to a greater extent. The presented results indicate the importance of the explicit treatment of electronic polarizability in molecular simulations of RNA, including in non-canonical regions.

Keywords: Drude polarizable force field; hairpin; RNA; CHARMM; dipole moment.

1. Introduction

All-atom molecular dynamics (MD) simulations have become and have repeatedly proven to be an essential tool in investigating biomolecular systems based on fundamental inter- and intra-atomic interactions.[1] To model biomolecular systems, MD simulations require an empirical potential energy function (i.e., empirical force field (FF)) to calculate the forces acting on the particles in the system, allowing for propagation of the dynamics of the system. Therefore, the quality of the simulations and thus predictions from MD simulations is heavily influenced by the underlying FF. In particular, the electrostatic part of the FFs plays a crucial role in simulating biomolecular systems, thereby strongly influencing the energetics and dynamics in the system. The ability to incorporate electrostatic interactions as accurately as possible has always been a priority for empirical FFs. The most widely used FFs for biomolecular simulations, namely AMBER,[2,3] CHARMM,[4–6] and OPLS,[7] are additive models that use fixed partial charges for each atom that have been developed to implicitly include polarization effects occurring in the aqueous phase.[8] However, such representation of electrostatic interactions has limited these methods' ability to adapt to diverse environments. For example, the polarity of the chemical environment may be very dynamic for an ion passing through ion channels,[9,10] for polar molecules on the interior of a bilayer,[11] and for the interatomic structure of water molecules in a confined space.[12] To improve the electrostatic interactions in a way that become more adaptive to such environments, polarizable FFs that treat electronic polarization explicitly have been developed.[13–16]

The classical Drude oscillator polarizable force field (Drude FF) is one of the models that include explicit treatment of electronic polarizability. Drude FF uses negatively charged auxiliary particles (Drude oscillators or particles) attached to all non-hydrogen atoms via harmonic springs (a.k.a. Charge-on-spring model[17]), which are used to simulate the induced polarization via displacement of Drude particles by responding to the local electric field. Details on the functional form of Drude FF can be found in Ref. 18. Briefly, Drude FF is an extension to CHARMM potential energy function as it maintains most of the functional form with an extension of the electrostatic energy term and addition of a self-polarization energy term. The electrostatic energy term is modified to incorporate the Drude-Drude and Drude-atom electrostatic interactions and can be seen in Eq. (1).

$$E_{\text{electrostatic}} = \frac{1}{4\pi D}\left(\sum_{i\neq j}\frac{q_i q_j}{\|r_i - r_j\|} + \sum_{i\neq j}\frac{q_{D,i} q_j}{\|r_{D,i} - r_j\|} + \sum_{i\neq j}\frac{q_{D,i} q_{D,j}}{\|r_{D,i} - r_{D,j}\|}\right). \tag{1}$$

In Eq. (1), r_i and r_j are positions of atoms i and j, and q_i, q_j, $q_{D,i}$, and $q_{D,j}$ are particle charges on atoms and corresponding Drude particles for atoms i and j. The addition of Drude particles to the system requires the addition of a self-polarization term given in Eq. (2),

$$E_{\text{self}} = \frac{1}{2}([K_{11}^{D}]d_1^2 + [K_{22}^{D}]d_2^2 + [K_{33}^{D}]d_3^2), \tag{2}$$

where d_1, d_2, and d_3 are the projections of the Drude particle-atom displacement vectors on the orthogonal axes defined based on the local molecular frame, and K^D is a tensor with off-diagonal elements set to zero. For the majority of non-hydrogen atoms (hydrogens do not carry explicit polarizability), the self-polarization is isotropic, with the anisotropy represented in Eq. (2) typically applied to hydrogen bond acceptors.

In combination, the treatment of electronic polarizability combined with the presence of lone pairs on hydrogen bond acceptors and the ability to include Lennard–Jones parameters on Drude particles[18] leads to improved treatment of the interactions between molecules. In the context of nucleic acids, the inclusion of the explicit treatment of electronic polarizability has shown improvements in the modeling with base flipping,[19] ion interactions,[20,21] the sensitivity of DNA structure to ion type,[21] and base-stacking.[22,23] To further investigate the potential utility of polarizable FFs, here we present an investigation on the influence of polarization on the structural changes, dipole moment distribution, and cation interactions in RNA hairpins systems by comparing MD simulations that using the Drude and CHARMM36 FFs. RNA hairpins are a particularly interesting test of a polarizable model given the problems encountered in simulations of this class of molecules using additive FFs.[24,25]

2. Methods

The RNA hairpin structures used in this study are GAAA (PDB: 1ZIF), CACG (PDB: 1ROQ), and UUCG (PDB: 2KOC). The crystal or NMR structures of these hairpins were obtained from the Protein Data Bank (PDB).[26] For the NMR structures, the first model in the PDB file was used to initiate the simulations. The cubic simulation box dimensions for GAAA, CACG, and UUCG systems were 47 Å, 62 Å, and 54 Å, respectively. The numbers of ions in MD simulations for GAAA and CACG were set to values determined to reproduce the experimental concentrations of 110 mM and 50 mM and yield an overall neutral system, which corresponds to 73 and 81 sodium ions (Na^+), and 64 and 69 chloride ions (Cl^-), respectively. The number of potassium ions (K^+) in the UUCG system was set to 13, which is determined to neutralize the negative charges on the RNA backbone. After initial configurations were prepared for each system, the systems were first minimized and equilibrated in CHARMM[6] with harmonic restraints on the non-hydrogen atoms with force constants 1 kcal/mol/Å^2 and 0.1 kcal/mol/Å^2, respectively. The equilibration simulations were run for 100 ps in the NPT ensemble with a timestep of 1 fs. CHARMM36 additive FF was used to model RNA and the ions.[5,27,28] Water molecules were modeled by the CHARMM-modified

TIP3P model.[29,30] Particle mesh Ewald (PME) method was applied for the calculation of electrostatic interactions with a real-space cutoff of 12 Å.[31] The lengths of all covalent bonds involving a hydrogen atom were constrained using the SHAKE algorithm.[32] The Lennard–Jones potential was smoothed to zero between 10 Å and 12 Å using a force switch. The production simulations were run in the NPT ensemble using the Langevin integrator (1 bar, 300°C) for 500 ns in the OpenMM simulation package[33,34] with a time step of 2 fs. A Monte Carlo (MC) anisotropic barostat was used to keep average pressure constant at 1 bar.

The equilibrated systems obtained from the CHARMM simulations were converted to the Drude polarizable model by adding Drude particles to all non-hydrogen atoms and lone pairs to all hydrogen bond acceptors using the Drude Prepper[35] in the CHARMM-GUI.[36] The TIP3P water model is also converted to polarizable SWM4-NDP water.[37] The Drude systems were again minimized and equilibrated using the same settings as with CHARMM36 as described above. Electrostatic interactions were treated using PME with a real-space cutoff of 12 Å.[31] The FF used in polarizable systems is the 2017 Drude RNA FF[38] with variations to account for over polarization of the Drude particles observed with that system (Sengul, Kognole, and MacKerell, work in progress). The FF parameters used in the present study may be obtained upon request. The production simulations were run in an NPT ensemble using the Langevin integrator (1 bar, 300°C) with a timestep of 1 fs for 500 ns using the OpenMM simulation package.[33,34] The "hard wall" feature is used for Drude particles that move more than 0.2 Å from their parent nuclei.[39]

Root Mean Square Deviation (RMSD) calculations were conducted by aligning the non-hydrogen backbone atoms of the helical regions of the MD hairpin structures and the corresponding experimental structures. Contact maps presented as heat maps represent the interactions between each nucleotide of the hairpins and cations based on a distance between the anionic phosphate oxygens and cations of less than 4.2 Å, which is used as the outer shell ion criterion. The contact maps were generated using the last 450 ns of the whole MD trajectory. The heat map values were calculated as the percentage of the MD trajectory that a cation exists as an outer shell ion. No direct contacts between the cations and the phosphates were observed in this study. CHARMM was used to calculate the dipole moment values for each base in each RNA hairpin separately. The dipole RMS fluctuation (RMSF) analyses were calculated over the whole MD trajectory, and the average dipole moment values were used as the reference dipole moment in the analyses.

3. Results and discussion

Three different RNA hairpin structures were selected to investigate the influence of polarizability on the structural dynamics of RNAs in an aqueous solution in the presence of monovalent cations. Bottaro *et al.*[40] conducted a clustering analysis to map all experimentally obtained hairpin structures according to their tetraloop conformations. Their investigations revealed that the most dominant hairpin structures are GNRA and UNCG type, where N can be any nucleotide and R can be either adenine (ADE) or guanine (GUA).[41] Accordingly, we have selected one hairpin structure from the GNRA family, namely GAAA (PDB: 1ZIF), and two hairpin structures from the YNMG family, namely CACG (PDB: 1ROQ) and UUCG (PDB: 2KOC). The YNMG family, where Y can either CYT or URA, N is any type, and M is either CYT or ADE, is an extension to UNCG family and includes the CNCG family.[42] The secondary structures of the selected hairpins can be seen in Fig. 1. The hairpins were simulated in aqueous solutions for 0.5 μs using system specific ionic concentrations. Identical systems were simulated for each hairpin using CHARMM36 and Drude FF, and identical analyses were conducted for each system.

The initial analysis involved the impact of polarization on the overall structure of the hairpins. Figure 2 shows the RMSDs of bases as a function of simulation time between simulated and experimental structures of the hairpins. The average RMSD and RMSF values are shown in Table 1. As can be seen from the figure and the table, RMSD values for the helical regions from both CHARMM36 and Drude simulations are similar for all systems. However, the RMSFs are systematically larger in the CHARMM36 simulations, as shown in

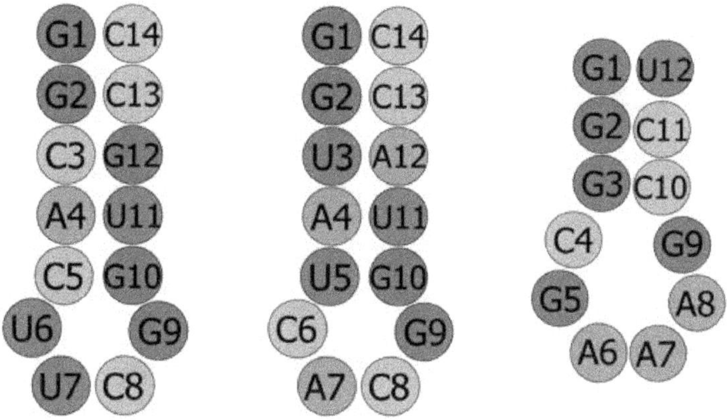

Fig. 1. (Color online) Secondary structure of the studied hairpins: UUCG (left), CACG (middle), and GAAA (right).

Table 1. Concerning the loop regions, the RMSDs and RMSFs are significantly smaller for Drude simulations than those with CHARMM36. These results indicate that the ability of the Drude FF to treat the variability of the base dipoles in different conformations is important for maintaining the structural integrity of the RNAs, particularly in the loop regions of the hairpins.

The ability of the Drude FF to maintain the overall integrity of the loop regions of the hairpins is presented in Fig. 3. From the overlays of the final frames from the Drude and CHARMM36 MD simulations with the experimental structures, it is evident that the overall conformation of the loops is significantly closer to experiment with the Drude FF. With UUCG and CACG, the distortion of the loop in the CHARMM36 FF is particularly evident with the conformation of the phosphodiester backbone significantly changed as well as the interactions between the individual bases.

The thermodynamic stability of UUCG hairpins has been shown to be strongly influenced by the interactions of bases in the loop region (URA6, CYT8, and GUA9).[24] More specifically, the URA6(O2′)-GUA9(O6) hydrogen bond is responsible for the exceptional stability of UUCG hairpin type,[43] and the dissociation of this signature hydrogen bond results in the destabilization of the hairpin.[44] Several studies have been conducted on the influence of the used FF on the

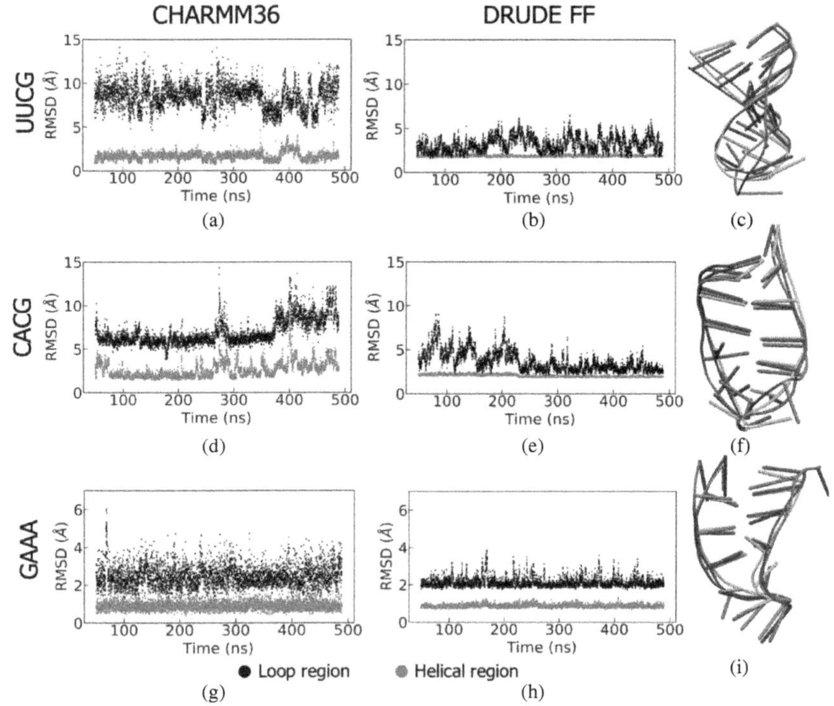

Fig. 2. (Color online) RMSDs of all heavy atoms in the backbone of the helical and loop regions of RNA hairpins (CHARMM36; a, d, g and Drude; b, e, h). The helical (magenta) and loop (black) regions were reported separately in the RMSD plots. Visualizations of RNA hairpin conformations (c, f, i) for the high-resolution NMR structures (green) and of the final snapshots of the Drude FF (blue) and CHARMM36 (red) MD simulations.

stability of UUCG hairpins. Banas *et al.*[25] showed that the signature hydrogen bond is lost in very short time scales in MD simulations conducted using three different versions of the AMBER nucleic acid FF (ff94, ff99, ff99bsc0).[2,45,46] Bergonzo *et al.*[44] reported structural RMSD increases in several replicas of UUCG simulations caused by the destabilization of the signature bond with an AMBER FF based on "ff99 + Peréz *et al.* $\alpha/\gamma+$ Zgarbová *et al.* χ_{OL3}." Similar destabilization issues were also reported for CHARMM36 FF.[25,44] and are again seen in the present study. However, our Drude FF results show increased stability for the signature hydrogen bond. As shown in Fig. 3, even after 500 ns simulations (Fig. 4(b)), the

Table 1. The RMSD and RMSF values for helical and loop regions of different RNA hairpin structures.

	Helical		Loop	
	RMSD	RMSF	RMSD	RMSF
		UUCG		
CHARMM36	1.69 ± 0.09	1.04	8.46 ± 0.21	5.19
DRUDE FF	1.88 ± 0.01	0.48	3.19 ± 0.26	0.73
		CACG		
CHARMM36	2.62 ± 0.56	2.37	6.90 ± 1.16	4.82
DRUDE FF	2.05 ± 0.12	0.72	3.76 ± 0.81	2.15
		GAAA		
CHARMM36	0.91 ± 0.01	1.31	2.51 ± 0.07	1.86
DRUDE FF	0.87 ± 0.02	0.75	2.17 ± 0.10	1.30

URA6(O2′)-GUA9(O6) hydrogen bond is conserved in Drude FF while the bond is immediately lost when the CHARMM36 simulation is started (Fig. 4(a)). To determine the reproducibility of this difference three replicates were run for each of the CHARMM36 and Drude FF simulations. The time series data for the replicate simulations, showing the Drude interactions to be stable while the CHARMM36 interaction is lost in all cases, is shown in Fig. S4. Additionally, we have analyzed several other non-canonical hydrogen bonds observed in the crystal structures of the loop regions of the hairpins. According to our analyses (Table 2) the Drude FF either produced more stable hydrogen bonds or behaved similarly compared to CHARMM36. The corresponding time series plots are shown in Figs. S5–S7. These results indicate that the ability of the Drude FF to maintain the conformations of the hairpins is due, in part, to the ability to better reproduce hydrogen bond interactions than the CHARMM36 FF. However, it does not appear that the Drude FF is systematically overstabilizing the hydrogen bonds as those in the GAAA hairpin are also stable in the CHARMM36 FF. This suggests that the improved maintenance of the hairpins and of selected hydrogen bonds is related to the explicit treatment of polarization as its influence on those properties depends on the explicit chemical environment.

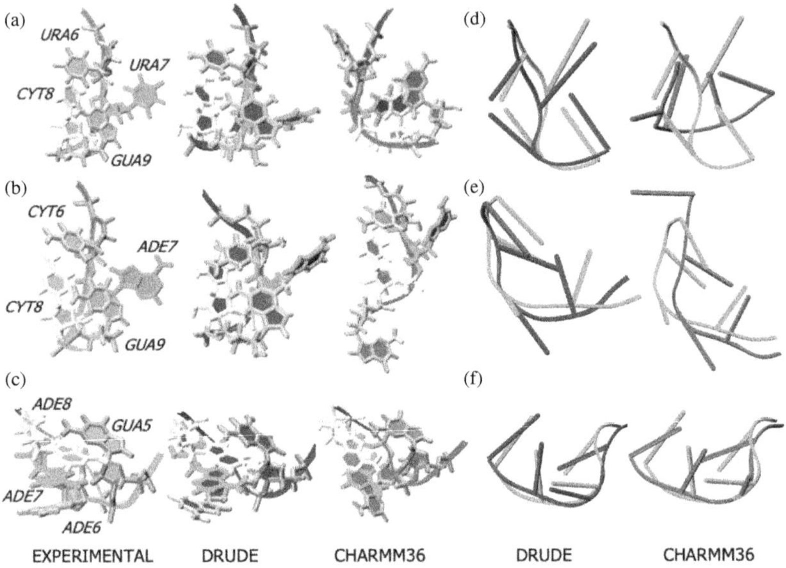

Fig. 3. (Color online) The MD snapshots of the loop regions of UUCG (a), CACG (b), and GAAA (c) from NMR (green), Drude FF (blue), and CHARMM36 (red). Overlaid stick representations of the experimental-Drude (left) and experimental-CHARMM36 (right) structures of the loop regions of UUCG (d), CACG (e), and GAAA (f). MD structures are from the final time frame of the MD simulations and structures were aligned based on the non-hydrogen atoms in the helical regions. The orientations of the different structure representations may differ to improve the visualization.

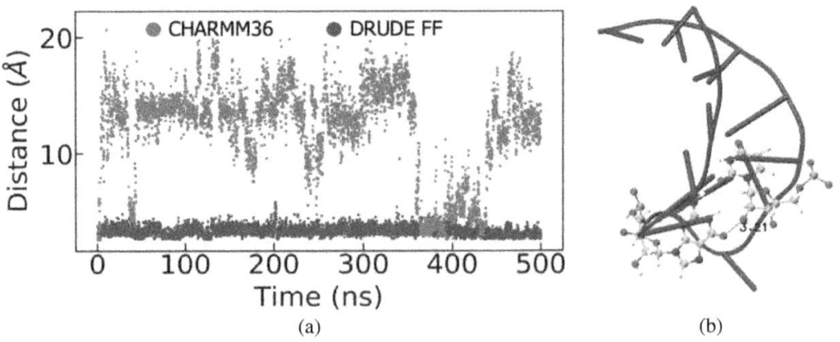

Fig. 4. (Color online) (a) The distance between $U_{L1}(O2')$ and $G_{L4}(O6)$ atoms as a function of simulation times, which corresponds to the signature hydrogen bond in the UUCG hairpin systems and (b) visualization of the signature bond in the final snapshot of the Drude FF MD simulation.

Table 2. The mean distances, with standard deviations, and RMS Fluctuation (RMSF) values for non-canonical hydrogen bonds observed in loop regions of different RNA hairpin structures. The experimental hydrogen bond distances are shown in the right column. Averages and standard deviations are based on five 100 ns blocks from the 500 ns simulations.

CHARMM36		DRUDE FF		
Mean	RMSF	Mean	RMSF	EXPERIMENTAL
		UUCG		
		URA6(O2)-GUA9(N1)		
10.88 ± 2.64	7.74	3.07 ± 0.03	0.05	2.77
		URA7(O2P)-CYT8(N4)		
10.43 ± 2.93	7.96	4.73 ± 2.16	3.11	2.99
		CACG		
		ADE7(O2P)-CYT8(N4)		
5.33 ± 4.13	0.70	2.90 ± 0.44	0.08	3.48
		CYT6(O2′)-GUA9(O6)		
17.37 ± 2.15	1.06	5.31 ± 0.45	0.91	2.54
		CYT6(O2)-GUA9(N1)		
15.84 ± 1.64	1.60	2.78 ± 0.11	0.03	3.09
		GAAA		
		GUA5(O2′)-ADE7(N7)		
3.92 ± 1.23	0.27	3.52 ± 1.05	1.05	2.37
		GUA5(O2′)-ADE7(N6)		
4.51 ± 1.71	1.21	4.18 ± 1.51	1.19	2.63
		GUA5(N2)-ADE8(O2P)		
2.98 ± 0.47	0.11	2.83 ± 0.13	0.01	2.87
		ADE6(O5′)-ADE6(C8)		
3.95 ± 0.83	0.23	3.51 ± 0.30	0.34	2.97

Another common FF artifact that results in inappropriate hairpin configurations is "ladder-like" structure formation in GAAA hairpins.[47] This occurred as a result of disproportionate accessibility of the high-anti region in AMBER nucleic acid FF,[25] an issue corrected in subsequent versions ff99χ_{YIL}[48] and ff99χ_{OL}.[49] Such conformations have never been observed with CHARMM36 RNA simulations, consistent with such conformations not being observed in this study with either the Drude or CHARMM36 FF.

Looking at the RMSD versus time for the separate regions of the hairpins, the CHARMM36 simulations show that helical and loop regions tend to influence each other's conformations. For example, UUCG and CACG hairpins undergo a structural change around 400 ns, with both the helical and loop regions showing high RMSD values (Figs. 2(a) and 2(d)). In contrast, for the Drude simulations, CACG helical region shows a minimal change in response to conformational changes happening in the loop regions in the first 250 ns of the simulations (Fig. 2(e)). This independent behavior between two different regions may be due to the additional stability of the individual helical and loop regions due to the presence of polarizability in the Drude FF, although additional investigations are needed to clarify this observation.

Figure 5 shows the contact maps between the anionic oxygens of the nucleotides of each hairpin and cations that populate RNA phosphate groups with outer hydration shell coordination. As shown in the figure, the cation distribution and the population of the cations around a hairpin are influenced by whether the FF is polarizable or non-polarizable. For the hairpins simulated using the Drude FF, the cation distribution is much more varied compared to the ones obtained using CHARMM36. In addition, the average time spent by any cation as an outer shell ion around each hairpin can be seen in Table 3. The hairpins that are simulated using Drude have a larger outer shell cation population around them compared to the ones simulated using CHARMM36. Furthermore, the standard deviation of population percentages (Table 3) is higher for Drude FF, which indicates that the ions can penetrate into and exit the outer shell hydration layer around the hairpins more easily. These results show the influence of polarizability on the interatomic dynamics between ions, RNA, and water molecules.

Analysis of the dipole moments of the bases was performed over all the individual bases for each system over the trajectories. Presented in Fig. 6 are the probability distributions of the different types of bases from all three hairpins over the 500 ns long MD trajectories. Analysis shows that the dipole moment fluctuations are larger for Drude systems compared to CHARMM36, as expected due to the explicit

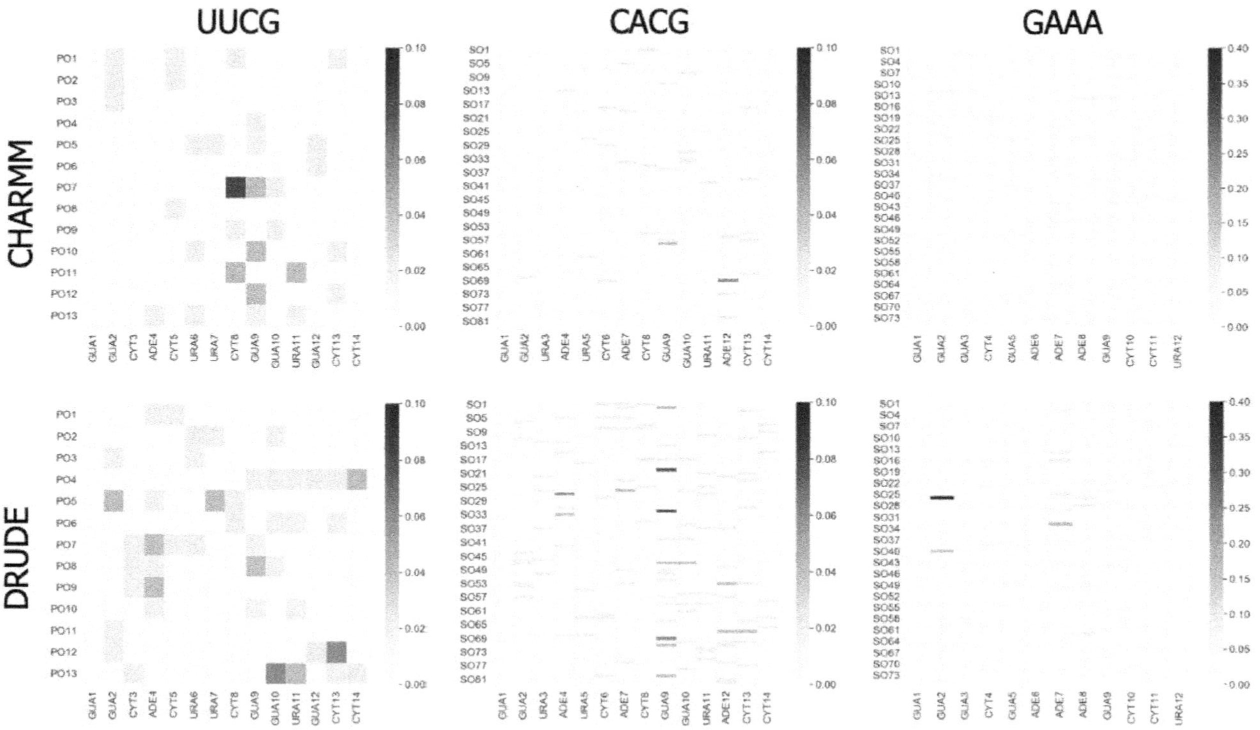

Fig. 5. (Color online) Contact maps between anionic phosphate oxygens of the RNA hairpin nucleotides and the cations present in the system. The maps show the time spent by cations around hairpin phosphate groups. Only outer shell cations were taken into consideration. The criterion used to assume a cation as an outer shell ion is given in the Method section. The heat map colors represent the percentage of the whole trajectory that a cation spent as an outer shell ion.

Table 3. The percent occupation of the cations around the anionic phosphate oxygens in the RNA hairpins as outer shell ions. The values are obtained by summation of percentage occupancy by all cations as outer shell ion throughout the full MD trajectory and therefore are unitless. The percentage occupancy is calculated as the time spent by cations as outer shell ion relative to the entire 500 ns long simulation time such that a value of 1 for a nucleotide represent full occupation throughout the entire simulation. For a hairpin composed of 14 nucleotides and assuming all nucleotides are fully occupied by outer shell ions, the percentage occupancy value should be 14. Average and standard deviations are based on five 100 ns blocks from the 500 ns simulations.

	UUCG	CACG	GAAA
CHARMM36	0.76 ± 0.18	0.70 ± 0.23	3.06 ± 0.46
DRUDE FF	1.10 ± 0.23	2.62 ± 0.87	4.72 ± 0.90

treatment of polarizability. With the CHARMM36 FF, variations of the dipole are solely associated with variations of the base geometries. Notably, the dipole moment distributions are at systematically larger values for the Drude model. This represents an improvement over CHARMM36, which tends to underestimate dipole moments compared to quantum mechanical calculations.[50] Notable, the median values of the base dipole for the Drude FF are systematically larger compared to the gas phase values for the bases. For example, the gas phase dipoles of the methylated and unmethylated GUA base are 6.76 and 6.21, respectively, significantly smaller than the peak of the distribution in Fig. 6(a). The corresponding gas phase values are 3.07/2.63 for ADE, 5.37/6.05 for CYT, and 4.72/4.59 for thymine (THY), where the X/Y are the methylated and unmethylated values, respectively. Thus, while the Drude electrostatic parameters were optimized to reproduce gas phase dipoles of the individual bases, in the condensed phase simulations, the dipoles sample systematically higher values. This indicates the importance of allowing the electron distributions of the bases to respond to the condensed phase environment in the Drude FF.

The average and RMSFs of the dipole moments for the individual bases in the three hairpins are shown on the secondary structures in

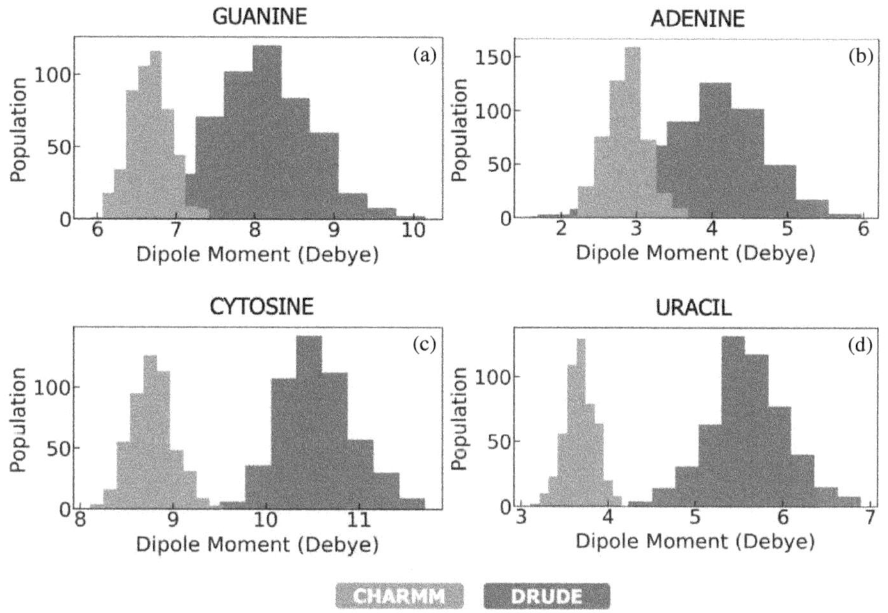

Fig. 6. (Color online) Dipole moment distribution over all the systems for the different types of the RNA bases obtained from the RNA hairpin simulations: (a) GUA, (b) ADE, (c) CYT and (d) URA. The distributions obtained from Drude FF simulations are represented with blue histograms, and from CHARMM36 simulations are represented with orange histograms.

Fig. 7. Results show that the Drude FF yields different dipole moments for the individual bases. While the base type is a dominant factor in determining the dipole values as expected, the magnitudes differ for the same base type, which is associated with the environment of the base. With the CYT bases, the mean values are ~10.6 in the helical regions with values of 10.3 in two of the three bases in the loop regions. With GUA bases, the range varies from 7.3 at the termini of GAAA to 8.1 in bases on CACG and GAAA. A low value of 3.2 occurs for one of the ADE bases in the GAAA loop with the remaining ADE values ranging from 3.8 to 4.2. An interesting example occurs with the URA7 of UUCG hairpin, which has a dipole moment of 6.7 that is significantly larger compared to other URA bases in all hairpins. As mentioned above, URA7 is the only nucleotide in the loop region of UUCG that does not directly

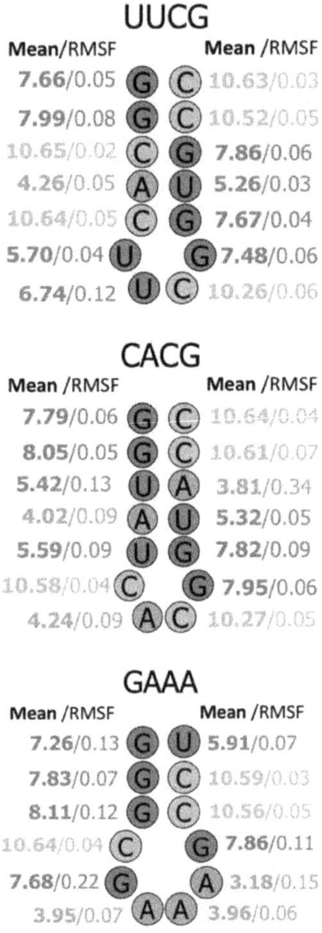

Fig. 7. (Color online) Average dipole moments and their RMS fluctuations for the individual bases from RNA hairpin simulations shown on secondary structures for different hairpins: UUCG (top), CACG (middle), GAAA (bottom).

contribute to the stability of the RNA structure through direct interactions with the rest of the RNA. Rather, this base is highly solvent exposed leading to the higher dipole moment. This higher dipole moment would favor the base maintaining a solvated conformation, limiting its interactions with the rest of the RNA. Indeed, this phenomenon may facilitate maintenance of the overall conformation of the loop by the Drude model. A similar phenomenon was observed in base flipping calculations with the Drude FF in DNA.[19] In that study, it was shown that the base dipole increased upon flipping and increased exposure to the aqueous environment. This shifted the equilibrium towards the open state as compared to the CHARMM36 FF leading to improved agreement with experimental NMR data. These results show that the explicit treatment of polarizability allows for local variability in dipole moments that impact the overall structural properties of the RNAs, leading to improved reproduction of the experimental geometries as compared to the additive CHARMM36 FF.

4. Conclusion

Polarizable FFs have been developed to improve the treatment of electrostatic interactions in MD simulations. Here, we present a comparative study on the influence of explicit treatment of polarizability on the modeling of RNA hairpins. We use a variation of the 2017 Drude RNA FF as the polarizable FF, which is known to show improved stability for canonical dsRNA structures as well as improvements in some hairpins, and CHARMM36 as the non-polarizable FF. We have selected three different RNA hairpin structures from the experimentally most observed hairpin types. The selected RNA hairpins were simulated in the aqueous phase in the presence of ions using both FFs. Cation-nucleotide contact map analysis, RMSD and RMSF analyses, and dipole distribution analysis for both Drude and CHARMM36 simulations were performed for each RNA hairpin system. Contact maps show that polarizability influences the distribution of cations around the hairpins, as Drude simulations show a higher population compared to the CHARMM36 simulations. The RMSD of helical regions shows similar values for both FFs, though the RMSFs are smaller with the Drude FF. In the loop regions both the RMSD and RMSF are smaller for Drude FF, showing that the FF yields loop conformations that are in excellent agreement with experimental structures. More specifically, the signature hydrogen bond in UUCG hairpins is conserved with Drude FF, which has been an issue for both CHARMM36 and

AMBER (ff94, ff99, ff99bsc0) simulations. The dipole moment distributions of the bases are systematically larger with Drude FF compared to CHARMM36, which is an improvement as these values tend to be underestimated compared to quantum mechanical calculations, and it indicates the importance of polarizability response of the bases to condensed phase environment. Notably, the average dipole moment values and their RMSFs in the Drude FF vary for the individual base type depending on their locations in the RNAs, showing that th that dipoles are responsive to their local environmental changes. For example, we reveal in this study that the base of the URA7, which is a loop nucleotide in the UUCG hairpin, has a larger dipole moment compared to other URA bases due to it being more solvent exposed, which is suggested to facilitate maintenance of the loop structure in UUCG.

Overall, the explicit treatment of electronic polarizability in the Drude FF allows the electronic structure of the nucleotides to respond to dynamic environmental changes happening around the RNA in the aqueous phase as seen variations in the dipole moments of the bases. This capability improves the stability of the structures of the UUCG, CACG and GAAA hairpins, which has been an issue with non-polarizable FFs.

Competing Interests

ADM is Co-founder and CSO of SilcsBio LLC.

Acknowledgments

The authors acknowledge funding support from the National Institutes of Health under Award No. GM131710. Appreciation to Dr. Abhishek Kognole for helpful discussions. M.Y.S., A.D.M. came up with the idea. The molecular dynamics simulations were run by M.Y.S. The analysis codes are written by M.Y.S. The analyses were conducted by M.Y.S. and A.D.M. The manuscript was written by M.Y.S. and A.D.M. All authors have given approval to the final version of the manuscript.

References

1. Dror, R. O.; Dirks, R. M.; Grossman, J. P.; Xu, H.; Shaw, D. E. Biomolecular Simulation: A Computational Microscope for Molecular Biology. *Annu. Rev. Biophys.* **2012**, *41*, 429–452.
2. Cornell, W. D.; Cieplak, P.; Bayly, C. I.; Gould, I. R.; Merz Jr. K. M.; Ferguson, D. M.; Spellmeyer, D. C.; Fox, T.; Caldwell, J. W.; Kollman, P. A. A Second Generation Force Field for the Simulation of Proteins, Nucleic Acids, and Organic Molecules. *J. Am. Chem. Soc.* **1995**, *117*, 5179–5197.
3. Cheatham III T. E.; Cieplak, P.; Kollman, P. A. A Modified Version of the Cornell *et al.* Force Field with Improved Sugar Pucker Phases and Helical Repeat. *J. Biomol. Struct. Dyn.* **1999**, *16* (4), 845–862.
4. Nicolas, F.; MacKerell Jr. A. D. All-Atom Empirical Force Field for Nucleic Acids: I. Parameter Optimization Based on Small Molecule and Condensed Phase Macromolecular Target Data. *J. Comput. Chem.* **2000**, *21* (2), 86–104.
5. Hart, K.; Foloppe, N.; Baker, C. M.; Denning, E. J.; Nilsson, L.; Mackerell Jr. A. D. Optimization of the CHARMM Additive Force Field for DNA: Improved Treatment of the BI/BII Conformational Equilibrium. *J. Chem. Theory Comput.* **2012**, *8* (1), 348–362.
6. Brooks, B. R.; Brooks III C. L.; Mackerell Jr. A. D.; Nilsson, L.; Petrella, R. J.; Roux, B.; Won, Y.; Archontis, G.; Bartels, C.; Boresch, S.; Caflisch, A.; Caves, L.; Cui, Q.; Dinner, A. R.; Feig, M.; Fischer, S.; Gao, J.; Hodoscek, M.; Im, W.; Kuczera, K.; Lazaridis, T.; Ma, J.; Ovchinnikov, V.; Paci, E.; Pastor, R. W.; Post, C. B.; Pu, J. Z.; Schaefer, M.; Tidor, B.; Venable, R. M.; Woodcock, H. L.; Wu, X.; Yang, W.; York, D. M.; Karplus, M. CHARMM: the Biomolecular Simulation Program. *J. Comput. Chem.* **2009**, *30* (10), 1545–1614.
7. Kaminski, G. A.; Friesner, R. A.; Tirado-Rives, J.; Jorgensen, W. L. Evaluation and Reparametrization of the OPLS-AA Force Field for Proteins via Comparison with Accurate Quantum Chemical Calculations on Peptides. *J. Phys. Chem. B* **2001**, *105*, 64744–66487.
8. Mackerell Jr. A. D. Empirical Force Fields for Biological Macromolecules: Overview and Issues. *J. Comput. Chem.* **2004**, *25* (13), 1584–1604.
9. Sandeep Patel, J. E. D.; Bauer, B. A. Exploring Ion Permeation Energetics in Gramicidin A Using Polarizable Charge Equilibration Force Fields. *J. Am. Chem. Soc.* **2009**, *131*, 13890–13891.
10. Allen, T. W.; Andersen, O. S.; Roux, B. Molecular Dynamics — Potential of Mean Force Calculations as a Tool for Understanding Ion Permeation and Selectivity in Narrow Channels. *Biophys. Chem.* **2006**, *124* (3), 251–267.
11. Vorobyov, I.; Bennett, W. F.; Tieleman, D. P.; Allen, T. W.; Noskov, S. The Role of Atomic Polarization in the Thermodynamics of Chloroform Partitioning to Lipid Bilayers. *J. Chem. Theory Comput.* **2012**, *8* (2), 618–628.
12. Fumagalli, L.; Esfandiar, A.; Fabregas, R.; Hu, S.; Ares, P.; Janardanan, A.; Yang, Q.; Radha, B.; Taniguchi, T.; Watanabe, K.; Gomila, G.; Novoselov, K. S.; Geim, A. K. Anomalously Low Dielectric Constant of Confined Water. *Science* **2018**, *360* (6395), 1339–1342.

13. Shi, Y.; Xia, Z.; Zhang, J.; Best, R.; Wu, C.; Ponder, J. W.; Ren, P. The Polarizable Atomic Multipole-based AMOEBA Force Field for Proteins. *J. Chem. Theory Comput.* **2013**, *9* (9), 4046–4063.
14. Rick, S. W.; Stuart, S. J.; Berne, B. J. Dynamical Fluctuating Charge Force Fields: Application to Liquid Water. *J. Chem. Phys.* **1994**, *101* (7), 6141.
15. Gao, W. X. A. J. Design of a Next Generation Force Field: The X-POL Potential. *J. Chem. Theory Comput.* **2007**, *3*, 1890–1900.
16. Donchev, A. G.; Ozrin, V. D.; Subbotin, M. V.; Tarasov, O. V.; Tarasov, V. I. A Quantum Mechanical Polarizable Force Field for Biomolecular Interactions. *Proc. Natl. Acad. Sci. USA* **2005**, *102* (22), 7829–7834.
17. Drude, P. M. C. R. M. R. A. *The Theory of Optics.* Longmans, Green, and Co., New York, 1902.
18. Lemkul, J. A.; Huang, J.; Roux, B.; MacKerell Jr. A. D. An Empirical Polarizable Force Field Based on the Classical Drude Oscillator Model: Development History and Recent Applications. *Chem. Rev.* **2016**, *116* (9), 4983–5013.
19. Lemkul, J. A.; Savelyev, A.; MacKerell Jr. A. D. Induced Polarization Influences the Fundamental Forces in DNA Base Flipping. *J. Phys. Chem. Lett.* **2014**, *5* (12), 2077–2083.
20. Gkionis, K.; Kruse, H.; Platts, J. A.; Mladek, A.; Koca, J.; Sponer, J. Ion Binding to Quadruplex DNA Stems. Comparison of MM and QM Descriptions Reveals Sizable Polarization Effects Not Included in Contemporary Simulations. *J. Chem. Theory Comput.* **2014**, *10* (3), 1326–1340.
21. Savelyev, A.; MacKerell Jr. A. D. Competition Among Li(+), Na(+), K(+), and Rb(+) Monovalent Ions for DNA in Molecular Dynamics Simulations Using the Additive CHARMM36 and Drude Polarizable Force Fields. *J. Phys. Chem. B* **2015**, *119* (12), 4428–4440.
22. Gresh, N.; Sponer, J. E.; Devereux, M.; Gkionis, K.; de Courcy, B.; Piquemal, J. P.; Sponer, J. Stacked and H-Bonded Cytosine Dimers. Analysis of the Intermolecular Interaction Energies by Parallel Quantum Chemistry and Polarizable Molecular Mechanics. *J. Phys. Chem. B* **2015**, *119* (30), 9477–9495.
23. Parker, T. M.; Sherrill, C. D. Assessment of Empirical Models versus High-Accuracy Ab Initio Methods for Nucleobase Stacking: Evaluating the Importance of Charge Penetration. *J. Chem. Theory Comput.* **2015**, *11* (9), 4197–4204.
24. Takeshi Sakata, H. H.; Oda, Y.; Tanaka, T.; Ikeharal, M.; Uesugi, S. Studies on the Structure and Stabilizing Factor of the CUUCGG Hairpin RNA Using Chemically Synthesized Oligonucleotides. *Nucl. Acids Res.* **1990**, *18* (13), 3831–3839.
25. Pavel Banas, D. H.; Zgarbova, M.; Jurecka, P.; Orozco, M.; Cheatham III T. E.; Sponer, J.; Otyepka, M. Performance of Molecular Mechanics Force Fields for RNA Simulations: Stability of UUCG and GNRA Hairpins. *J. Chem. Theory Comput.* **2010**, *6*, 3836–3849.
26. Berman, H. M.; Westbrook, J.; Feng, Z.; Gilliland, G.; Bhat, T. N.; Weissig, H.; Shindyalov, I. N.; Bourne, P. E. The Protein Data Bank. *Nucl. Acids Res.* **2000**, *28* (1), 235–242.

27. Denning, E. J.; Priyakumar, U. D.; Nilsson, L.; Mackerell Jr. A. D. Impact of 2′-Hydroxyl Sampling on the Conformational Properties of RNA: Update of the CHARMM All-Atom Additive Force Field for RNA. *J. Comput. Chem.* **2011**, *32* (9), 1929–1943.
28. Beglov, D.; Roux, B. Finite Representation of an Infinite Bulk System: Solvent Boundary Potential for Computer Simulations. *J. Chem. Phys.* **1994**, *100* (12), 9050–9063.
29. Jorgensen, W. L.; Chandrasekhar, J.; Madura, J. D.; Impey, R. W.; Klein, M. L. Comparison of Simple Potential Functions for Simulating Liquid Water. *J. Chem. Phys.* **1983**, *79* (2), 926–935.
30. Durell, S. R.; Brooks, B. R.; Ben-Naim, A. Solvent-Induced Forces between Two Hydrophilic Groups. *J. Phys. Chem.* **1994**, *98*, 2198–2202.
31. Essmann, U.; Perera, L.; Berkowitz, M. L.; Darden, T.; Lee, H.; Pedersen, L. G. A Smooth Particle Mesh Ewald Method. *J. Chem. Phys.* **1995**, *103* (19), 8577–8593.
32. Ryckaert, J.-P.; Ciccotti, G.; Berendsen, H. J. C. Numerical Integration of the Cartesian Equations of Motion of a System with Constraints: Molecular Dynamics of *n*-Alkanes. *J. Comput. Phys.* **1977**, *23*, 327–341.
33. Izaguirre, J. A.; Sweet, C. R.; Pande, V. S. Multiscale Dynamics of Macromolecules Using Normal Mode Langevin. *Pac. Symp. Biocomput.* **2015**, *2010*, 240–251.
34. Eastman, P.; Swails, J.; Chodera, J. D.; McGibbon, R. T.; Zhao, Y.; Beauchamp, K. A.; Wang, L. P.; Simmonett, A. C.; Harrigan, M. P.; Stern, C. D.; Wiewiora, R. P.; Brooks, B. R.; Pande, V. S. OpenMM 7: Rapid Development of High Performance Algorithms for Molecular Dynamics. *PLoS Comput. Biol.* **2017**, *13* (7), e1005659.
35. Kognole, A. A.; Lee, J.; Park, S. J.; Jo, S.; Chatterjee, P.; Lemkul, J. A.; Huang, J.; MacKerell Jr., A. D.; Im, W. CHARMM-GUI Drude Prepper for Molecular Dynamics Simulation Using the Classical Drude Polarizable Force Field. Manuscript submitted for publication.
36. Jo, S.; Kim, T.; Iyer, V. G.; Im, W. CHARMM-GUI: A Web-Based Graphical User Interface for CHARMM. *J. Comput. Chem.* **2008**, *29* (11), 1859–1865.
37. Lamoureux, G.; Harder, E.; Vorobyov, I. V.; Roux, B.; MacKerell, A. D. A Polarizable Model of Water for Molecular Dynamics Simulations of Biomolecules. *Chem. Phys. Lett.* **2006**, *418* (1–3), 245–249.
38. Lemkul, J. A.; MacKerell Jr. A. D. Polarizable Force Field for RNA Based on the Classical Drude Oscillator. *J. Comput. Chem.* **2018**, *39* (32), 2624–2646.
39. Li, H.; Chowdhary, J.; Huang, L.; He, X.; MacKerell Jr. A. D.; Roux, B. Drude Polarizable Force Field for Molecular Dynamics Simulations of Saturated and Unsaturated Zwitterionic Lipids. *J. Chem. Theory Comput.* **2017**, *13* (9), 4535–4552.
40. Bottaro, S.; Lindorff-Larsen, K. Mapping the Universe of RNA Tetraloop Folds. *Biophys. J.* **2017**, *113* (2), 257–267.
41. Woese, C. R.; Winker, S.; Gutell, R. R. Architecture of Ribosomal RNA: Constraints on the Sequence of "Tetra-Loop". *Proc. Natl. Acad. Sci. USA* **1990**, *87*, 8467–8471.

42. Du, Z.; Yu, J.; Andino, R.; James, T. L. Extending the Family of UNCG-like Tetraloop Motifs: NMR Structure of a CACG Tetraloop from Coxsackievirus B3. *Biochemistry* **2003**, *42*, 4373–4383.
43. Kollman, J. L. M. A. P. A. Theoretical Studies of an Exceptionally Stable RNA Tetraloop: Observation of Convergence from an Incorrect NMR Structure to the Correct One Using Unrestrained Molecular Dynamics. *J. Mol. Biol.* **1997**, *270*, 436–450.
44. Bergonzo, C.; Henriksen, N. M.; Roe, D. R.; Cheatham III T. E. Highly Sampled Tetranucleotide and Tetraloop Motifs Enable Evaluation of Common RNA Force Fields. *RNA* **2015**, *21* (9), 1578–1590.
45. Perez, A.; Marchan, I.; Svozil, D.; Sponer, J.; Cheatham III T. E.; Laughton, C. A.; Orozco, M. Refinement of the AMBER Force Field for Nucleic Acids: Improving the Description of Alpha/Gamma Conformers. *Biophys. J.* **2007**, *92* (11), 3817–3829.
46. Wang, J.; Cieplak, P.; Kollman, P. A. How Well Does a Restrained Electrostatic Potential (RESP) Model Perform in Calculating Conformational Energies of Organic and Biological Molecules? *J. Comput. Chem.* **2000**, *21* (12), 1049–1074.
47. Mlynsky, V.; Banáš, P.; Hollas, D.; Reblova, K.; Walter, N. G.; Sponer, J.; Otyepka, M. Extensive Molecular Dynamics Simulations Showing That Canonical G8 and Protonated A38H+ Forms Are Most Consistent with Crystal Structures of Hairpin Ribozyme. *J. Phys. Chem. B* **2010**, *114*, 6642–6652.
48. Yildirim, I.; Stern, H. A.; Kennedy, S. D.; Tubbs, J. D.; Turner, D. H. Reparameterization of RNA? Torsion Parameters for the AMBER Force Field and Comparison to NMR Spectra for Cytidine and Uridine. *J. Chem. Theory Computa.* **2010**, *6*, 1520–1531.
49. Zgarbova, M.; Otyepka, M.; Sponer, J.; Mladek, A.; Banas, P.; Cheatham III T. E.; Jurecka, P. Refinement of the Cornell *et al.* Nucleic Acids Force Field Based on Reference Quantum Chemical Calculations of Glycosidic Torsion Profiles. *J. Chem. Theory Comput.* **2011**, *7* (9), 2886–2902.
50. Lemkul, J. A.; MacKerell Jr. A. D. Polarizable Force Field for DNA Based on the Classical Drude Oscillator: I. Refinement Using Quantum Mechanical Base Stacking and Conformational Energetics. *J. Chem. Theory Comput.* **2017**, *13* (5), 2053–2071.

Part II
Novel Methods in Computational Biophysics and Chemistry and their Applications to Biological Problems

https://doi.org/10.1142/9789819808199_0004

Chapter 4

Ab-*initio* Binding of Barnase–Barstar with DelPhiForce Steered Molecular Dynamics (DFMD) Approach*

Mahesh Koirala and Emil Alexov†

Department of Physics and Astronomy
Clemson University, Clemson, SC 29634, USA
†*ealexov@g.clemson.edu*

Receptor–ligand interactions are involved in various biological processes, therefore understanding the binding mechanism and ability to predict the binding mode are essential for many biological investigations. While many computational methods exist to predict the 3D structure of the corresponding complex provided the knowledge of the monomers, here we use the newly developed DelPhiForce steered Molecular Dynamics (DFMD) approach to model the binding of barstar to barnase to demonstrate that first-principles methods are also capable of modeling the binding. Essential component of DFMD approach is enhancing the role of long-range electrostatic interactions to provide guiding force of the monomers toward their correct binding orientation and position. Thus, it is demonstrated that the DFMD can successfully dock barstar to barnase even if the initial positions and orientations of both are completely different from the correct ones. Thus, the electrostatics provides orientational guidance along with pulling force to deliver the ligand in close proximity to the receptor.

Keywords: DFMD; protein–protein interactions; electrostatic interactions; steered molecular dynamics simulations; barnase–barstar; electrostatic forces.

---

*To cite this article, please refer to its earlier version published in *J. Theor. Comput. Chem.* **2020**, *19* (4), 2050016, DOI: 10.1142/S0219633620500169
†Corresponding author.

1. Introduction

Protein–protein interactions (PPIs) are essential for numerous biological processes such as cell-to-cell interactions, metabolic control and development.[1] This importance is demonstrated by the fact that deviations of the wild-type patterns of PPIs and the corresponding biophysical characteristics of protein complexes were shown to be associated with diseases.[2–9] Upon binding, a protein–protein complex is formed and depending on the participating monomers the complexes are classified as homodimers, heterodimers, enzyme-inhibitor, and antibody-protein complexes,[10] just to mention some. Furthermore, these complexes are characterized according to shape and size,[11] complementarity between interfaces[12,13] and residue interface propensities[14,15] including hydrogen bonding[16] and secondary structure.[17] Among the driving forces causing complex formation, the electrostatic interactions are the only long-range interactions and thus play a primary role in guiding monomers toward the binding. The electrostatic energies and forces were shown to play a major role in different molecular phenomena such as PPIs, ion binding,[8,18–20] protein folding and stability,[21,22] protein–DNA/RNA interactions,[23–25] and protein–microtubule binding.[3,26–28]

Modeling PPIs is a two-fold problem: predicting the binding pose and revealing the binding trajectory. There are many approaches for predicting binding pose, including homology modeling[29] and docking,[30] while much less attention has been paid to predicting the binding pathway. The last is very important for cases involving binding that requires large conformational changes involving backbone motion[31] and opening of the binding pocket.[32]

The technique capable in principle of addressing both, the binding trajectory and the binding pose, is molecular dynamics (MD) simulations of PPIs.[33] Essential component in these MD simulations is the treatment of water phase: either with explicit solvent waters[34] or with Generalized Born (GB) implicit model.[35,36] Explicit solvent model is computationally too expensive when large complexes are involved, and monomers are separated at large distances. Alternative is the GB model, which is much less computationally demanding. However, the default setup of GB model involves cut-offs for the calculations of both Born radii and pair-wise energies. So, if the distance between receptor and ligand is larger than the cut-off distance, the default setup of GB model will result in omitting contributions of distant atoms, thus affecting the modeling.[37] Increasing the cut-offs in MD setup presumably should solve this issue, however it results in dramatical slowdown of the calculations and in addition, still does not model electrostatic interactions correctly, as demonstrated in Ref. 37.

To address such a deficiency, in our previous work, we introduced a hybrid method, the DelPhiForce steered MD (DFMD), that takes advantage of the accurate calculations of electrostatic forces at each atom of the ligand and ports these forces into steered MD.[37] It was demonstrated that DFMD correctly predicts the binding trajectory (the entrance of the binding pocket) and binding pose of a small ligand (spermidine) bound to a receptor (spermine synthase).[37] However, DFMD has never been tested on PPIs. Here, we extend the applicability of DFMD by applying it to model the binding of barstar to barnase. We show that DFMD is capable of guiding barstar–barnase complex formation irrelevant of the starting positions and orientations of separated monomers.

2. Materials and methods

2.1. *Delphi*

Delphi is a popular software that solves the Poisson–Boltzmann equation (PBE), which uses finite-difference method to compute electrostatic potential distribution throughout the modeling box.[38–40]

2.2. *DelPhiForce*

DelPhiForce (http://compbio.clemson.edu/delphi-force) is available within Delphi package to calculate the electric field, forces and energy of a two-molecular system. In our study, it is used to compute the electric force between receptor and ligand. The receptor's atoms are charged using amber force field[41] and the ligand's atoms are kept neutral. Then Delphi uses finite-difference approach to solve the PBE and computes the electrostatic potential at each atom of ligands due to charge in the receptor atoms, and from there to deliver the atomic electrostatic forces.[42,43]

2.3. *DFMD simulations*

We used DFMD [37] method to study the binding of ligand to receptor, i.e. the barstar binding to the barnase. The DFMD method uses an approach that combines long-range electrostatic force via DelPhi-Force and NAMD MD simulation package.[44] The ligand (barstar) was offset 30 Å away from its bound position to ensure that the only nonzero force is the long-range electrostatic force. Then the electrostatic force on each atom of barstar was calculated with DelPhiForce.[42,43] The forces are then given to steered MD module of NAMD.[44] Simulations were done with Generalized Born implicit solvent (GBIS) model.[45] The ion concentration was 0.15 M, solvent dielectric constant was 80 and the temperature was maintained at 300 K using a Langevin thermostat. In addition, cut-off was taken at 18 Å and periodic boundary conditions were applied with a cubical box of 150 Å. Constant pulling forces were applied in the simulations where the direction and magnitude of the steered forces were calculated by DelPhiForce.[42,43,45] The steered force was recalculated and updated every 500 steps of the simulations. The simulations were done using the default value of steered electrostatic force range: lower limit: 0.2 $\text{kcal}\cdot\text{mol}^{-1}\cdot\text{Å}^{-1}$ and upper limit: 1.18 $\text{kcal}\cdot\text{mol}^{-1}\cdot\text{Å}^{-1}$.

2.4. *Selection and preparation of barnase and barstar monomers*

The barnase–barstar complex structure was taken from Protein Data Bank (PDB ID: 1BRS)[46] and we selected chain "C" as barnase and chain "F" as barstar. The PDB file had two mutations, C80A and C82A, with respect to wild type and these mutations have been mutated back to wild-type residues. The reverse mutations were performed using UCSF Chimera[47] where the rotameric state with highest occupancy in Dunbrach library of rotamers is utilized.

The distance between the centers of mass (CoMs) of barnase and barstar complex is calculated as 23.39 Å using VMD.[48] The CoM of barnase is positioned at the origin of the reference frame and barstar is rotated around the axes passing through the CoM of barnase so that the CoM of barstar resides on the X-axis (Fig. 1). The starting positions for each simulation are obtained by offsetting barstar 30 Å away from its bound position ($\Delta D = 30$ Å) along the X-axis. To generate alternative unbound orientations for barnase and barstar, 64 initial structures with different relative orientations of barstar and barnase were created. The first set was generated keeping the barnase orientation fixed, while barstar was rotated around the X- and Z-axes passing through its CoM in increments of 90°. Then the same

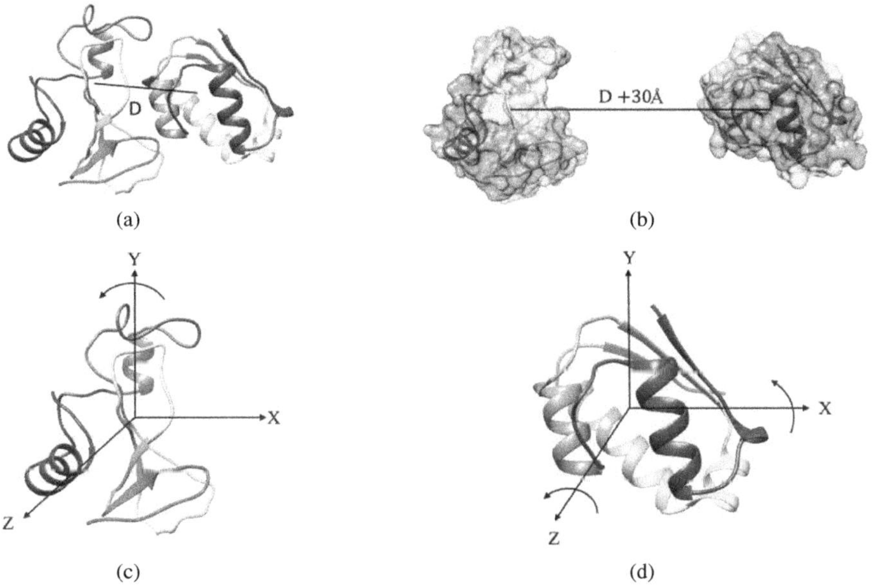

Fig. 1. Selection and manipulations of barnase and barstar monomers: (a) Experimental structure of barnase–barstar complex with the CoM distance denoted by D (left: barnase and right: barstar); (b) barnase–barstar complex separated by $\Delta D = 30$ Å between the centers of mass; (c) different initial orientations of barnase created by rotating barnase about the Y-axis through its center of mass with the step increments of 90°; and (d) different initial orientations of barstar created by rotating barstar about the X- and Z-axes though its center of mass with the step increments of 90°.

procedure is repeated for four orientations of barnase rotating it around the Y-axis passing through its CoM [Figs. 1(a)–1(d)].

2.5. *Ligand RMSD and interface RMSD*

Ligand RMSD (L-RMSD) is the RMSD of ligand obtained through the superimposition of the receptor docking model to its crystallographic structure and then computing the RMSD of the ligand.[49] The interface RMSD (I-RMSD) is the backbone RMSD of interface residues of the ligand.[50]

3. Results

Sixty-four independent DFMD simulations were carried out starting from different positions and orientations of both barnase and barstar. The success of the modeling was monitored via several quantities, mentioned in Sec. 2: L-RMSD, I-RMSD and the offset from crystallographic bound position (ΔD). Here, we provide the results of the simulations by showing several showcase examples and then statistics of the entire set of simulations.

3.1. *Showcase examples*

It is not anticipated that each simulation will result in successful binding, because the simulation time is relatively short (5 ns) and the binding process may involve transient binding at wrong places before the correct binding mode is found. Therefore, here we present two successful cases (Fig. 2), simply to illustrate that DFMD can bind correctly the relatively large proteins even when the initial positions and orientations are completely away from the crystallographic ones (the same simulation without DelPhiForce assistance resulted in no binding). The first case is DFMD simulation with initial positions and orientations of monomers as follows: Barnase is fixed in its crystallographic position while barstar is moved away by $\Delta D =$ 30 Å along the X-axis and rotated by 90° around the X-axis and 270° around the Z-axis (see Fig. 1 for more detail). In the second case, the

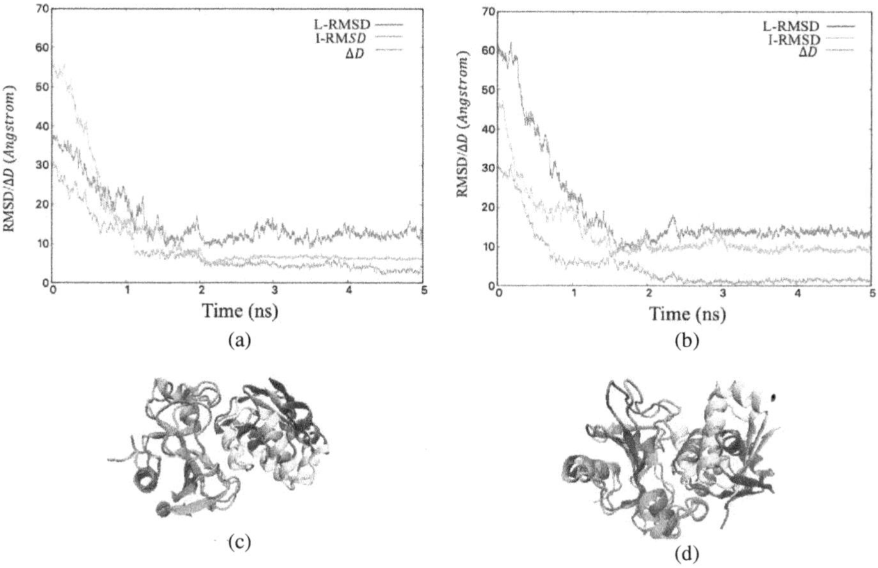

Fig. 2. L-RMSD, ΔD and I-RMSD as a function of simulation time: (a) L-RMSD, I-RMSD and ΔD of barnase–barstar at different time steps during the simulation with barnase fixed and barstar being rotated 90° around the X-axis and 270° around the Z-axis passing through the CoM; (b) L-RMSD, I-RMSD and ΔD at different time steps during the simulation with barnase being rotated 90° around the Y-axis and barstar being rotated 270° around the Z-axis passing through the CoM; and (c) and (d) final positions of barnase–barstar at the last step of simulation compared to experimental structure when the simulation is performed with initial orientations explained for (a) and (b), respectively: barnase colored green and barstar colored purple, while the experimental structures are colored red (barnase) and yellow (barstar).

initial positions and orientations of monomers are as follows: Barnase is rotated by 90° along the Y-axis and barstar is moved away along the X-axis by $\Delta D = 30$Å and rotated by 270° along the Z-axis. Both cases clearly represent difficult binding initial conditions, especially the second case where both binding interfaces are in completely wrong orientations. Nevertheless, 5-ns DFMD simulations are quite successful as shown in Fig. 2.

The results show that L-RMSD approaches the 10Å limit, which is the cut-off used in Critical Assessment of Predicted Interactions (CAPRI) to indicate acceptable predictions.[49] It is important to observe that once the near-binding mode is achieved, the L-RMSD does not change much and the system is quite stable [Figs. 2(a) and 2(b)]. The assessment of the success of simulation is also done via monitoring ΔD. Indeed, one sees that ΔD goes quickly to near zero, indicating the binding partners are situated at almost the same distance as seen in the crystallographic structure [Figs. 2(a) and 2(b)]. Furthermore, we monitored I-RMSD for both cases [Figs. 2(a) and 2(b)]. Similar observations as above can be made — the system quickly reaches a binding pose that is very similar to the crystallographic one and stays stable during the rest of the simulations. The reason why the system does not exactly reach the crystallographic conformation is the internal flexibility of both the receptor and the ligand. They sample conformations that are several Å RMSD away from their crystallographic structures, an observation that has been reported by other researchers as well.[51] The final positions of barnase–barstar at the end of simulation compared to their native structures are also shown for guidance of the eye [Figs. 2(c) and 2(d)].

3.2. *Statistical analysis*

Statistical analysis shows that most of the simulations are successful despite the short simulation time. To illustrate the time evolutions of the quantities (ΔD, L-RMSD and I-RMSD) that are used in this work to assess the success of simulations, we select several time windows from the DFMD simulations (time windows 2–3, 3–4 and 4–5 ns). Each time window consists of 4000 frames. One anticipates that longer simulations will bring the ligand closer to the crystallographic binding position. Results are shown in Fig. 3. Analyzing the offset distance, ΔD [Fig. 3(a)], one sees that the number of cases with very small ΔD is greater for the time windows of 3–4 ns and 4–5 ns compared with the 2–3-ns time window. This indicates that indeed as the time of simulations progresses the DFMD brings the partners together close to the experimental positions. Furthermore, Fig. 3(b) shows the time evolutions of L-RMSD in these three time windows separately. Here the trend is not as clear as it is for ΔD. Practically, there is no difference of the number of cases with low L-RMSD

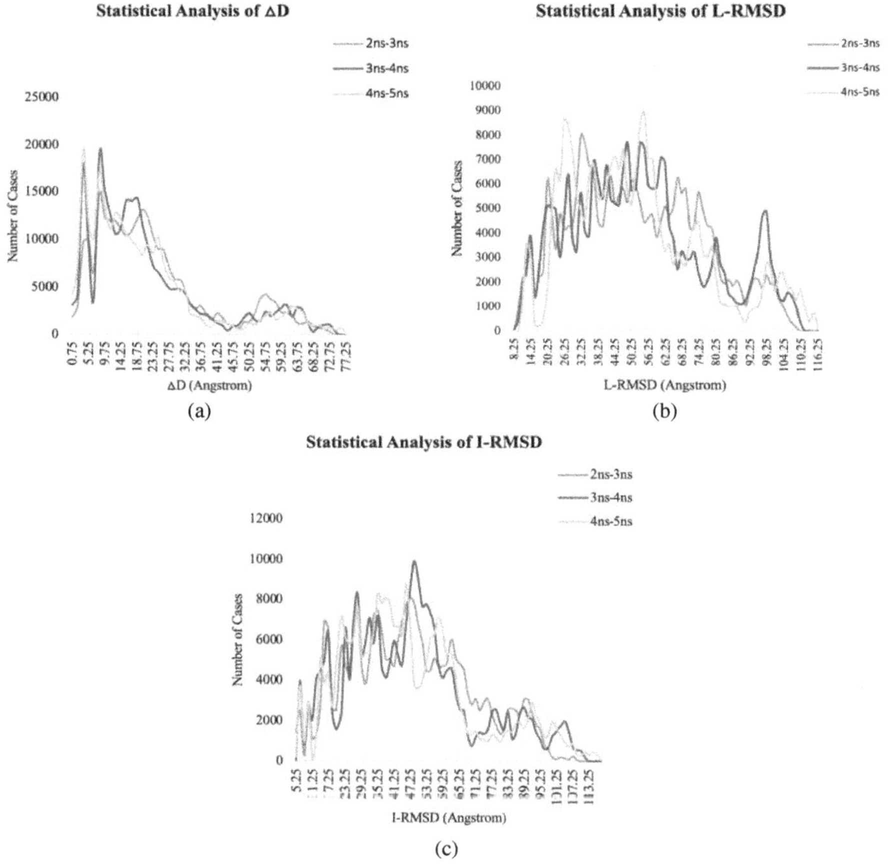

Fig. 3. Statistical analysis of the results based on ΔD, L-RMSD and I-RMSD during different time windows of the simulations: (a) analysis of ΔD; (b) analysis of L-RMSD; and (c) analysis of I-RMSD.

between the three time windows. However, there is a tendency in the 60–90Å range of L-RMSD that longer simulations provide better outcome, i.e. less cases being with such large L-RMSD. It should be made clear that L-RMSD has three components: L-RMSD due to (1) offset distance, (2) wrong orientation of the ligand and (3) conformational changes of the ligand. In addition, due to the fact that the receptor also undergoes conformational changes during the simulations, the structural superimposition of crystallographic 3D structure of the receptor and the corresponding snapshot is contributing to L-RMSD as well. Finally, the results about I-RMSD are shown in Fig. 3(c). Here, we see both effects mentioned above: an enrichment of case with low I-RMSD for the time windows of 3–4 ns and 4–5 ns compared with the case in the time window of 2–3 ns; and at the same time fewer cases in the regions 60–90 Å. The arguments about the factors contributing to I-RMSD are the same as outlined for L-RMSD.

Overall, the statistical analysis indicates that DFMD facilitates the binding and attempts to bring the apartments to their crystallographic positions in the complex. However, none of the 64 simulations resulted in perfect match of the binding mode of barnase–barstar complex as seen in the crystallographic structure, i.e. $\Delta D = 0$ Å, $\mathrm{L-RMSD} = 0$ Å and I-RMSD $= 0$ Å. One plausible reason for that is in our protocol the structures are not rigid and they sample various conformations (away from the crystallographic 3D structures) and thus even when they are sampling the crystallographic binding mode, still the corresponding ΔD, L-RMSD and I-RMSD will be different from zero.

3.3. *Role of electrostatics*

The work was done on diverse sets of initial orientations and positions of both barnase and its ligand barstar. We did not observe particular preference of successful binding with respect to the corresponding initial conditions. In some cases, such as case 2 in Sec. 3.1, the partners were positioned and oriented in completely wrong way (by wrong, we mean positions and orientations completely different from the crystallographic ones), and still DFMD was able to correctly bring them together very close to the crystallographic binding mode. What is the reason for that? Figure 4 illustrates two cases of initial conditions, an easy case where the barstar is simply moved away from its crystallographic position [Fig. 4(a)] and a difficult case, where barstar is moved away and rotated as well [Fig. 4(b)]. In the first case [Fig. 4(a)] the binding interfaces are still facing each other and presumably the recognition should be easier. Indeed, as it can

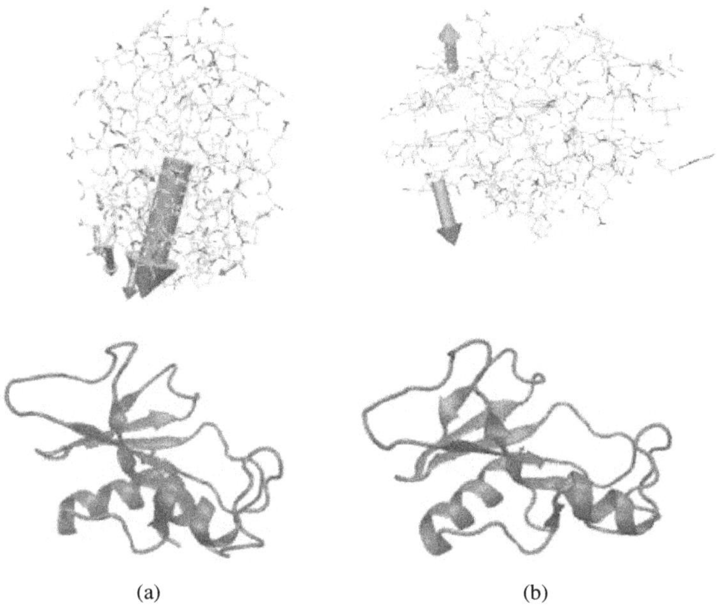

Fig. 4. Electrostatic forces generated by barnase (cartoon representation) to barstar placed 30 Å away from its bound position. Orange arrows represent forces acting on each residue, while the green arrow is the total resultant force. The length of the arrow reflects calculated magnitude of the force. The side chains of residues of barstar are represented as lines, of which blue and red ones denote positively charged residue and negatively charged residue, respectively: (a) Initial crystallographic structure of barnase–barstar separated by 30 Å from its bound position; and (b) electrostatic torque formed when keeping barnase fixed and barstar is rotated around the X-axis by 90° and around the Z-axis by 270° and separated by 30 Å from its bound position.

be seen in Fig. 4(a), the resulting electrostatic force coming from barnase and acting on barstar is pointing toward the receptor, thus contributing to the association process. In the second case [Fig. 4(b)], the interfaces are not facing each other and if barstar is simply pulled toward the barnase, the binding will be totally wrong. However, Fig. 4(b) indicates that electrostatics plays a crucial role by providing a torque that makes barstar to rotate and thus to adopt correct orientation of the binding interface. Such an electrostatic torque effect was shown to be common for various binding partners.[43]

3.4. *Roles of conformational flexibility and longer simulation time*

It was mentioned above that none of the simulations resulted in a perfect binding mode. The corresponding quantities, ΔD, L-RMSD and I-RMSD, were found to be different from zero in all 64 runs. To probe what is the plausible reason for that, we carried out traditional MD simulations (without DelPhiForce) starting from the crystallographic structure of barnase–barstar complex. The results of 5-ns simulations are shown in Fig. 5(a). It can be seen that ΔD, L-RMSD and I-RMSD are different from zero, confirming our initial thought that conformational dynamics associated with MD results in deviation from crystallographic structure and thus one cannot expect that DFMD modeling will reproduce the binding mode seen in X-ray experiment.

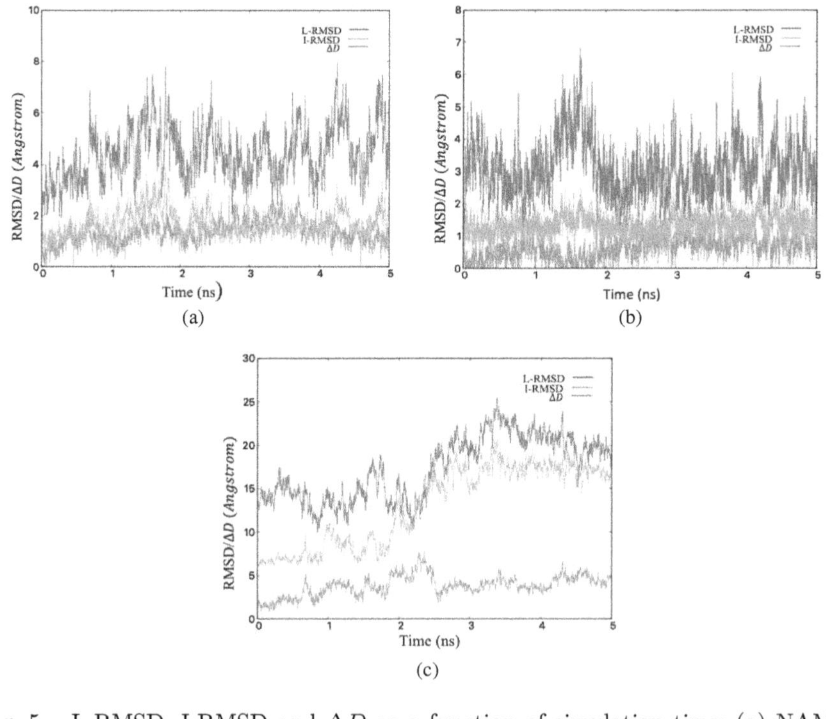

Fig. 5. L-RMSD, I-RMSD and ΔD as a function of simulation time: (a) NAMD simulation of bound crystal structure; (b) DFMD simulation of bound crystal structure; and (c) NAMD simulation of last frame of successful DFMD simulation shown in Fig. 2(b).

Furthermore, we tested if the DelPhiForce could be causing problems when monomers are in bound state. For this purpose, we carried out DFMD simulations starting from the X-ray structure of the complex [Fig. 5(b)]. One can compare Figs. 5(a) and 5(b) to observe that there is no significant difference. This indicates that DFMD in bound state delivers results similar to traditional MD and does not cause artifacts.

Finally, one may expect that once the near-binding mode is reached, one can do traditional MD simulations to refine the binding mode. To quickly probe this, we took a particular case of binding mode resulting in small ΔD, L-RMSD and I-RMSD [taken from the last frame of the simulations shown in Fig. 2(b)] and subjected it to additional 5-ns NAMD. Results are shown in Fig. 5(c). Unfortunately, no improvement was observed.

4. Conclusion

This work showed that DFMD is capable of modeling protein–protein binding in case of barnase–barstar. Successful binding was observed independently of the initial positions, indicating that the protocol is not biased toward easy cases when the partners are simply separated by a particular distance but still maintain the correct interface orientation. Thus, DFMD not only facilitates the association of the partners, but also assists them to adopt correct mutual orientation. The DFMD approach is fast and computationally less expensive than the traditional MD-based approaches to model receptor–ligand association. Furthermore, the traditional MD was unable to generate any binding mode. Thus, DFMD is an efficient tool to model protein–protein binding especially if one has no knowledge of how to position and orient the binding partners prior to the simulations.

5. Acknowledgments

The work was supported by grants from NIH, Grant Nos. R01GM093937, R01GM125639 and P20GM121342. We thank Shailesh Kumar Panday for useful discussions and proofreading the manuscript.

References

1. Szklarczyk D *et al.*, STRING v10: protein–protein interaction networks, integrated over the tree of life, *Nucleic Acids Res* **43**:D447–D452, 2015.
2. Kuzmanov U, Emili A, Protein-protein interaction networks: probing disease mechanisms using model systems, *Genome Med* **5**:37, 2013.
3. Li L *et al.*, Forces and disease: Electrostatic force differences caused by mutations in kinesin motor domains can distinguish between disease-causing and non-disease-causing mutations, *Sci Rep* **7**:8237, 2017.
4. Peng Y *et al.*, Revealing the effects of missense mutations causing Snyder-Robinson syndrome on the stability and dimerization of spermine synthase, *Int J Mol Sci* **17**:77, 2016.
5. Peng Y, Alexov E, Investigating the linkage between disease-causing amino acid variants and their effect on protein stability and binding, *Proteins* **84**: 232–239, 2016.
6. Petukh M, Kucukkal TG, Alexov E, On human disease-causing amino acid variants: statistical study of sequence and structural patterns, *Hum Mutat* **36**: 524–534, 2015.
7. Kucukkal TG *et al.*, Structural and physico-chemical effects of disease and non-disease nsSNPs on proteins, *Curr Opin Struct Biol* **32**: 18–24, 2015.
8. Petukh M *et al.*, Chronic beryllium disease: revealing the role of beryllium ion and small peptides binding to HLA-DP2, *PLoS ONE* **9**:e111604, 2014.
9. Nishi H *et al.*, Cancer missense mutations alter binding properties of proteins and their interaction networks, *PLoS ONE* **8**:e66273, 2013.
10. Jones S, Thornton JM, Principles of protein-protein interactions, *Proc Natl Acad Sci USA* **93**: 13–20, 1996.
11. Chothia C, The nature of the accessible and buried surfaces in proteins, *J Mol Biol* **105**: 1–12, 1976.
12. Li Y, Zhang X, Cao D, The role of shape complementarity in the protein-protein interactions, *Sci Rep* **3**:3271, 2013.
13. Skrabanek L *et al.*, Computational prediction of protein–protein interactions, *Mol Biotechnol* **38**: 1–17, 2008.
14. Yan C *et al.*, Characterization of protein–protein interfaces, *Protein J* **27**: 59–70, 2008.
15. Jones S, Thornton JM, Analysis of protein-protein interaction sites using surface patches, *J Mol Biol* **272**: 121–132, 1997.
16. Teng S *et al.*, Modeling effects of human single nucleotide polymorphisms on protein-protein interactions, *Biophys J* **96**: 2178–2188, 2009.
17. Jochim AL, Arora PS, Assessment of helical interfaces in protein–protein interactions, *Mol BioSyst* **5**: 924–926, 2009.
18. Petukh M, Zhang M, Alexov E, Statistical investigation of surface bound ions and further development of BION server to include pH and salt dependence, *J Comput Chem* **36**: 2381–2393, 2015.
19. Petukh M, Kimmet T, Alexov E, BION web server: predicting non-specifically bound surface ions, *Bioinformatics* **29**: 805–806, 2013.

20. Petukh M *et al.*, Predicting nonspecific ion binding using DelPhi, *Biophys J* **102**: 2885–2893, 2012.
21. Takano K *et al.*, An X-linked channelopathy with cardiomegaly due to a CLIC2 mutation enhancing ryanodine receptor channel activity, *Hum Mol Genet* **21**: 4497–4507, 2012.
22. Witham S *et al.*, A missense mutation in CLIC2 associated with intellectual disability is predicted by *in silico* modeling to affect protein stability and dynamics, *Proteins* **79**: 2444–2454, 2011.
23. Peng Y, Alexov E, Computational investigation of proton transfer, pKa shifts and pH-optimum of protein-DNA and protein-RNA complexes, *Proteins* **85**: 282–295, 2017.
24. Yang Y *et al.*, Binding analysis of methyl-CpG binding domain of MeCP2 and Rett syndrome mutations, *ACS Chem Biol* **11**: 2706–2715, 2016.
25. Kucukkal TG *et al.*, Impact of Rett syndrome mutations on MeCP2 MBD stability, *Biochemistry* **54**: 6357–6368, 2015.
26. Li L, Alper J, Alexov E, Cytoplasmic dynein binding, run length, and velocity are guided by long-range electrostatic interactions, *Sci Rep* **6**:31523, 2016.
27. Li L, Alper J, Alexov E, Multiscale method for modeling binding phenomena involving large objects: application to kinesin motor domains motion along microtubules, *Sci Rep* **6**:23249, 2016.
28. Tajielyato N *et al.*, E-hooks provide guidance and a soft landing for the microtubule binding domain of dynein, *Sci Rep* **8**:13266, 2018.
29. Schwede T *et al.*, SWISS-MODEL: an automated protein homology-modeling server, *Nucleic Acids Res* **31**: 3381–3385, 2003.
30. Lengauer T, Rarey M, Computational methods for biomolecular docking, *Curr Opin Struct Biol* **6**: 402–406, 1996.
31. Abagyan R, Totrov M, Kuznetsov D, ICM: A new method for protein modeling and design: applications to docking and structure prediction from the distorted native conformation, *J Comput Chem* **15**: 488–506, 1994.
32. Bottegoni G *et al.*, A new method for ligand docking to flexible receptors by dual alanine scanning and refinement (SCARE), *J Comput-Aided Mol Des* **22**: 311–325, 2008.
33. Salsbury, JrFR, Molecular dynamics simulations of protein dynamics and their relevance to drug discovery, *Curr Opin Pharmacol* **10**: 738–744, 2010.
34. Duan L *et al.*, Accelerated molecular dynamics simulation for helical proteins folding in explicit water, *Front Chem* **7**:540, 2019.
35. Zhang LY *et al.*, Solvent models for protein–ligand binding: Comparison of implicit solvent Poisson and surface generalized Born models with explicit solvent simulations, *J Comput Chem* **22**: 591–607, 2001.
36. Mongan J *et al.*, Generalized Born model with a simple, robust molecular volume correction, *J Chem Theory Comput* **3**: 156–169, 2007.
37. Peng Y *et al.*, DFMD: Fast and effective DelPhiForce steered molecular dynamics approach to model ligand approach toward a receptor: application to spermine synthase enzyme, *Front Mol Biosci* **6**:74, 2019.
38. Li L *et al.*, DelPhi: a comprehensive suite for DelPhi software and associated resources, *BMC Biophys* **5**:9, 2012.

39. Li C *et al.*, DelPhi suite: New developments and review of functionalities, *J Comput Chem* **40**: 2502–2508, 2019.
40. Panday SK *et al.*, Modeling electrostatics in molecular biology: A tutorial of DelPhi and associated resources [Article v1. 0], *Living J Comput Mol Sci* **1**:10841, 2019.
41. Wang J *et al.*, Development and testing of a general amber force field, *J Comput Chem* **25**: 1157–1174, 2004.
42. Li L *et al.*, DelPhiForce web server: electrostatic forces and energy calculations and visualization, *Bioinformatics* **33**: 3661–3663, 2017.
43. Li L, Chakravorty A, Alexov E, DelPhiForce, a tool for electrostatic force calculations: Applications to macromolecular binding, *J Comput Chem* **38**: 584–593, 2017.
44. Phillips JC *et al.*, Scalable molecular dynamics with NAMD, *J Comput Chem* **26**: 1781–1802, 2005.
45. Tanner DE *et al.*, Parallel generalized Born implicit solvent calculations with NAMD, *J Chem Theory Comput* **7**: 3635–3642, 2011.
46. Buckle AM, Schreiber G, Fersht AR, Protein-protein recognition: Crystal structural analysis of a barnase-barstar complex at 2.0-.ANG. resolution, *Biochemistry* **33**: 8878–8889, 1994.
47. Pettersen EF *et al.*, UCSF Chimera — A visualization system for exploratory research and analysis, *J Comput Chem* **25**: 1605–1612, 2004.
48. Humphrey W, Dalke A, Schulten K, VMD: visual molecular dynamics, *J Mol Graph* **14**: 33–38, 1996.
49. Janin J *et al.*, CAPRI: a Critical Assessment of PRedicted Interactions, *Proteins* **52**: 2–9, 2003.
50. Moreira IS *et al.*, SpotOn: high accuracy identification of protein-protein interface hot-spots, *Sci Rep* **7**:8007, 2017.
51. Agrawal P *et al.*, Benchmarking of different molecular docking methods for protein-peptide docking, *BMC Bioinformatics* **19**:426, 2019.

https://doi.org/10.1142/9789819808199_0005

Chapter 5

The Accuracy of Force Fields on the Simulation of Intrinsically Disordered Proteins: A Benchmark Test on the Human p53 Tumor Suppressor*

Shangbo Ning, Jun Liu, Na Liu† and Dazhong Yan‡

School of Biology and
Pharmceutical Pharmaceutical Engineering
Wuhan Polytechnic University
Wuhan 430023, P. R. China
†*liuna@whpu.edu.cn*
‡*yandz6808@163.com*

Intrinsically disordered proteins (IDPs) are a class of proteins without stable three-dimensional structures under physiological conditions. IDPs exhibit high dynamic nature and could be described by structural ensembles. As one of the most widely used tools, molecular dynamics (MD) simulation could provide the atomic descriptions of the structural ensemble of IDPs. However, the accuracy of the MD simulation largely depends on the accuracy of the force field. In this paper, we compared the structural ensembles of the activation domain 1 (AD1) in p53 tumor suppressor obtained from the widely used force fields, AMBER99SB-ILDN, CHARMM27, CHARMM36m with different water models. The results show that CHARMM36m generates more extended conformations than other force fields, while CHARMM27 prefers to sample the α-helical structure. Moreover, the chemical shifts obtained by CHARMM36m are the closest to the experimental measurements. These results indicate that the CHARMM36m force field performs best in characterizing the structure properties of p53 AD1. Water models are also critical to describe the structural ensemble of IDPs.

---

*To cite this article, please refer to its earlier version published in *J. Theor. Comput. Chem.* **2020**, *19* (4), 2050011, DOI: 10.1142/S021963362050011X

†,‡Corresponding authors.

TIP4P water model can obtain more extended conformations and produce more local helical conformations than the TIP3P model in our simulation. In addition, we also compare the chemical shifts predicted by different chemical shift predicting programs with experimental measurements, the results show that SHIFTX2 obtains the best performance in the chemical shifts prediction.

Keywords: Intrinsically disordered proteins; molecular dynamics simulation; force fields; chemical shifts.

1. Introduction

Intrinsically disordered proteins (IDPs) or intrinsically disordered regions (IDRs) are abundant in eukaryote and play significant roles in many biological processes.[1] They are also associated with many human diseases such as cancer, cardiovascular disease and neurodegenerative diseases.[2] IDPs lack well-defined three-dimensional structures in physiological conditions, the structural ensembles of IDPs contain a large number of widely different conformations and these conformations are in rapid interconversion.[3] The enrichment of polar and charged amino acids makes IDPs have a larger interaction surface area and the conformational flexibility allows IDPs to adopt different conformations. These characteristics make IDPs play essential roles in protein–protein interactions.[4] Thus, it is necessary to characterize the structural properties of IDPs to better understand their structure-function relations. The high conformational flexibility of IDPs makes it is difficult to be characterized by experimental techniques. Molecular dynamics (MD) simulations are currently a promising tool for studying the atomic structural properties for proteins. The accuracy of the empirical force fields is critical for studying the structure, stability, folding, and function of proteins.[5–7] Most classic force fields are developed for folded proteins which are composed of well-defined three-dimensional structures. The accuracy of these force fields in simulating IDPs still needs further evaluation and verification.[8–10]

Many experimental technologies, including the small-angle X-ray scattering (SAXS) and nuclear magnetic resonance (NMR), can only provide some ensemble-averaged descriptions of IDPs. SAXS provides long-range structural correlations information related to protein shape. J-coupling provides information on the dihedral angle, while NOE implies the formation of local turn and helix. The chemical shifts reflect local secondary structure, while the residual dipole couplings (RDCs) provide information about local and long-range interactions.[4,9] Chemical shift is an easy measured and reproducible parameter, which could be used in the identification, annotation and location of the individual atoms.[11,12] Chemical shifts are also widely employed as the benchmark to estimate the accuracy of the MD simulations. Nowadays, many tools have been developed to predict the chemical shifts of ordered proteins, such as SHIFTX2, SPARTA+, PPM, etc.[13] However, the accuracy of these programs for disordered proteins is still unknown.

p53, a tumor suppressor, which is activated as a transcriptional regulator in response to DNA damage, leading to the termination of cell cycle and apoptosis. p53 has also been discovered to regulate many processes related to cell differentiation, autophagy and maintain cellular gene integrity.[14,15] It is shown that almost 50% of human cancers are related to p53 mutations.[16] The full-length p53 consists of 393 residues, in which both the N-terminal transactivation domain (TAD) and the C-terminal domain (CTD) are IDRs. TAD domain (residues 1–63) contains two subdomains, AD1 (residues 1–40) and AD2 (residues 43–63).[17] These two subdomains interact with various proteins, such as CBP/p300, CSN5/Jab1, Mdm2, RPA, TFIID, and TFIIH, to perform their functions.[18,19] AD1 and AD2 usually form stable α-helix structure upon binding to the target protein.[17] Due to the important role of TAD, extensive experimental studies have been carried out on this protein. High-resolution NMR structures and chemical shifts were produced.[17,20–22] Hence, the TAD region of p53 would be a good model to evaluate the accuracy of the force field.

In this work, the performance of three commonly used force fields, AMBER99SB-ILDN, CHARMM36m and CHARMM27, was verified on p53 AD1. These force fields have been modified for a better balance between different secondary structures. For instance, the amino acid side-chain torsion potentials were modified in AMBER99SB-ILDN force field, which has a better balance in sampling helix and coil structures.[23] The CMAP potential was added to CHARMM22 in producing CHARMM27.[24,25] CHARMM36m is a newly developed force field for simulating both IDPs and folded proteins with the backbone CMAP potential and specific salt bridge interactions were refined. We analyzed the structural characteristics of the AD1 region to evaluate the accuracy of these force fields. Moreover, three chemical shift prediction programs were also employed to assess their performance in predicting the chemical shifts of IDPs. Our results show that CHARMM36m performs best in sampling p53 AD1, the structure ensemble generated by CHARMM36m more in line with experimental measurements. And SHIFTX2 software performs better than the other software in predicting the chemical shifts of IDPs.

2. Materials and methods

2.1. *Simulation details*

The initial structure of p53 was derived from the NMR structure (PDB ID: 2L14).[22] Since only the chemical shifts of residue 13–37 in the AD1 region were obtained in NMR experiments,[21] a truncated p53 polypeptide (13-37 aa) was constructed through UCSF Chimera (the truncated p53 polypeptide is shown in Fig. 1).[26] The processed protein was placed in a water solvent box with periodic boundary conditions. The box size was about 5.02 nm × 6.32 nm × 6.13 nm and totally 19,000 atoms. The system was neutralized by adding Na^+ and Cl^-. The concentration of NaCl is 50 mM, which is consistent with the experimental environment.[21]

Totally four simulations with different force fields and water models — AMBER99SB-ILDN (TIP3P),[23] AMBER99SB-ILDN (TIP4P), CHARMM27 (TIP3P),[25] and CHARM36m (TIP3P)[27] were carried out.[28] All simulations were performed by AMBER16.[29] For the CHARMM force field systems, the systems were constructed with GROMACS-5.0.4,[30] and then converted to AMBER topology files for simulation with ParmED (http://github.com/ParmEd/ParmEd).

PLSQETFSDLWKLLPENNVLSPLPS

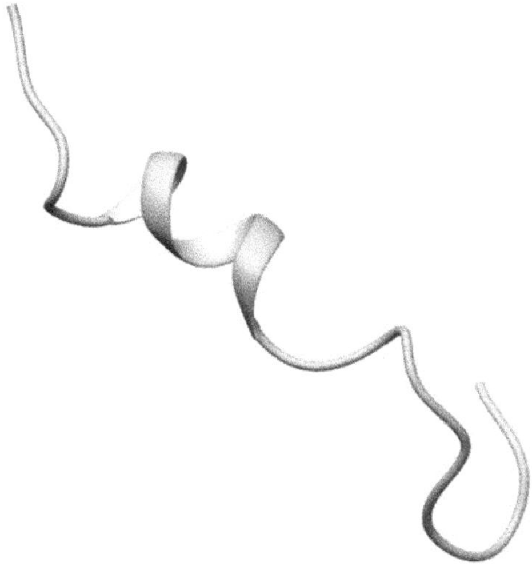

Fig. 1. The initial structure and sequence of simulated p53 AD1 and the red box in the sequence indicates the α-helical structure.

In simulations, long-range electrostatic interactions were calculated using the Particle Mesh Ewald (PME) method[31] with a van der Waals cut-off radius of 10 Å. The time step for integrating the equation of motion was set as 2 fs, and each system experienced a total of 40,000 steps of energy minimization in two steps. In the simulation, the temperature was controlled to 300K by the Nosé–Hoover thermostat method.[32] The NPT process used semi-isotropic coupling to maintain the pressure in all directions, and the relaxation time was 0.5 ps. The Parrinello–Rahman barostat[33] was employed to maintain pressure at 1.01×10^5 Pa and the characteristic frequency was 2.0 ps. All bond lengths were constrained by the LINCS algorithm.[34] Each system was simulated for 2 μs and the subsequent analyses were performed using the last 1 μs trajectories.

2.2. *Analysis*

Secondary structure of the protein was analyzed by STRIDE method. Cluster analyses for the simulations were performed in Gromacs5.0.4

with the GROMOS cluster method,[35] with a cutoff of 0.5 nm. SHIFTX,[36] SHIFTX2[11] and SPARTA+[37] were employed to calculate the chemical shifts of N and H_N for each simulation. The secondary chemical shifts were obtained by subtracting the value in the random coil structure according to the literature.[38]

3. Results and discussion

3.1. *The convergence of simulations*

To estimate the convergence of the simulations, the RMSD between simulated and experimental chemical shifts as a function of time for different force fields were analyzed firstly (Fig. 2). For CHARMM36m and AMBER99SB-ILDN (TIP3P), the RMSD of N chemical shifts clearly decreases during the first 1 μs. No substantial changes are observed in the simulations of CHARMM27 and AMBER99SB-ILDN (TIP4P) during the whole simulation time. A clear decrease for the RMSD of H_N chemical shifts was observed during the first 1 μs for all force fields. In addition, backbone atoms RMSD between the snapshots and the NMR p53 AD1 structure were also calculated along the simulation time (shown in Fig. S1). The initial structure of p53 AD1 used in this work is derived from the complex structure in which residues 19–25 form stable helical structure. The RMSD fluctuation is larger during the whole simulation under AMBER99SB-ILDN (TIP4P) and CHARMM36m, especially in the case of CHARMM36m, which indicates that more disordered conformations are explored by these force fields. In all the simulated trajectories, the fluctuation range of RMSD tended to be stable after 1 μs. Considering all of the above information, only the last 1 μs of each simulation were considered in the subsequent analyses.

3.2. *Structure dimensions of p53 AD1 under different force fields*

Previous studies have shown that the isolated p53 AD1 region is an IDR.[39,40] It is generally considered that IDPs possess an extended structure and higher average R_g due to its intrinsic flexibility and the

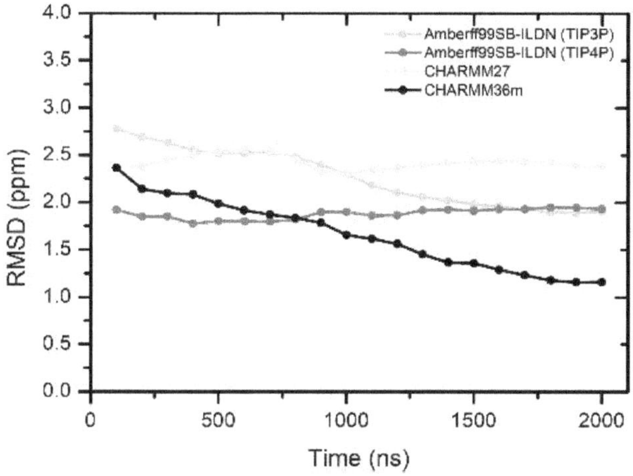

(a)

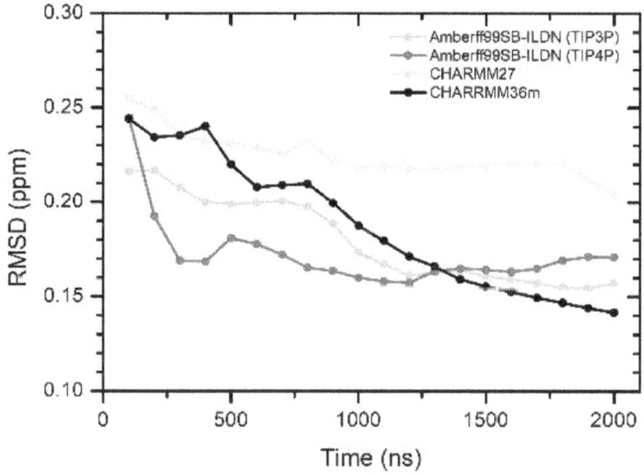

(b)

Fig. 2. RMSD between simulated and experimental chemical shifts as a function of time for (A) N chemical shifts, (B) H_N chemical shifts for AMBER99SB-ILDN-TIP3P (cyan), AMBER99SB-ILDN-TIP4P (magenta), CHARMM27 (yellow), CHARMM36m (black).

abundance of polar and charged residues.[41] The distribution of the radius of gyration (R_g) was analyzed to characterize the average dimensions of the p53 AD1 conformational ensembles obtained by different force fields. The results are plotted in Fig. 3. We observe that

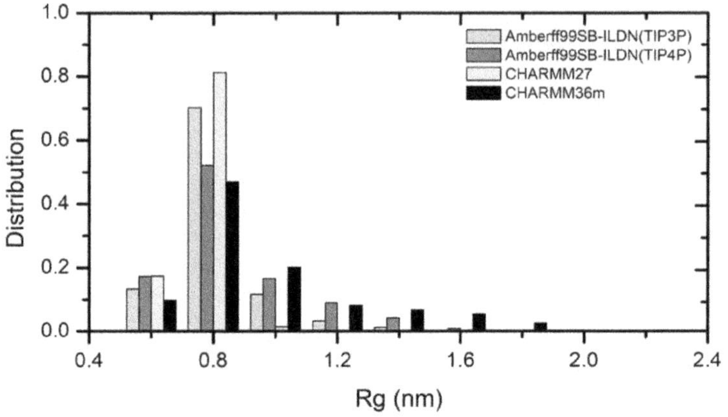

Fig. 3. The distribution of R_g for AMBER99SB-ILDN-TIP3P (cyan), AMBER99S-B-ILDN-TIP4P (magenta), CHARMM27 (yellow), CHARMM36m (black).

the R_g peaks are all located around 0.8 nm under these four force fields. However, the conformation ensembles generated by CHARMM36m and AMBER99SB-ILDN (TIP4P) have a wider distribution of R_g values, especially for CHARMM36m. The conformation ensemble sampled by CHARMM36m includes the conformations with R_g larger than 1.8 nm, which indicates that more extended conformations are sampled under this force field. SAXS experiment indicated that the R_g of full-length p53 (1-91) NTD was approximately 2.7 nm.[42] In our simulations, the p53 AD1 only contains residues 13–37 that leads to the far lower R_g than the full-length p53 NTD. And this comparison supports that the conformations ensemble in the CHARMM36m is more rational than other force fields.

Robustelli *et al.*[43] assessed the accuracy of several force fields by simulating disordered and ordered proteins. They discovered that CHARMM36m (TIP3P) produces larger average R_g value for small disordered proteins (< 60 residues). Ouyang *et al.*[44] studied the conformational features of p53 AD2 with different force fields and found that many flexible conformations were generated by CHARMM36m (TIP3P) and AMBER99SB-ILDN (TIP3P), while CHARMM27 obtained more compact conformations. Similar situations also occurred in our simulation. As shown in Fig. 3,

CHARMM36m (TIP3P) produces the largest average R_g and almost no conformations with R_g larger than 0.8 nm are explored by CHARMM27 force field. But in our simulation conformations with lower R_g are also generated under AMBER99SB-ILDN (TIP3P). Our results show that the R_g produced by AMBER99SB-ILDN (TIP4P) is better than AMBER99SB-ILDN (TIP3P). This indicates that water dispersion interactions have obvious effects on the dimension of conformations. We also analyzed the variation of R_g as a function of the simulation time (as shown in Fig. S2). We can observe that CHARMM36m and AMBER99SB-ILDN (TIP4P) could explore larger R_g conformations in a short simulation time, while the R_g values of conformations obtained by CHARMM27 and AMBER99SB-ILDN (TIP3P) are less than 1.2 nm during the whole simulation. Overall, the conformation ensemble of p53 AD1 generated by CHARMM36m(TIP3P) and AMBER99SB-ILDN (TIP4P) are more extended than the other force fields. And they could sample extended conformations in a short simulation time.

3.3. *Structure properties of p53 AD1*

The secondary structure property of IDPs is often essential for signal transduction. Many disease-related mutations in IDPs have been predicted to alter the secondary structure.[45] The p53 TAD mutant K24N associated with cancer has been shown to decrease the helicity of residues 18–27.[46] Therefore, it is important to assess the quality of current force fields in capturing the residual secondary structure of IDPs. We analyzed different secondary structure proportions in different simulations. As can be seen in Fig. 4(a), the secondary structure proportions largely depend on the chosen force field. The coil and turn possess the largest fraction in all the simulations except under CHARMM27 force field. The fraction of random coil reaches almost 60%, and the β-turn structure possesses 30% in the simulation of CHARMM36m, while almost no α-helix is detected. In the ensemble sampled with AMBER99SB-ILDN, more β-turn structures and little 3_{10}-helix are detected. The Amber family is likely to generate more 3_{10}-helix than other force fields. The secondary

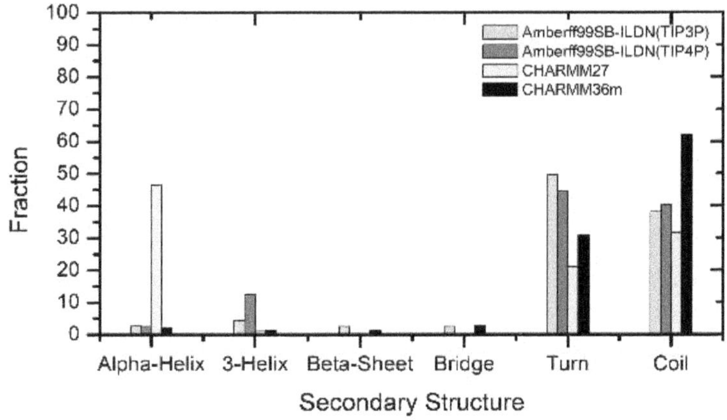

(a)

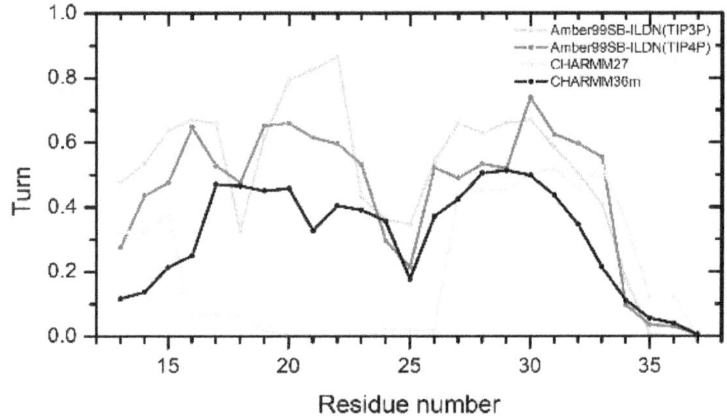

(b)

Fig. 4. (a) The average fraction of different secondary structures in p53 AD1 (13-37) for different simulations. (b) Probability of β-turn structure of each residue by different force fields.

structure property obtained with CHARMM27 is distinct from other force fields, where a much higher probability of α-helix structure is obtained, coming up to 45%. The features displayed in CHARMM36m and AMBER99SB-ILDN are consistent with the previous speculations that there is no tendency to form stable α-helix or β-sheet for free state of p53 (14–28).[47] Robustelli *et al.*[43] proposed that AMBER99SB-ILDN (TIP4P-D) strongly destabilizes helical conformations and this combination performs well for disordered proteins with no residual secondary structure such as Aβ40 and α-synuclein. However, many β-turn and little 3-helix secondary structure were sampled in our system with AMBER99SB-ILDN (TIP4P). This difference may ascribe to the different systems and distinct water models which imply that the careful selection of water models is required in the simulations of IDPs.

NMR studies found that residues 19–25 tend to form two β-turn in the truncated p53 (17–29) and this region would form α-helical structure when binding to MDM2 domain of CBP protein with a coupled folding and binding process.[22,47–49] These residues play an important role in the transactivation process and the coupled folding and binding process.[46,50] Here, we analyzed the β-turn propensity of each amino acid based on the simulations under different force fields and water models. As shown in Fig. 4(b), residues 19–25 and residues 26–33 have a great probability forming β-turn under AMBER99SB-ILDN and CHARMM36m force fields. The β-turn structure proportion obtained by AMBER99SB-ILDN is slightly higher than CHARMM36m. In the conformation ensemble obtained with the CHARMM27, only the C-terminus of p53 AD1 has a certain probability of forming turn structure, and the N-terminal of p53 AD1 maintains helical structure during the whole simulation.

Residues 40–45, 49–54 in p53 AD2 exhibit β-turn structure under experimental measurements,[48,51] but Ouyang *et al.*[44] discovered that partly α-helical structure of these regions was sampled under AMBER99SB-ILDN and Amber03 fore fields. While in our simulation the probability of α-helix obtained by the AMBER99SB-ILDN force field is very low. This implies that the secondary structure property of proteins not only relates to the employed force field but also the properties of the residues in the sequence, residues 40–54 in p53 AD2 contains more charged amino acids than residues 17–29 in p53 AD1.

Cluster analysis was also performed for visualizing the structural details. Top five clusters for these force fields are shown in Fig. 5. The total percentages of top five clusters for AMBER99SB-ILDN (TIP3P), AMBER99SB-ILDN (TIP4P), CHARMM27, and CHARMM36m are 61.83%, 73.11%, 100%, and 55.20%, respectively. The first five

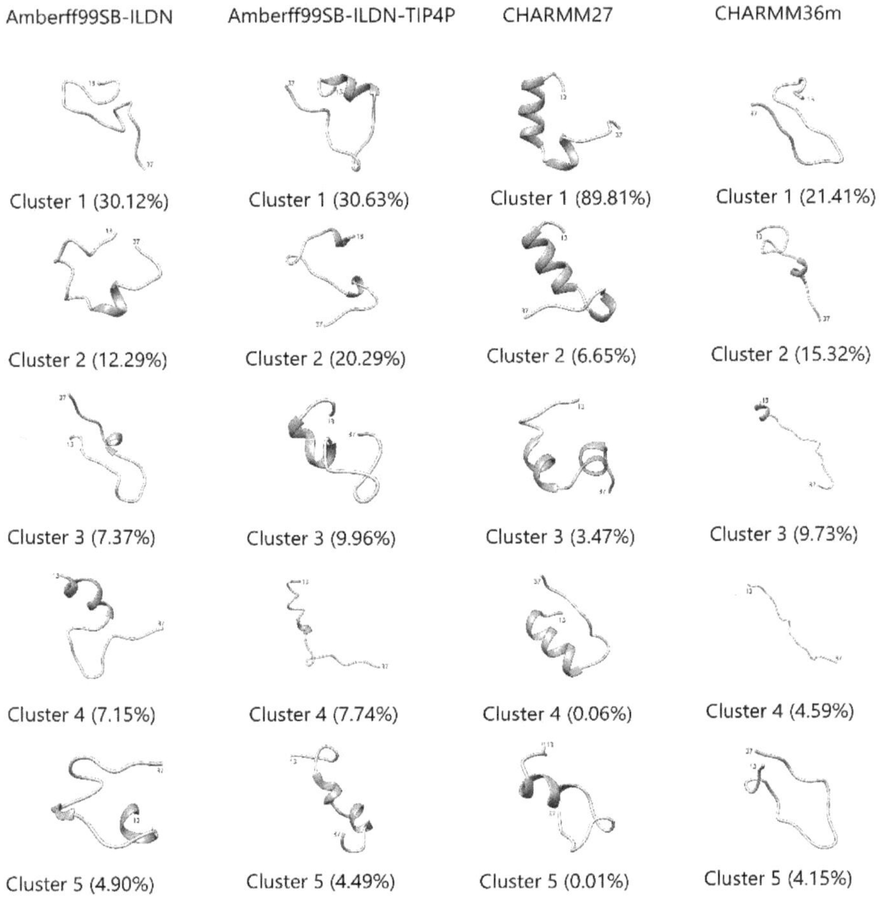

Fig. 5. Representative structures and the proportions of the top five clusters in different simulations. The orange color indicates the α-helical structure, while the yellow indicates the β-turn structure.

clusters in the AMBER99SB-ILDN and CHARMM36m force fields only represent part of the conformation space, which indicates that they explored a more heterogeneous conformational space. The top five clusters of CHARMM27 represent the entire conformation space and the proportion of the first cluster reaching 89.81%, which is more consistent with the conformational characteristics of globular proteins. All the five clusters of CHARMM27 contain helical structure, while clusters generated by CHARMM36m mainly contain random coil and a localized β-turn structure. The conformations obtained with AMBER99SB-ILDN are more heterogeneous, the clusters contain α-helix, β-turn and random coil structures. The proportion of α-helix conformation is higher when AMBER99SB-ILDN combined with the TIP4P water model. Somavarapu *et al.*[52] used different water models to simulate Aβ40 and found that the simulation with TIP3P water model tended to obtain more compact, helical protein structures. This is partly inconsistent with our results, which means that the accuracy of the water model is also highly relevant to the sequence of the system.

3.4. *Comparison with experimental data*

NMR chemical shift is sensitive to the local chemical environment formed by the local structure of proteins. Some excellent chemical shift predicting programs, including SHIFTX, SHIFTX2, SPRATA+, CamShift, etc., have been developed for ordered proteins. SHIFTX uses a unique semi-empirical method to calculate chemical shifts, while SHIFTX2 combines machine learning methods with sequence alignment-based methods to calculate protein chemical shifts for backbone and side-chain atoms.[11] SPARTA+ employs a well-trained neural network algorithm to increase the efficiency of chemical shift prediction based on the known structures.[37] Although the results of structured proteins predicted through these programs are often agreed well with experimental data, the prediction accuracy for intrinsically disordered proteins is still unknown.

To evaluate the performance of different force fields and chemical shift calculated programs, we calculated chemical shifts of N and H_N with different programs and compared with the experimental data.[21] The RMSD between simulated and experimental data for N and H_N chemical shifts in all residues (13–37 aa) is shown in Fig. 6. From Fig. 6(a), we can see that the accuracy of the predicted chemical shifts also depends on the selection of the force fields. The CHARMM27 produces the largest RMSD, while CHARMM36m gives the smallest RMSD. In the case of AMBER99SB-ILDN, TIP3P water model is better than TIP4P water model.

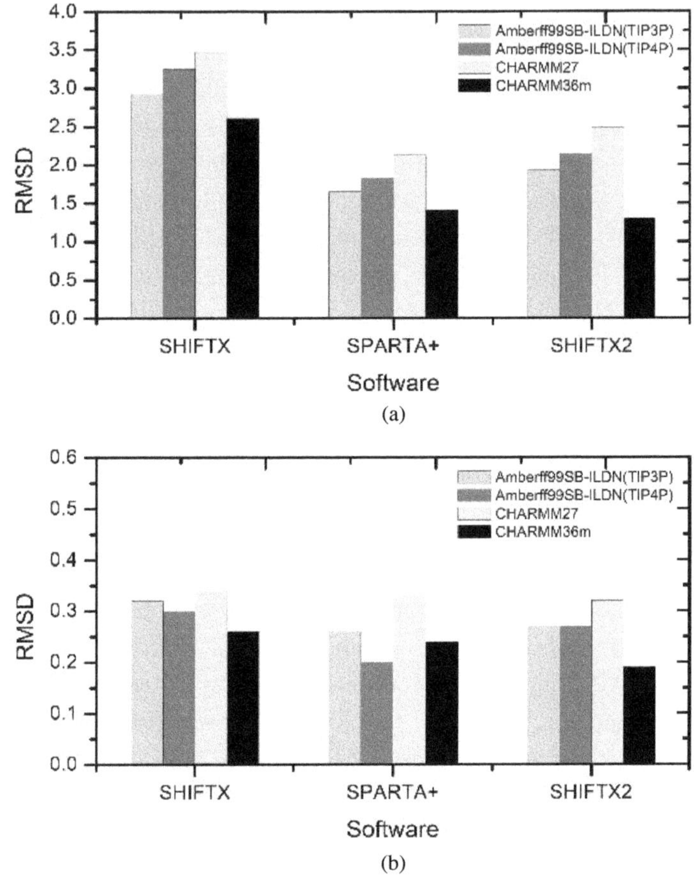

Fig. 6. (a) The RMSD for all residues between simulated and experimental ones for N chemical shifts. (b) The RMSD for all residues between simulated and experimental ones for H_N chemical shifts for these four force fields. Amber99SB-ILDN-TIP3P (cyan), AMBER99SB-ILDN-TIP4P (magenta), CHARMM27 (yellow), CHARMM36m (black).

In Fig. 6(a), the RMSD value of SHIFTX2 and SPARTA+ are close, which indicates that these two programs have similar precision, both are more accurate than SHIFTX. The accuracy of SHIFTX2 is slightly higher than SPARTA+, the RMSD of N Chemical shifts calculated by SHIFTX2 is 1.3, lower than that calculated by SPARTA+ (RMSD = 1.41). The same conclusion also obtained by the RMSD of H_N in Fig. 6(b) (0.19 and 0.24 for SHIFTX2 and

SPARTA+, respectively). In a word, when CHARMM36m combined with SHIFTX2, the chemical shifts are most close to experimental measurements. SHIFTX2 is not only suitable for predicting the chemical shifts of ordered proteins but also for IDPs.

4. Conclusion

The accurate description of the structural feature is important for the application of MD simulations in IDPs. In this paper, three empirical force fields — CHARMM27, AMBER99SB-ILDN, CHARMM36m and two types of water models were employed to explore the structural ensemble of the isolated p53 AD1 domain to assess the quality of these force fields.

Our results indicate that the conformation characteristics of p53 AD1 predicted by MD simulations are strongly related to the force fields. We observe that CHARMM36m produces the structure ensemble with a wide R_g distribution in a short time and CHARMM36m shows the best agreement with experimental chemical shifts. And a low proportion of α-helix and β-turn structure but a higher random coil proportion is produced by CHARMM36m. Thus, CHARMM36m is more suitable for simulating of the proteins with lower secondary structure contents like p53 AD1. The CHARMM27 force field overestimates the α-helix propensity. From the results of cluster analysis, the clusters obtained by AMBER99SB-ILDN are more heterogeneous than the other two force fields. The β-turn structure proportion generated by AMBER99SB-ILDN force field is the highest among three tested force fields. Combined with the chemical shifts results, the accuracy of this force field is better than CHARMM27 but lower than CHARMM36m. Overall, CHARMM36m is more suitable for the simulation of p53 AD1 than other force fields. Different water models also affect the conformation characteristic of p53 AD1. TIP4P can obtain conformation ensemble with larger R_g values, and this water model produces more local helical conformations than the TIP3P water model in our system. We also compared the prediction accuracy of the chemical shift calculation tools SHIFTX, SHIFTX2 and SPRATA+. By comparing with experimental values, SHIFTX2 gets the best performance, which is followed by SPRATA+. Overall, our work provides instructive reference for further studies of p53 AD1 and this work is also conductive to the selection of chemical shifts calculation tool for IDPs.

Acknowledgments

This work was supported by the National Natural Science Foundation of China (31270112) and the Wuhan Polytechnic University initial Scientific Research Fund (2018RZ13).

References

1. van der Lee R, Buljan M, Lang B, Weatheritt RJ, Daughdrill GW, Dunker AK, Fuxreiter M, Gough J, Gsponer J, Jones DT, Kim PM, Kriwacki RW, Oldfield CJ, Pappu RV, Tompa P, Uversky VN, Wright PE, Babu MM, Classification of intrinsically disordered regions and proteins, *Chem Rev* **114**:6589–6631, 2014.
2. Uversky VN, Dave V, Iakoucheva LM, Malaney P, Metallo SJ, Pathak RR, Joerger AC, Pathological unfoldomics of uncontrolled chaos: Intrinsically disordered proteins and human diseases, *Chem Rev* **114**:6844–6879, 2014.
3. Uversky VN, p53 proteoforms and intrinsic disorder: An illustration of the protein structure-function continuum concept, *Int J Mol Sci* **17**:1874–1910, 2016.
4. Huang J, MacKerell AD, Force field development and simulations of intrinsically disordered proteins, *Curr Opin Struct Biol* **48**:40–48, 2018.
5. Whitford PC, Noel JK, Gosavi S, Schug A, Sanbonmatsu KY, Onuchic JN, An all-atom structure-based potential for proteins: Bridging minimal models with all-atom empirical forcefields, *Proteins* **75**:430–441, 2009.
6. Best RB, Hummer G, Optimized molecular dynamics force fields applied to the helix-coil transition of polypeptides, *J Phys Chem B* **113**:9004–9015, 2009.
7. Piana S, Lindorff-Larsen K, Shaw DE, How robust are protein folding simulations with respect to force field parameterization?, *Biophys J* **100**:L47–L49, 2011.
8. Moore SJ, Sonar K, Bharadwaj P, Deplazes E, Mancera RL, Characterisation of the structure and oligomerisation of islet amyloid polypeptides (IAPP): A review of molecular dynamics simulation studies, *Molecules* **23**:2142–2178, 2018.
9. Palazzesi F, Prakash MK, Bonomi M, Barducci A, Accuracy of current all-atom force-fields in modeling protein disordered states, *J Chem Theory Comput* **11**:2–7, 2015.
10. Su X, Wang K, Liu N, Chen J, Li Y, Duan M, All-atom structure ensembles of islet amyloid polypeptides determined by enhanced sampling and experiment data restraints, *Proteins* **87**:541–550, 2019.

11. Han B, Liu Y, Ginzinger SW, Wishart DS, SHIFTX2: Significantly improved protein chemical shift prediction, *J Biomol NMR* **50**:43–57, 2011.
12. Wishart DS, Case DA, Use of chemical shifts in macromolecular structure determination, *Method Enzymol* **338**:3, 2001.
13. Pantelopulos GA, Mukherjee S, Voelz VA, Microsecond simulations of mdm2 and its complex with p53 yield insight into force field accuracy and conformational dynamics, *Proteins* **83**:1665–1676, 2015.
14. Bode AM, Dong Z, Post-translational modification of p53 in tumorigenesis, *Nat Rev Cancer* **4**:793–805, 2004.
15. Vousden KH, Lane DP, p53 in health and disease, *Nat Rev Mol Cell Biol* **8**:275–283, 2007.
16. Greenblatt MS, Bennett WP, Hollstein M, Harris CC, Mutations in the p53 tumor suppressor gene: Clues to cancer etiology and molecular pathogenesis, *Cancer Res* **54**:4855–4878, 1994.
17. Ferreon JC, Lee CW, Arai M, Martinez-Yamout MA, Dyson HJ, Wright PE, Cooperative regulation of p53 by modulation of ternary complex formation with CBP/p300 and HDM2, *Proc Natl Acad Sci USA* **106**:6591–6596, 2009.
18. Soussi T, The p53 tumor suppressor gene: From molecular biology to clinical investigation, *Ann N Y Acad Sci* **910**:121–137; discussion 137–129, 2000.
19. Lee CW, Arai M, Martinez-Yamout MA, Dyson HJ, Wright PE, Mapping the interactions of the p53 transactivation domain with the KIX domain of CBP, *Biochemistry* **48**:2115–2124, 2009.
20. Rajagopalan S, Andreeva A, Rutherford TJ, Fersht AR, Mapping the physical and functional interactions between the tumor suppressors p53 and BRCA2, *Proc Natl Acad Sci USA* **107**:8587–8592, 2010.
21. Arai M, Ferreon JC, Wright PE, Quantitative analysis of multisite protein-ligand interactions by NMR: Binding of intrinsically disordered p53 transactivation subdomains with the TAZ2 domain of CBP, *J Am Chem Soc* **134**:3792–3803, 2012.
22. Lee CW, Martinez-Yamout MA, Dyson HJ, Wright PE, Structure of the p53 transactivation domain in complex with the nuclear receptor coactivator binding domain of CREB binding protein, *Biochemistry* **49**:9964–9971, 2010.
23. Lindorff-Larsen K, Piana S, Palmo K, Maragakis P, Klepeis JL, Dror RO, Shaw DE, Improved side-chain torsion potentials for the Amber ff99SB protein force field, *Proteins* **78**:1950–1958, 2010.
24. MacKerell AD, Bashford D, Bellott M, Dunbrack RL, Evanseck JD, Field MJ, Fischer S, Gao J, Guo H, Ha S, Joseph-McCarthy D, Kuchnir L, Kuczera K, Lau FT, Mattos C, Michnick S, Ngo T, Nguyen DT, Prodhom B, Reiher WE, Roux B, Schlenkrich M, Smith JC, Stote R, Straub J, Watanabe M, Wiorkiewicz-Kuczera J, Yin D, Karplus M, All-atom empirical potential for molecular modeling and dynamics studies of proteins, *J Phys Chem B* **102**:3586–3616, 1998.
25. Mackerell AD, Jr., Feig M, Brooks CL, 3rd, Extending the treatment of backbone energetics in protein force fields: Limitations of gas-phase quantum mechanics in reproducing protein conformational distributions in molecular dynamics simulations, *J Comput Chem* **25**:1400–1415, 2004.

26. Pettersen EF, Goddard TD, Huang CC, Couch GS, Greenblatt DM, Meng EC, Ferrin TE, UCSF Chimera–a visualization system for exploratory research and analysis, *J Comput Chem* **25**:1605–1612, 2004.
27. Huang J, Rauscher S, Nawrocki G, Ran T, Feig M, de Groot BL, Grubmuller H, MacKerell AD, Jr., CHARMM36m: An improved force field for folded and intrinsically disordered proteins, *Nat Methods* **14**:71–73, 2017.
28. Jorgensen W, Chandrasekhar J, Madura J, Impey R, Klein M, Comparison of simple potential functions for simulating liquid water, *J Chem Phys* **79**:926–935, 1983.
29. Case DA, Cheatham TE, 3rd, Darden T, Gohlke H, Luo R, Merz KM, Jr., Onufriev A, Simmerling C, Wang B, Woods RJ, The Amber biomolecular simulation programs, *J Comput Chem* **26**:1668–1688, 2005.
30. Hess B, Kutzner C, van der Spoel D, Lindahl E, GROMACS 4: Algorithms for highly efficient, load-balanced, and scalable molecular simulation, *J Chem Theory Comput* **4**:435–447, 2008.
31. Essmann U, Perera L, Berkowitz ML, Darden T, Lee H, Pedersen LG, A smooth particle mesh Ewald method, *J Chem Phys* **103**:8577–8570, 1995.
32. Nose S, A unified formulation of the constant temperature molecular-dynamics methods, *J Chem Phys* **81**:511–519, 1984.
33. Parrinello M, Rahman A, Polymorphic transitions in single-crystals — A new molecular-dynamics method, *J Appl Phys* **52**:7182–7190, 1981.
34. Hess B, Bekker H, Berendsen HJC, Fraaije JGEM, LINCS: A linear constraint solver for molecular simulations, *J Chem Theory Comput* **18**:1463–1472, 1997.
35. Daura X, Gademann K, Jaun B, Seebach D, van Gunsteren WF, Mark AE, Peptide folding: When simulation meets experiment, *Angew Chem Int Ed* **38**:236–240, 1999.
36. Neal S, Nip AM, Zhang H, Wishart DS, Rapid and accurate calculation of protein 1H, 13C and 15N chemical shifts, *J Biomol NMR* **26**:215–240, 2003.
37. Shen Y, Bax A, SPARTA+: A modest improvement in empirical NMR chemical shift prediction by means of an artificial neural network, *J Biomol NMR* **48**:13–22, 2010.
38. Wishart DS, Bigam CG, Holm A, Hodges RS, Sykes BD, 1H, 13C and 15N random coil NMR chemical shifts of the common amino acids. I. Investigations of nearest-neighbor effects, *J Biomol NMR* **5**:67–81, 1995.
39. Ayed A, Mulder FA, Yi GS, Lu Y, Kay LE, Arrowsmith CH, Latent and active p53 are identical in conformation, *Nat Struct Biol* **8**:756–760, 2001.
40. Dawson R, Müller L, Dehner A, Klein C, Kessler H, Buchner J, The N-terminal domain of p53 is natively unfolded, *J Mol Biol* **332**(5):1131–1141, 2003.
41. Das RK, Pappu RV, Conformations of intrinsically disordered proteins are influenced by linear sequence distributions of oppositely charged residues, *Proc Natl Acad Sci USA* **110**:13392–13397, 2013.
42. Huang F, Rajagopalan S, Settanni G, Marsh RJ, Armoogum DA, Nicolaou N, Bain AJ, Lerner E, Haas E, Ying L, Fersht AR, Multiple conformations of full-length p53 detected with single-molecule fluorescence resonance energy transfer, *Proc Natl Acad Sci USA* **106**:20758–20763, 2009.

43. Robustelli P, Piana S, Shaw DE, Developing a molecular dynamics force field for both folded and disordered protein states, *Proc Natl Acad Sci* **115**:E4758, 2018.
44. Ouyang Y, Zhao L, Zhang Z, Characterization of the structural ensembles of p53 TAD2 by molecular dynamics simulations with different force fields, *Phys Chem Chem Phys* **20**:8676–8684, 2018.
45. Vacic V, Markwick PR, Oldfield CJ, Zhao X, Haynes C, Uversky VN, Iakoucheva LM, Disease-associated mutations disrupt functionally important regions of intrinsic protein disorder, *PLoS Comput Biol* **8**:e1002709, 2012.
46. Zhan YA, Wu H, Powell AT, Daughdrill GW, Ytreberg FM, Impact of the K24N mutation on the transactivation domain of p53 and its binding to murine double-minute clone 2, *Proteins* **81**:1738–1747, 2013.
47. Botuyan MV, Momand J, Chen Y, Solution conformation of an essential region of the p53 transactivation domain, *Fold Des* **2**:331–342, 1997.
48. Lee H, Mok KH, Muhandiram R, Park KH, Suk JE, Kim DH, Chang J, Sung YC, Choi KY, Han KH, Local structural elements in the mostly unstructured transcriptional activation domain of human p53, *J Biol Chem* **275**:29426–29432, 2000.
49. Kussie PH, Gorina S, Marechal V, Elenbaas B, Moreau J, Levine AJ, Pavletich NP, Structure of the MDM2 oncoprotein bound to the p53 tumor suppressor transactivation domain, *Science* **274**:948–953, 1996.
50. Harris SL, Levine AJ, The p53 pathway: Positive and negative feedback loops, *Oncogene* **24**:2899–2908, 2005.
51. Chi SW, Lee SH, Kim DH, Ahn MJ, Kim JS, Woo JY, Torizawa T, Kainosho M, Han KH, Structural details on mdm2-p53 interaction, *J Biol Chem* **280**:38795–38802, 2005.
52. Somavarapu AK, Kepp KP, The dependence of amyloid-beta dynamics on protein force fields and water models, *Chemphyschem* **16**:3278–3289, 2015.

https://doi.org/10.1142/9789819808199_0006

Chapter 6

Changes in Structure and Flexibility of p53 TAD2 Upon Binding to p300 Taz2*

Tongtong Li[†], Amy O. Stevens[†], Laura I. Gil Pineda[†,‡], Shenghan Song[†], Christabel A. Ameyaw Baah[†] and Yi He[†,§]

[†]*Department of Chemistry & Chemical Biology*
The University of New Mexico, Albuquerque
New Mexico 87131, United States
[‡]*Department of Chemical Sciences*
Universidad Icesi, Cali 760031, Colombia
[§]*yihe@unm.edu*

p53 is a transcription factor with intrinsically disordered regions that plays an essential role in many cellular processes. As a tumor suppressor, the dysfunction of p53 causes various cancers. p53 can be activated by binding with cofactors in the cell due to stresses or DNA damages. The N-terminal transactivation domain (TAD) of p53 can regulate cell apoptosis by interacting with its binding partners, such as the transcriptional adaptor zinc-binding 2 (Taz2) domain of p300, and cofactors, i.e. cyclic-AMP response element-binding protein (CREB)-binding protein (CBP). The experimentally solved structure of p300 Taz2 and p53 TAD2 has provided insights into the interactions that potentially lead to the structural changes of p53 TAD2 and stabilize the complex. To explore the structural changes as well as the residues that lead to such changes from an isolated state to a bound state in p53 TAD2, we used all-atom molecular dynamics (MD) to simulate two different systems: (1) the p300 Taz2–p53 TAD2 complex and (2) isolated p53 TAD2 (residues 35–59). Although still largely unstructured, residues across the p53 TAD2 contribute significantly to

*To cite this article, please refer to its earlier version published in *J. Theor. Comput. Chem.* **2020**, *19* (4), 2040007, DOI: 10.1142/S0219633620400076
§Corresponding author.

stabilizing the binding between p300 Taz2 and p53 TAD2. Our results suggest that the binding affinity of the p300 Taz2–p53 TAD2 complex originates from hydrophobic and electrostatic interactions. The results are in agreement with previous reports and experimental data. By comparing the two simulated systems, our results not only demonstrate the structural changes of p53 TAD2 after binding with Taz2 but also identify the key residues leading to such changes. We also identify the critical residues that can provide insight into the interaction network between p300 Taz2 and p53 TAD2.

Keywords: Molecular dynamics; IDP; p53 TAD2; p300 Taz2; structural transition.

1. Introduction

Protein–protein interactions are an essential part of biological processes in all organisms. This process requires complementary sizes, charges and hydrophilic/hydrophobic properties between the two binding partners. In general, several critical sites contribute to the stability of the complex and facilitate the binding process. The interaction of intrinsically disordered proteins (IDPs) with other proteins is notably unique considering IDPs do not possess specific structures in their free state. An in-depth understanding of the interactions between IDPs and their binding partners in the functional biological complexes is the key to accelerating the development of drugs[1] that mimic signal factors.

p53 factor consists of three domains,[2] including a DNA binding domain, a tetramerization domain and an N-terminal transactivation domain (TAD). The p53 N-terminal transactivation region is composed of two TADs, TAD1 and TAD2.[3] Both domains can activate the transactivation process independently.[4] As a tumor suppressor, p53 plays an essential role in numerous signaling pathways relating to the cell division cycle and DNA damage.[5] The disruption of p53 causes many diseases, such as cancer,[6–8] fatty liver disease,[9,10] and Alzheimer's disease.[11] In typical cells, Murine Double Minute 2 (MDM2) can bind with the TAD of p53 to mediate p53 degradation.[12] However, p53 will be activated to respond to stresses by binding with transcriptional cofactor-cyclic-AMP response element-binding protein (CREB)-binding protein (CBP) and p300 in cells due to stresses or DNA damage. The 3D structures of the corresponding domains which interact with p53 in CBP and p300 are similar.[13] p53 can bind to four domains of p300. Among the four domains, the Taz2 domain binds with the p53 TAD with the greatest affinity.[14] Taz2 domain competitively binds with the p53 TAD, protecting p53 from degradation by MDM2. The Taz2 domain includes four helices ($\alpha 1$, $\alpha 2$, $\alpha 3$, and $\alpha 4$ helices) and three zinc finger motifs stabilizing the helical structures.[15] The $\alpha 1$, $\alpha 2$ and $\alpha 3$ helices form a hydrophobic pocket which often accepts hydrophobic residues from its binding partners, such as TAD.[16] It was suggested that both the TAD1 and TAD2 domains could form a stable amphipathic α-helix upon binding with Taz2 through hydrophobic and electrostatic interactions.[17] The Taz2–p53 complex regulates the transcription process and induces subsequent cellular activities.

To date, the structures of p53 complexes with different binding partners have been determined using multiple methods, such as NMR spectroscopy,[14,18,19] small-angle X-ray scattering (SAXS),[20] computational simulation,[2,21] and other methods.[3,22] Residual dipolar couplings (RDCs) from NMR and SAXS have been measured to analyze isolated TAD, TAD in tetrameric p53 and the p53 complex with its cofactor. The findings indicate that the isolated TAD has a similar conformation to TAD in the context of p53 due to the stiff proline-rich region (PPR) between TAD and core domain.[20] With these results, p53 can be simplified to the isolated TAD when studying the binding of TAD with other proteins. The exchange broadening of the NMR spectra of the complexes creates a great challenge when analyzing the data. A recent study proposed a method to solve this problem using fusion proteins in the study of the p53–Taz2 complex.[19]

In recent years, many reports have studied the interaction networks in the Taz2–p53 complex.[16,18] Interactions between the critical residues provide key information on the interaction network between p300 Taz2 and p53 TAD2 in the complex conformation. In the p300 Taz2–p53 TAD1 complex,[19] the hydrophobic residues (Phe19, Leu22, W23, Leu25, and Leu26) of p53 TAD1 can form a hydrophobic core with residues (Thr1812, Leu1823, Ile1824, Ala1825, Phe1843) of CBP Taz2. Mutational analyses[16] also demonstrated that residues Phe19, Leu22 and Leu25 form hydrophobic interactions with the groove formed by the $\alpha1$, $\alpha2$ and $\alpha3$ helices of p300 Taz2. The aromatic ring of Phe19 contacts with side chains of Arg1737 and Lys1760. Additionally, the positive charge on Lys1760 can be stabilized by the π electrons of Phe19. Leu22 contacts with Val1764 and Met1761, while Leu25 contacts with Ile1781 and Pro1780. Several N-terminal residues of TAD1 stabilize the complex by electrostatic interactions with charged or polar residues of Taz2. This was confirmed by NMR data as the chemical shifts of residues 10, 11, 14, and 15 change largely. NMR data[19] indicated that hydrophobic and electrostatic interactions increase the affinity between Taz2 and p53 TAD. In the p300 Taz2–p53 TAD2 complex, the hydrophobic residues (M44, L45) can bind with the hydrophobic surface of Taz2. Additionally, the hydrophobic surface of Taz2 is electropositive, forming additional electrostatic contacts with acidic p53 residues (D41, D42 and E51). I50, W53 and F54 can form a hydrophobic core in the binding pocket of Taz2. It has been reported that two mutants[14] of p53 (L22Q/W23S, W53Q/F54S) have a much lower affinity with Taz2 when compared with the wild-type p53 structure. The NMR spectra and mutational analysis based on isothermal titration calorimetry (ITC)[18] indicated that hydrophobic interactions and salt bridge bonds between Taz2 and p53 TAD contribute to the stability of the complex. Interestingly, the TAD1 and TAD2 subunits interact with the same interface of Taz2 but bind in a perpendicular orientation, due to the flexibility of TAD. Post-translational modification is a common way to mediate the functions of proteins. Phosphorylation of p53 at specific sites (Ser15 and Thr18 in TAD1) can increase its binding affinity with Taz2, whereas phosphorylation of residues in TAD2 (Ser46 and Thr55) does not affect the binding affinity.[3] A recent report[1] compared the binding affinity of TAD1 with Taz2 domain and MDM2 and analyzed the influence of phosphorylation on affinity. They found that TAD1 can bind with MDM2 with a higher binding affinity due to different residues anchoring in the corresponding binding pocket. Phosphorylation at particular sites of TAD1 can increase its binding affinity with Taz2, however decrease when binding with MDM2. This is because phosphorylation increases the electrostatic interaction in Taz2–TAD1 by introducing negative charge to TAD1.

The structure and interaction networks of the Taz2–p53 complex are essential to understanding the driving forces behind binding. Although some reports are centered on the Taz2–p53 TAD1 complex, there is an urgent need to explore the inherent binding mechanisms of Taz2–p53 TAD2 with extensive molecular dynamics (MD) simulations. In this paper, we focus on p53 TAD2 (residues L35-G59) and p300 Taz2 (residues A1723-K1812, including four mutations: C1738A, C1746A, C1789A, and C1790A). An early report suggested that the mutant not only has similar behaviors as wild-type Taz2 of p300 but also has almost identical structures.[3] Taz2 of CBP has similar properties with its mutant (from cysteine to alanine residues).[15] To analyze the critical residues and their interactions in the binding pocket to provide new insights to the binding structural basis, we used all-atom MD simulations to simulate the dynamics of Taz2–p53 TAD2 (residues 35–59) complex and isolated p53 (residues 35–59).

2. Methods

Two sets of simulations were carried out. The first set of simulations were used to investigate the dynamics of the p300 Taz2–p53 TAD2 complex. The experimentally determined p300 Taz2–p53 TAD2 complex structure (PDB ID: 2MZD)[18] was used as the starting structure for this set of simulations. The second set of simulations was used to investigate the dynamics of the isolated N-terminal TAD of p53 TAD2. The p53 structure extracted from the previously used complex structure was used as the starting structure. The starting structure for the p300 Taz2–p53 TAD2 complex and isolated p53 TAD2 is shown in Fig. 1. All the simulations were carried out using the CHARMM36m force field with explicit solvents and the Groningen Machine for Chemical Simulations (GROMACS) package,[23–25] version 2018.3. The protein was placed in a cubic box with the

(a) (b)

Fig. 1. (Color online) Starting structures of Taz2–p53 (35–59) complex and free p53 TAD2 (35–59) simulations. (a) Starting structure of the p300 Taz2 (green) and p53 TAD2 (red) complex. (b) Starting structure of free p53 TAD2 simulations (colored in rainbow). The structure of p53 TAD2 used here is the same as in the complex but with Taz2 removed. The structure shown above in panel B is rotated 90° compared to the p53 TAD2 shown in panel A.

corresponding water model and counter ions (Na^+ or Cl^-) to neutralize the whole system at 300 K. Long-range electrostatics were calculated using the particle-mesh Ewald (PME) algorithm.[26,27] Periodic boundary conditions were applied in all directions. Each system of protein, water and counter ions was prepared using CHARMM-GUI,[28,29] which generates a series of GROMACS inputs for subsequent MD simulations.

To generate equilibrated starting structures for the MD simulations, steepest-descent minimization was carried out, followed by a 1-ns MD equilibrium simulation with a time step of 1 fs, to heat the whole system from 1 K to the desired temperature (300 K). All bonds with hydrogen atoms are converted to constraints with the algorithm LINear Constraint Solver (LINCS),[30] using the default parameters of the GROMACS package. The equilibrated structures obtained from the above steps were used for subsequent production runs. A Nose–Hoover temperature thermostat[31,32] was used to maintain the temperature. The time step was 2 fs, and snapshots were taken every 100 ps. A cubic water box size of 110 Å was employed and run for a total of 12 μs per system, including two 6 μs long MD trajectories.

3. Results and discussion

3.1. *Helical propensity of p53 TAD2 in the complex and in isolated state*

Based on the starting structures of the complex and isolated p53 TAD2 simulations, p53 TAD2 in both the bound and unbound states contains an α-helix that spans residues 47–55. To investigate the stability of the helical regions in p53 TAD2, we performed DSSP[33,34] analysis for the combined trajectory of the p300 Taz2–p53 TAD2 complex as well as the combined trajectory of the isolated p53 TAD2 simulations.

As shown in Fig. 2, p53 residues 47–55 have helical preferences in the isolated environment but form a stable amphipathic α-helix upon binding with Taz2. This agrees with the previous report by Miller Jenkins *et al.*[18] and demonstrates that the binding between p300 Taz2 and p53 TAD2 enhances the stability of the helix formed by residue 47–55. The N-terminus and C-terminus of TAD2 show no distinct helical structures, illustrating that they are flexible and potentially extended. This flexibility helps TAD to recognize a diverse array of binding partners. In the isolated p53 TAD2, the α-helix is transient. However, when bound to Taz2, the frequency of α-helix increases almost four-folds, indicating the conformation of the α-helix is more stable compared to the isolated fragment. The secondary structure of

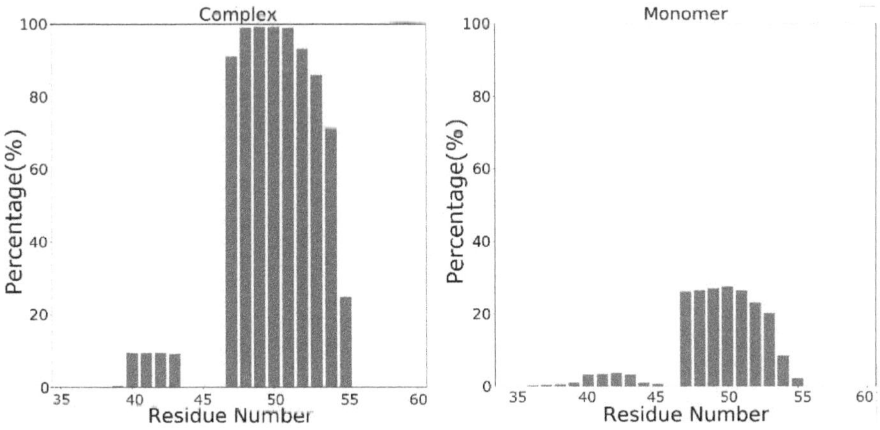

Fig. 2. Helical propensity analyses of p53 TAD2 in the complex (left) and in isolation (right).

p53 TAD2 is in agreement with experimental data.[18] The N-terminal residues 40–43 can form a more dynamic α-helix compared with residues 47–55. The stability of this helix also increased in the complex, due to the formation of contacts between the N-terminal residues and p300 Taz2. However, even bound with p300 Taz2, the frequency of N-terminal α-helix is less than 10%.

To explore the flexibility of the N- and C-terminal of p53 TAD2 region, RMSF analysis was used. The RMSF plot (Fig. 3) displays the fluctuation of residues in different regions. As shown in Fig. 2, p53 TAD2 is composed of the α-helix in the middle region and disordered structures in the N-terminus and C-terminus. The fluctuation in the disordered terminal regions is clearly larger than that in the helical region. When bound with p300 Taz2, the fluctuation of p53 TAD2 decreases distinctively. The RMSF plot indicates the stability of the helix structure increases greatly upon binding with p300 Taz2. Furthermore, the fluctuation of terminal regions also decreases,

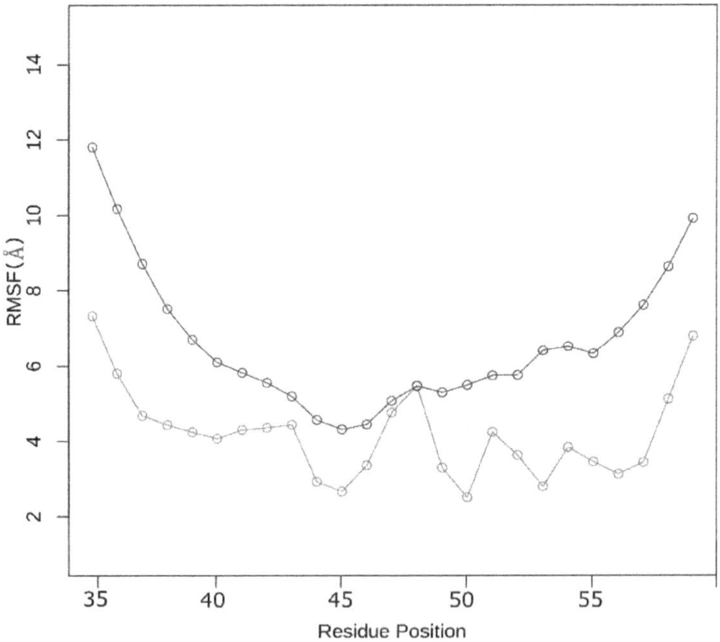

Fig. 3. (Color online) RMSF plot of p53 TAD2 (35–59). p53 TAD2 (35–59) in the complex is colored in red and isolated p53 TAD2 (35–59) in blue.

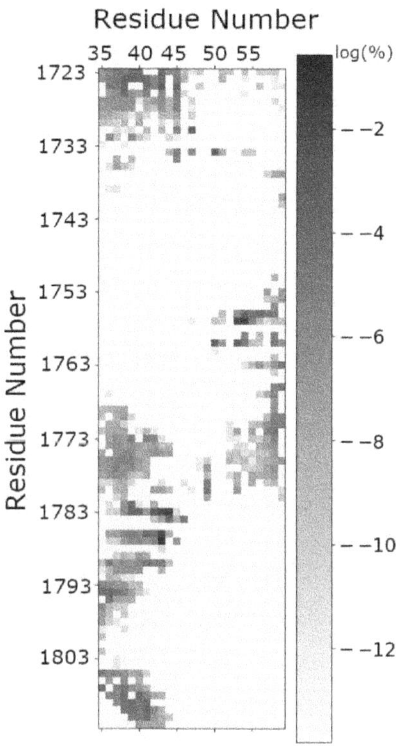

Fig. 4. (Color online) Contact map of the Taz2–p53 TAD2 (35–59) complex. The X-axis indicates the residue number of the 25-residue (35–59) N-terminal p53 TAD2, and the Y-axis is the p300 Taz2 (residues 1723–1812). The color of the scale bar represents the intermolecular forces between the residues of TAD2 and Taz2, the darker color indicating stronger interaction.

indicating the flexibility of the disordered residues is restrained due to the interactions between the flexible residues and p300 Taz2 (more details in Fig. 4).

3.2. *Critical residues and interaction network in the p300 Taz–p53 TAD2 complex*

It was proposed earlier that in the p300 Taz2–p53 TAD2 complex, p53 TAD2 forms a short amphipathic α-helix containing hydrophobic and acid residues. Taz2 forms a hydrophobic and electropositive binding pocket.[18] Therefore, p53 TAD2 binds with p300 Taz2 through hydrophobic and electrostatic interactions. In the contact map (shown in Fig. 4), it is clear that the preferred interaction regions are quite different between the N-terminus and C-terminus (including the helical part, residues 47–55) of p53 TAD2. The stable α-helix (residues 47–55) in p53 TAD2 tends to interact with p300 Taz2 residues 1756–1781. It should be noted that the residues 35–44 of the N-terminal p53 TAD2 form interactions with more than 50% of the p300 Taz2 protein and play a significant role in stabilizing the p300 Taz2–p53 TAD2 complex. The interactions with p300 Taz2 restrain the flexibility of disordered regions and improve the stability of the secondary structure of p53 TAD2.

The amphipathic α-helix in p53 TAD2 can interact with p300 Taz2 through hydrophobic and electrostatic interactions. It has been shown that mutations of conserved Trp53 and Phe54[14] induce weakened binding with Taz2. According to the contact map, Trp53 forms hydrophobic interactions with Pro1756 and the nonpolar portion of Lys1760. Phe54 forms hydrophobic interactions with Pro1756. The hydrogen of indole group (Trp53) and hydrogens of phenyl ring (Phe54) carry partially positive charge, forming strong interactions with the hydroxyl oxygen of Ser1757. The interactions between the side chains of Trp53 and Phe54 and the hydrophobic pocket of Taz2 limit the flexibility of the C-terminus of the helix in p53 TAD2. Moreover, Ile50 of p53 TAD2 contacts with nonpolar portions of Ser1734 and Lys1760 of p300 Taz2, which further immobilizes the middle region of the helix in p53 TAD2. Pro47 forms hydrophobic interactions with the nonpolar portion of Arg1731, preventing the disassociation of the N-terminus of the helix in p53 TAD2. The hydrophobic residues (Trp53, Phe54, Ile50, and Pro47) stabilize the helix conformation in the p300 Taz2–p53 TAD2 complex by interacting with nonpolar residues or the nonpolar portions of polar residues in p300 Taz2. The acidic residues in the α-helix of TAD2 form electrostatic interactions with the positive binding pocket of Taz2. For example, Glu51 contacts with polar Ser1734 and electropositive Lys1760.

The N-terminal residues of p53 TAD2 can bind with a large range of residues in p300 Taz2. Leu43 binds with Ala1787, Ile1786 and the

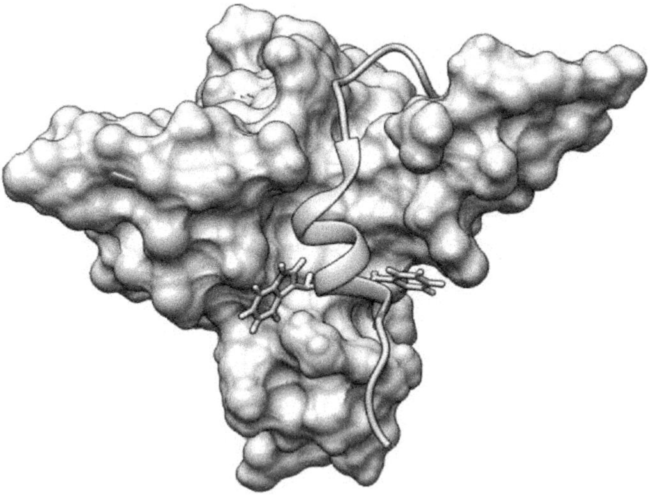

Fig. 5. Structure of the Taz2–p53 (35–59) complex.

nonpolar portion of Lys1783. Met44 binds with the nonpolar portion of Lys1783. The acidic Asp42 forms the salt bridge with electropositive Lys1783.

The conserved residues (Trp53 and Phe54, shown in Fig. 5) play significant roles in the formation of p300 Taz2–p53 TAD2 complex. The binding affinity[14,18] of the p300 Taz2–p53 TAD2 complex is a result of both hydrophobic interactions between nonpolar residues of p53 TAD2 and the hydrophobic binding site of p300 Taz2 as well as electrostatic interactions between acidic residues of p53 TAD2 and polar or basic residues of p300 Taz2.

4. Conclusion

With MD simulations, we analyzed the changes in structure and flexibility of p53 TAD2 upon binding to the Taz2 domain of p300. The helix frequency plot and RMSF plot indicate that the stability of helices in the p53 TAD2 increases greatly upon binding with p300 Taz2. This originates from (1) the hydrophobic and electrostatic interactions between helix in p53 TAD2 and hydrophobic binding pocket of p300 Taz2 and (2) interactions between flexible N-terminus in p53 TAD2 and broad regions in p300 Taz2. Both the amphipathic helix and the flexible N-terminus provide contributions to the formation of the p300 Taz2–p53 TAD2 complex. In the interaction networks, several critical residues[14] (Trp53 and Phe54) in p53 TAD2 interact with the hydrophobic interface formed by $\alpha 1$, $\alpha 2$ and $\alpha 3$ helices of Taz2. The electropositive binding pocket interacts with acidic residues in TAD2, further stabilizing the complex. The structural analyses of the Taz2–p53 TAD2 complex can provide insights into the binding structural analysis of p53 TAD2 with its cofactors and ultimately facilitate drug designs to regulate critical signaling pathways.

References

1. Kannan S, Partridge AW, Lane DP, Verma CS, The dual interactions of p53 with MDM2 and p300: Implications for the design of MDM2 inhibitors, *Int J Mol Sci* **20**:5996(1–21), 2019.
2. Ithuralde RE, Turjanski AG, Phosphorylation regulates the bound structure of an intrinsically disordered protein: The p53-TAZ2 case, *PLoS One* **11**:1–17 (2016).
3. Jenkins LMM *et al.*, Two distinct motifs within the p53 transactivation domain bind to the taz2 domain of p300 and are differentially affected by phosphorylation, *Biochemistry* **48**:1244–1255, 2009.
4. Candau R *et al.*, Two tandem and independent sub-activation domains in the amino terminus of p53 require the adaptor complex for activity, *Oncogene* **15**:807–816, 1997.
5. Green DR, Kroemer G, Cytoplasmic functions of the tumour suppressor p53, *Nature* **458**:1127–1130, 2009.
6. Vogelstein B, Lane D, Levine AJ, Surfing p53 network, *Nature* **408**:307–310, 2000.
7. Rosenfeldt MT *et al.*, P53 status determines the role of autophagy in pancreatic tumour development, *Nature* **504**:296–300, 2013.
8. Zenz T *et al.*, TP53 mutation and survival in chronic lymphocytic leukemia, *J Clin Oncol* **28**:4473–4479, 2010.
9. Zhou H *et al.*, Effects of melatonin on fatty liver disease: The role of NR4A1/DNA-PKcs/p53 pathway, mitochondrial fission, and mitophagy, *J Pineal Res* **64**:1–19, 2018.
10. Castro RE *et al.*, MiR-34a/SIRT1/p53 is suppressed by ursodeoxycholic acid in the rat liver and activated by disease severity in human non-alcoholic fatty liver disease, *J Hepatol* **58**:119–125, 2013.
11. Jazvinšćak Jembrek M, Slade N, Hof PR, Šimić G, The interactions of p53 with tau and Aß as potential therapeutic targets for Alzheimer's disease, *Prog Neurobiol* **168**:104–127, 2018.

12. Haupt Y, Maya R, Kazaz A, Oren M, Mdm2 promotes the rapid degradation of p53, *Nature* **387**:296–299, 1997.
13. Papaleo E, Casiraghi N, Arrigoni A, Ranzani V, Atomistic insights into the regulatory mechanisms mediated by post-translational modifications: Molecular dynamics investigations, *Curr Phys Chem* **2**:344–362, 2012.
14. Teufel DP, Freund SM, Bycroft M, Fersht AR, Four domains of p300 each bind tightly to a sequence spanning both transactivation subdomains of p53, *Proc Natl Acad Sci USA* **104**:7009–7014, 2007.
15. De Guzman RN, Liu HY, Martinez-Yamout M, Dyson HJ, Wright PE, Solution structure of the TAZ2 (CH3) domain of the transcriptional adaptor protein CBP, *J Mol Biol* **303**:243–253, 2000.
16. Feng H *et al.*, Structural basis for p300 Taz2-p53 TAD1 binding and modulation by phosphorylation, *Structure* **17**:202–210, 2009.
17. Lee H *et al.*, Local structural elements in the mostly unstructured transcriptional activation domain of human p53, *J Biol Chem* **275**:29426–29432, 2000.
18. Miller Jenkins LM, Feng H, Durell SR, Tagad HD, Mazur SJ, Tropea JE, Bai Y, Appella E, Characterization of the p300 Taz2-p53 TAD2 complex and comparison with the p300 Taz2-p53 TAD1 complex, *Biochemistry* **54**:2001–2010, 2015.
19. Krois AS, Ferreon JC, Martinez-Yamout MA, Dyson HJ, Wright PE, Recognition of the disordered p53 transactivation domain by the transcriptional adapter zinc finger domains of CREB-binding protein, *Proc Natl Acad Sci USA* **113**:E1853–E1862, 2016.
20. Wells M *et al.*, Structure of tumor suppressor p53 and its intrinsically disordered N-terminal transactivation domain, *Proc Natl Acad Sci USA* **105**:5762–5767, 2008.
21. Saglam AS, Wang DW, Zwier MC, Chong LT, Flexibility vs preorganization: Direct comparison of binding kinetics for a disordered peptide and its exact preorganized analogues, *J Phys Chem B* **121**:10046–10054, 2017.
22. Coactivator T, Sengupta J, Tumor suppressor p53-mediated structural reorganization of the transcriptional coactivator p300, *Biochemistry* **58**:3434–3443, 2019.
23. Berendsen HJC, van der Spoel D, van Drunen R, GROMACS: A message-passing parallel molecular dynamics implementation, *Comput Phys Commun* **91**:43–56, 1995.
24. Pall S, Abraham MJ, Kutzner C, Hess B, Lindahl E, Tackling exascale software challenges in molecular dynamics simulations with GROMACS, in *Lecture Notes in Computer Science (including subseries Lecture Notes in Artificial Intelligence and Lecture Notes in Bioinformatics)*, Springer, Vol. 8759, pp. 3–27, 2015.
25. Abraham MJ *et al.*, Gromacs: High performance molecular simulations through multi-level parallelism from laptops to supercomputers, *SoftwareX* **1–2**:19–25, 2015.
26. Darden T, York D, Pedersen L, Particle mesh Ewald: An N·log(N) method for Ewald sums in large systems, *J Chem Phys* **98**:10089, 1993.

27. Essmann U *et al.*, A smooth particle mesh Ewald method, *J Chem Phys* **103**:8577–8593, 1995.
28. Jo S, Kim T, Iyer VG, Im W, CHARMM-GUI: A web-based graphical user interface for CHARMM, *J Comput Chem* **29**:1859–1865, 2008.
29. Lee J *et al.*, CHARMM-GUI input generator for NAMD, GROMACS, AMBER, OpenMM, and CHARMM/OpenMM simulations using the CHARMM36 additive force field, *J Chem Theory Comput* **12**:405–413, 2016.
30. Hess B, Bekker H, Berendsen HJC, Fraaije JGEM, LINCS: A linear constraint solver for molecular simulations, *J Comput Chem* **18**:1463–1472, 1997.
31. Nose S, A unified formulation of the constant temperature molecular dynamics methods, *J Chem Phys* **81**:511, 1984.
32. Hoover WG, Canonical dynamics: Equilibrium phase-space distributions, *Phys Rev A* **31**:1695–1697, 1985.
33. Kabsch W, Sander C, Dictionary of protein secondary structure: Pattern recognition of hydrogen-bonded and geometrical features, *Biopolymers* **22**:2577–2637, 1983.
34. Touw WG *et al.*, A series of PDB-related databanks for everyday needs, *Nucleic Acids Res* **43**:D364–D368, 2015.

Part III
Computational Biophysics and Chemistry Methods to Predict pKa of Ionizable Groups in Proteins, RNAs, DNAs, and Small Molecules

https://doi.org/10.1142/9789819808199_0007

Chapter 7

Computing Protein pKas Using the TABI Poisson–Boltzmann Solver*

Jiahui Chen[†,‖], Jingzhen Hu[‡,**], Yongjia Xu[§,††], Robert Krasny[¶,‡‡] and Weihua Geng[§,§§]

[†]*Department of Mathematics, Michigan State University, East Lansing, MI 48824, USA*
[‡]*Department of Mathematics, Duke University, Durham, NC 27710, USA*
[§]*Department of Mathematics, Southern Methodist University, Dallas, TX 75275, USA*
[¶]*Department of Mathematics, University of Michigan, Ann Arbor, MI 48109, USA*
[‖]*chenj159@msu.edu*
[**]*jingzhen.hu@duke.edu*
[††]*yongjiax@smu.edu*
[‡‡]*krasny@umich.edu*
[§§]*wgeng@smu.edu*

A common approach to computing protein pKas uses a continuum dielectric model in which the protein is a low dielectric medium with embedded atomic point charges, the solvent is a high dielectric medium with a Boltzmann distribution of ionic charges, and the pKa is related to the electrostatic free energy which is obtained by solving the Poisson–Boltzmann equation. Starting from the model pKa for a titrating residue, the method obtains the intrinsic pKa and then computes the protonation

*To cite this article, please refer to its earlier version published in *J. Comput. Biophys. Chem.* **2021**, *20* (2), 175–187, DOI: 10.1142/S2737416520420065
§§Corresponding author.

probability for a given pH including site–site interactions. This approach assumes that acid dissociation does not affect protein conformation aside from adding or deleting charges at titratable sites. In this work, we demonstrate our treecode-accelerated boundary integral (TABI) solver for the relevant electrostatic calculations. The pKa computing procedure is enclosed in a convenient Python wrapper which is publicly available at the corresponding author's website. Predicted results are compared with experimental pKas for several proteins. Among ongoing efforts to improve protein pKa calculations, the advantage of TABI is that it reduces the numerical errors in the electrostatic calculations so that attention can be focused on modeling assumptions.

Keywords: Protein pKa; acid dissociation constant; continuum dielectric model; Poisson–Boltzmann equation; boundary integral equation; treecode.

1. Introduction

Proton transfer between the ionizable amino acids of a solvated protein and the surrounding aqueous solvent has an important effect on pH-dependent processes such as enzymatic activity[1] and protein–ligand binding.[2] In the ideal case of an acid with a single ionizable group, the pKa can be measured by simple acid–base titration. The acid dissociation reaction, $HA \rightleftharpoons A^- + H^+$, has equilibrium constant defined by

$$\mathrm{Ka} = \frac{[A^-][H^+]}{[HA]}, \tag{1}$$

where $[HA]$, $[A^-]$, $[H^+]$ are the equilibrium concentrations of the protonated acid, its deprotonated conjugate, and free protons. The Henderson–Hasselbalch equation,

$$\mathrm{pH} = \mathrm{pKa} + \log_{10}\frac{[A^-]}{[HA]}, \tag{2}$$

relates the solvent $\mathrm{pH} = -\log_{10}[H^+]$ with the acid $\mathrm{pKa} = -\log_{10}\mathrm{Ka}$ and shows that $\mathrm{pH} = \mathrm{pKa}$ when $[HA] = [A^-]$, i.e. when the acid is equally likely to be protonated or deprotonated.

Proteins however have multiple ionizable groups that titrate at different rates depending on the solution pH, and measuring the pKas of the various sites is more difficult. One approach to determining protein pKas uses chemical shifts measured experimentally by NMR spectroscopy,[3] Alternatively, protein pKas can be predicted theoretically,[4] and methods reported in the literature employ continuum dielectric models,[5–7] Monte Carlo methods,[8] molecular dynamics and QM/MM simulations,[9,10] and empirical relationships.[11,12]

This work focuses on a continuum dielectric model in which the protein pKas are given by electrostatic free energy shifts, where the solvation free energy is obtained by solving the Poisson–Boltzmann (PB) equation.[5,13,14] The protein is represented by a low dielectric cavity with embedded atomic partial charges, and the surrounding aqueous solvent is a high dielectric medium with dissolved ions.[15,16] The dielectric interface is most commonly assumed to be the molecular surface (or solvent-excluded surface).[17] The model assumes that protonation/deprotonation has negligible effect on the protein conformation, and the electrostatic free energy shifts arise entirely from changes in the protonation state of the ionizable groups. Under this assumption, the same protein structure is used for all protonation states, and the pKa computation requires solving the PB equation many times with different charge distributions to obtain the electrostatic free energy shifts. This calls for an accurate and efficient numerical method to handle the challenges of PB simulations which arise due to the irregular geometry of the molecular surface, the jump in dielectric constant across the surface, the singularity of the atomic partial charges representing the protein, and the infinite domain on which the problem is posed.

Among several numerical PB solvers developed in the literature, the treecode-accelerated boundary integral (TABI) method is chosen for the pKa computations in this work. The boundary integral formulation used in TABI analytically satisfies the dielectric jump conditions and accounts for the singular charges and infinite domain, while the irregular geometry of the molecular surface is resolved by triangulation. Compared to alternative grid-based PB solvers, the TABI solver is particularly useful in treating the complex charge distributions necessary in pKa calculations. Our numerical results show that TABI can provide accurate electrostatic free energies and protein pKas even with a relatively coarse triangulation. A Python wrapper was written to pipeline the entire pKa computing procedure using TABI and is publicly available.

The rest of this article is organized as follows. Section 2 presents the continuum dielectric model based on the PB equation. Section 3 presents the boundary integral form of the PB equation and the TABI solver. Section 4 reviews how protein pKas are computed from statistical mechanics. Section 5 describes the pKa calculating process and its implementation in a Python wrapper. Section 6 reports results in computing pKas for two sample proteins. The article ends with a conclusion section summarizing the main points.

2. Continuum dielectric model

Figure 1 illustrates the model showing the protein domain Ω_p, solvent domain Ω_s, and the interface Γ between Ω_p and Ω_s, which in this work is taken to be the molecular surface (or solvent excluded surface).[18] Partial charges q_k are located at atoms $\mathbf{y}_k \in \Omega_p$ representing the protein for $k = 1 : N_c$, while a Boltzmann charge distribution in Ω_s

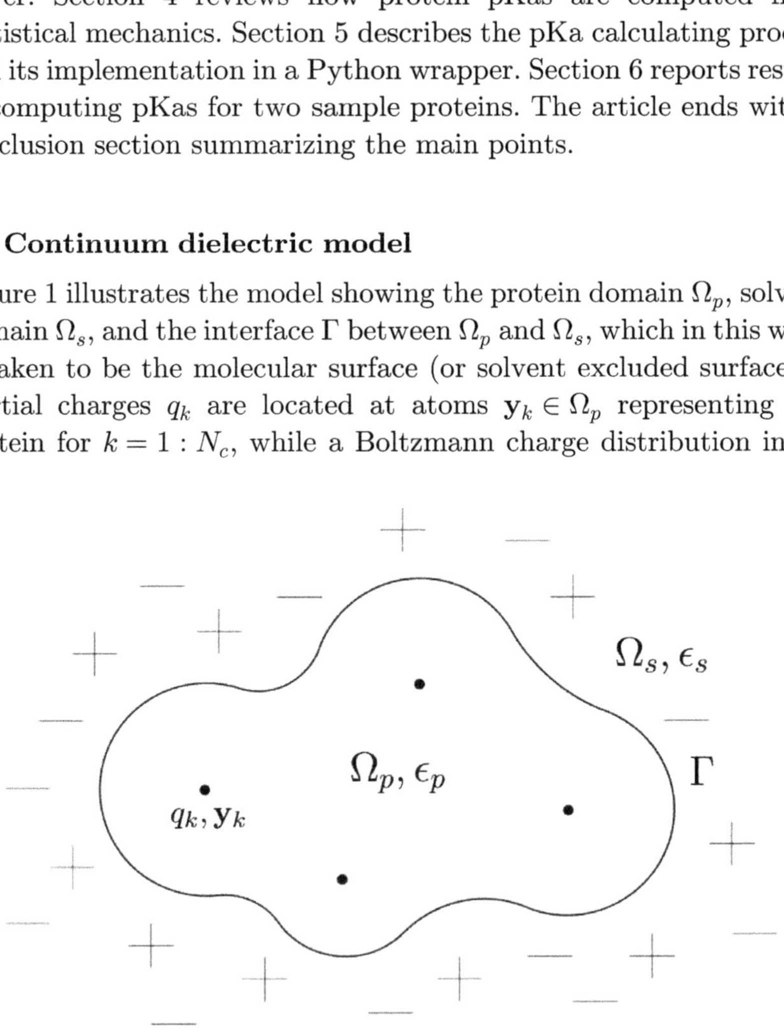

Fig. 1. Dielectric continuum model of solvated protein, protein domain Ω_p with dielectric constant ϵ_p and partial charges q_k at atom locations $\mathbf{y}_k$, $k = 1 : N_c$, solvent domain Ω_s with dielectric constant ϵ_s and dissolved ions (+/−), molecular surface Γ.

represents the dissolved ions. Applying Gauss's law to the charge distributions in Ω_p, Ω_s, and taking the limit of weak ionic strength leads to the linear PB equation for the electrostatic potential,

$$-\nabla \cdot (\epsilon(\mathbf{x})\nabla\phi(\mathbf{x})) + \bar{\kappa}^2(\mathbf{x})\phi(\mathbf{x}) = 4\pi \sum_{k=1}^{N_c} q_k \delta(\mathbf{x} - \mathbf{y}_k), \quad \mathbf{x} \in \Omega_p, \Omega_s. \tag{3}$$

The potential and electric displacement are continuous across the interface,

$$\phi_p(\mathbf{x}) = \phi_s(\mathbf{x}), \quad \epsilon_p \frac{\partial \phi_p(\mathbf{x})}{\partial n} = \epsilon_s \frac{\partial \phi_s(\mathbf{x})}{\partial n}, \quad \mathbf{x} \in \Gamma, \tag{4}$$

where the subscripts indicate limiting values from either domain, and the potential satisfies the far-field boundary condition, $\phi(\mathbf{x}) \to 0$ as $|\mathbf{x}| \to \infty$. The dielectric constant and screening parameter are taken to be piecewise constant,

$$\epsilon(\mathbf{x}) = \begin{cases} \epsilon_p, & \mathbf{x} \in \Omega_p \\ \epsilon_s, & \mathbf{x} \in \Omega_s, \end{cases} \quad \bar{\kappa}^2(\mathbf{x}) = \begin{cases} 0, & \mathbf{x} \in \Omega_p \\ \epsilon_s \kappa^2, & \mathbf{x} \in \Omega_s \end{cases}, \tag{5}$$

where κ is the inverse Debye length determined by the ion concentration I in the solvent (see Refs. 19 and 20 for details on the definition and units of these parameters). We consider several values for the protein dielectric constant ϵ_p, with solvent dielectric constant $\epsilon_s = 80$ and inverse Debye length $\kappa = 0.1257\,\text{Å}^{-1}$ corresponding to ion concentration $I = 0.15\,\text{M}$.

The PB equation (3) has analytic solutions only for simple interface geometries such as a plane or sphere, and for general molecular surfaces, it is solved numerically using methods described in several comprehensive reviews.[21,22] Numerical solution of the PB equation faces several challenges, in particular the protein molecular surface is geometrically complex, the dielectric constant is discontinuous across the molecular surface, the protein is represented by singular atomic partial charges, and the domain is unbounded. Numerical methods fall into two main categories: *grid-based methods* discretize the equation directly on a volumetric grid using finite-difference or finite-element approximations,[23–30] while *boundary element methods* solve an equivalent integral equation on the triangulated molecular surface.[31–38] Boundary element methods effectively address the challenges mentioned above, but they result in dense linear systems which can be expensive to solve by iterative methods such as GMRES.[39] In this work, we employ a treecode to compute the dense matrix–vector products required in GMRES iteration, thereby reducing the operation count from $O(N^2)$ to $O(N \log N)$, where N is the number of faces in the triangulation of the molecular surface. The resulting TABI solver[35] is described in the next section.

3. TABI solver

This section describes the boundary integral form of the PB equation,[33] then the discretization of the integral equations, then the treecode algorithm for fast matrix–vector product and finally the choice of the MSMS density parameter for triangulating the molecular surface.

3.1. *Boundary integral form of PB equation*

Applying Green's second identity to Eq. (3), the electrostatic potential in each domain can be expressed in terms of single-layer and double-layer potentials on the molecular surface[33]

$$\phi(\mathbf{x}) = \int_\Gamma \left[G_0(\mathbf{x},\mathbf{y}) \frac{\partial \phi(\mathbf{y})}{\partial n} - \frac{\partial G_0(\mathbf{x},\mathbf{y})}{\partial n} \phi(\mathbf{y}) \right] dS_{\mathbf{y}} + \sum_{k=1}^{N_c} q_k G_0(\mathbf{x},\mathbf{y}_k), \quad \mathbf{x} \in \Omega_p, \tag{6a}$$

$$\phi(\mathbf{x}) = \int_\Gamma \left[-G_\kappa(\mathbf{x},\mathbf{y}) \frac{\partial \phi(\mathbf{y})}{\partial n} + \frac{\partial G_\kappa(\mathbf{x},\mathbf{y})}{\partial n} \phi(\mathbf{y}) \right] dS_{\mathbf{y}}, \quad \mathbf{x} \in \Omega_s, \tag{6b}$$

where the Coulomb potential and screened Coulomb potential are

$$G_0(\mathbf{x},\mathbf{y}) = \frac{1}{4\pi|\mathbf{x}-\mathbf{y}|}, \quad G_\kappa(\mathbf{x},\mathbf{y}) = \frac{e^{-\kappa|\mathbf{x}-\mathbf{y}|}}{4\pi|\mathbf{x}-\mathbf{y}|}. \tag{7}$$

The singular charges in the protein domain are represented by the Coulomb potential source term in Eq. (6a), and the far-field boundary condition, $\phi(\mathbf{x}) \to 0$ as $|\mathbf{x}| \to \infty$, is analytically satisfied.

Enforcing the interface conditions in Eq. (4) yields a set of boundary integral equations relating the surface potential ϕ_1 and its normal derivative $\partial\phi_1/\partial n$ on Γ,

$$\frac{1}{2}(1+\varepsilon)\phi_1(\mathbf{x}) = \int_\Gamma \left[K_1(\mathbf{x},\mathbf{y})\frac{\partial\phi_1(\mathbf{y})}{\partial n} + K_2(\mathbf{x},\mathbf{y})\phi_1(\mathbf{y})\right] dS_\mathbf{y} + S_1(\mathbf{x}), \quad \mathbf{x}\in\Gamma, \tag{8a}$$

$$\frac{1}{2}(1+\varepsilon^{-1})\frac{\partial\phi_1(\mathbf{x})}{\partial n} = \int_\Gamma \left[K_3(\mathbf{x},\mathbf{y})\frac{\partial\phi_1(\mathbf{y})}{\partial n} + K_4(\mathbf{x},\mathbf{y})\phi_1(\mathbf{y})\right] dS_\mathbf{y} + S_2(\mathbf{x}), \quad \mathbf{x}\in\Gamma, \tag{8b}$$

where $\varepsilon = \varepsilon_2/\varepsilon_1$ is the ratio of dielectric constants. The kernels $K_{1,2,3,4}$ and source terms $S_{1,2}$ are linear combinations of G_0, G_k and their first- and second-order normal derivatives.[33,35] Note that Eqs. (8a) and (8b) are coupled integral equations of the second kind, and together with the properties of the kernels, this ensures that they are well-conditioned. The electrostatic free energy of the solvated protein is

$$\Delta G = \frac{1}{2}\sum_{\substack{j,k=1\\ j\neq k}}^{N_c} \frac{q_j q_k}{|\mathbf{y}_j - \mathbf{y}_k|} + \frac{1}{2}\sum_{k=1}^{N_c} q_k \phi_{\text{reac}}(\mathbf{y}_k), \tag{9}$$

where the first term on the right is the Coulomb energy and the second term is the electrostatic solvation energy, and the reaction potential at the kth charge site is

$$\phi_{\text{reac}}(\mathbf{y}_k) = \int_\Gamma \left[K_1(\mathbf{y}_k,\mathbf{y})\frac{\partial\phi_1(\mathbf{y})}{\partial n} + K_2(\mathbf{y}_k,\mathbf{y})\phi_1(\mathbf{y})\right] dS_\mathbf{y}. \tag{10}$$

3.2. *Discretization of boundary integral equations*

The molecular surface Γ is triangulated using MSMS,[18] where the density parameter d is the number of vertices per Å^2 of surface area. Figure 2 shows two examples with a relatively coarse density $d = 2$,

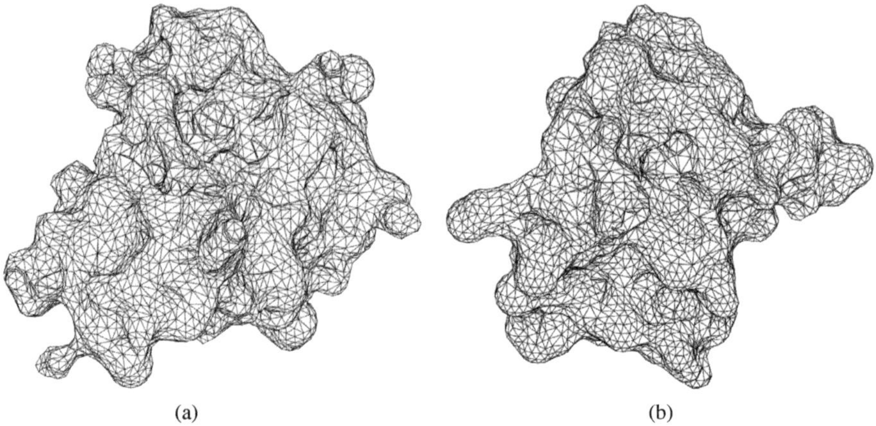

Fig. 2. Triangulation of protein molecular surfaces using MSMS[18] with density $d = 2$: (a) bovine pancreatic trypsin inhibitor (BPTI, PDB ID 4pti) and (b) turkey ovomucoid third domain (OMTKY3, PDB ID 2ovo).

(a) BPTI, (b) OMTKY3. Letting $\mathbf{x}_i$, $i = 1 : N$ denote the triangle centroids, the integrals in Eqs. (8a) and (8b) are discretized by a centroid collocation boundary element method,

$$\frac{1}{2}(1+\varepsilon)\phi_1(\mathbf{x}_i) = \sum_{\substack{j=1 \\ j\neq i}}^{N} \left[K_1(\mathbf{x}_i,\mathbf{x}_j)\frac{\partial\phi_1(\mathbf{x}_j)}{\partial n} + K_2(\mathbf{x}_i,\mathbf{x}_j)\phi_1(\mathbf{x}_j)\right]A_j + S_1(\mathbf{x}_i), \quad (11a)$$

$$\frac{1}{2}(1+\varepsilon^{-1})\frac{\partial\phi_1(\mathbf{x}_i)}{\partial n} = \sum_{\substack{j=1 \\ j\neq i}}^{N} \left[K_3(\mathbf{x}_i,\mathbf{x}_j)\frac{\partial\phi_1(\mathbf{x}_j)}{\partial n} + K_4(\mathbf{x}_i,\mathbf{x}_j)\phi_1(\mathbf{x}_j)\right]A_j + S_2(\mathbf{x}_i), \quad (11b)$$

where A_j is the triangle area, and the $j = i$ term in the sums is omitted to avoid the kernel singularity. Equation (10) is discretized in a similar way to obtain the reaction potential. Equations (11a) and (11b) form a $2N \times 2N$ linear system for the values of the surface potential $\phi_1(\mathbf{x}_i)$ and its normal derivative $\partial\phi_1(\mathbf{x}_i)/\partial n$ at the triangle centroids. The system is solved by GMRES iteration which requires a matrix–vector product at each step.[39] Since the matrix is dense, computing the product by direct summation requires $O(N^2)$ operations, which is prohibitively expensive when N is large. The next subsection describes the treecode algorithm used to reduce the cost of the matrix–vector product.

3.3. *Treecode acceleration of matrix–vector product*

We summarize the treecode algorithm and refer to previous work for more detail.[40,41] The matrix–vector product arising from Eqs. (11a) and (11b) consists of terms in the form of charged particle interactions,

$$V_i = \sum_{\substack{j=1 \\ j \neq i}}^{N} K(\mathbf{x}_i, \mathbf{x}_j) q_j, \quad i = 1, \ldots, N, \tag{12}$$

where V_i is the potential at a target particle $\mathbf{x}_i$, K is one of the kernels, and q_j is a charge associated with the source particle $\mathbf{x}_j$. To evaluate the potentials V_i rapidly, the particles $\mathbf{x}_i$ are divided into a hierarchy of clusters having a tree structure as depicted in Fig. 3(a). Then V_i is evaluated as a sum of particle–cluster interactions,

$$V_i \approx \sum_{C \in N_i} \sum_{\mathbf{x}_j \in C} K(\mathbf{x}_i, \mathbf{x}_j) q_j + \sum_{C \in F_i} \sum_{||\mathbf{k}||=0}^{p} a_{\mathbf{k}}(\mathbf{x}_i, \mathbf{x}_c) m_{\mathbf{k}}(C), \tag{13}$$

where C denotes a cluster, and N_i, F_i denote the near-field and far-field clusters of particle $\mathbf{x}_i$. Figure 3(b) depicts a single particle–cluster interaction. The first term on the right-hand side of Eq. (13) is a direct sum for particles $\mathbf{x}_j$ near $\mathbf{x}_i$, while the second term is a pth order Cartesian Taylor approximation about the cluster center $\mathbf{x}_c$, for clusters that are well-separated from $\mathbf{x}_i$, where $a_{\mathbf{k}}(\mathbf{x}_i, \mathbf{x}_c)$ are the Taylor coefficients and $m_{\mathbf{k}}(C)$ are the cluster moments.[41]

A particle $\mathbf{x}_i$ and cluster C are *well-separated* if the multipole acceptance criterion (MAC) is satisfied, $r_c/R \leq \theta$, where r_c is the cluster radius, $R = |\mathbf{x}_i - \mathbf{x}_c|$ is the particle–cluster distance and θ is a user-specified parameter.[40] The treecode cycles through the particles $\mathbf{x}_i$

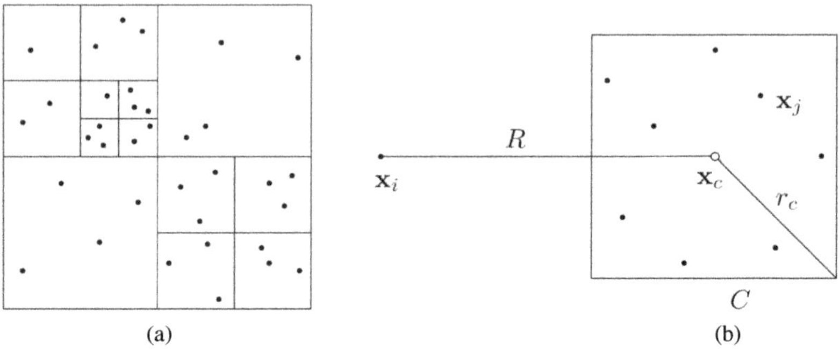

Fig. 3. Treecode schematic in 2D: (a) tree structure with four levels of particle clusters; (b) interaction between particle $\mathbf{x}_i$ and cluster $C = \{\mathbf{x}_j\}$, cluster center $\mathbf{x}_c$, cluster radius r_c, particle-cluster distance $R = |\mathbf{x}_i - \mathbf{x}_c|$.

and interacts with the clusters in the tree; if the MAC is satisfied, then the Taylor approximation is used; otherwise, the child clusters are checked, unless the cluster is a leaf at which point direct summation is used. The accuracy of the treecode is controlled by the Taylor approximation order p and MAC parameter θ. Using the treecode, the operation count for the matrix–vector product is reduced to $O(N \log N)$; the factor N is the number of particles $\mathbf{x}_i$, and the factor $\log N$ is the number of levels in the tree. This completes the description of the TABI solver.[35]

3.4. *Choice of MSMS density*

The MSMS density d is critical in determining the accuracy of the TABI solver. To demonstrate this, we computed the electrostatic free energy ΔG for two proteins, bovine pancreatic trypsin inhibitor (BPTI, PDB ID: 4pti)[42] and turkey ovomucoid third domain (OMTKY3, PDB ID: 2ovo).[43] The treecode used MAC parameter $\theta = 0.8$ and order $p = 3$, which ensures that the treecode approximation error is less than the discretization error in the boundary element method. We use PDB2PQR[44] to assign partial charges to the PDB protein structures based on the PARSE force field.[45] Table 1 presents results for MSMS density $d = 5, 10, 20, 40$ and protein dielectric constant $\epsilon_p = 4, 20$; the number displayed for $d = 5, 10, 20$ is the difference in ΔG from the $d = 40$ value in units of kcal/mol, followed

Table 1. Electrostatic free energy ΔG in Eq. (9) computed by TABI solver, (a) protein BPTI, (b) protein OMTKY3, protein dielectric constant $\epsilon_p = 4, 20$, MSMS density $d = 5, 10, 20, 40$ vertices/Å^2, the number displayed for $d = 5, 10, 20$ is the difference in ΔG from the $d = 40$ value in units of kcal/mol, relative error (%) is computed using $d = 40$ value as reference, run time in seconds.

	ΔG (kcal/mol)		Relative error (%)		Run time (s)	
d	$\epsilon_p = 4$	$\epsilon_p = 20$	$\epsilon_p = 4$	$\epsilon_p = 20$	$\epsilon_p = 4$	$\epsilon_p = 20$
(a) BPTI						
5	−17.3	−2.6	0.29	0.22	47.8	43.5
10	−6.0	−0.9	0.10	0.08	106.4	91.3
20	−1.8	−0.3	0.03	0.02	277.9	205.5
40	−5888.5	−1162.9	—	—	816.9	461.5
(b) OMTKY3						
5	−11.2	−1.6	0.26	0.19	56.2	83.7
10	−3.9	−0.6	0.09	0.07	100.2	79.2
20	−1.2	−0.2	0.03	0.02	194.4	169.4
40	−4326.8	−852.6	—	—	468.2	381.8

by the relative percent error using the $d = 40$ value as the reference. The last column in Table 1 shows the run time in seconds using a MacBook Pro with 2.2 GHz i7 Intel Core and 16 GB 1600 MHz DDR3 memory; these results used block diagonal preconditioning to reduce the number of GMRES iterations.[46] The ΔG values vary significantly with the protein dielectric constant ϵ_p, but for each choice of ϵ_p, the computed ΔG converges to the reference value as the MSMS density d increases. With density $d = 10$, the relative error is 0.1% or less, and the run time is less than 2 min; this density value is used in the pKa computations below.

4. Protein pKa from free energy shifts

The equilibrium constant of the acid dissociation reaction is related to the Gibbs free energy,

$$\Delta G = -RT \ln \mathrm{Ka} = RT \ln 10 \cdot \mathrm{pKa}, \tag{14}$$

where $R = 8.31\,\mathrm{J/(mol\ K)}$ is the gas constant and $T = 298\,\mathrm{K}$ is the temperature in this work, and this enables protein pKas to be

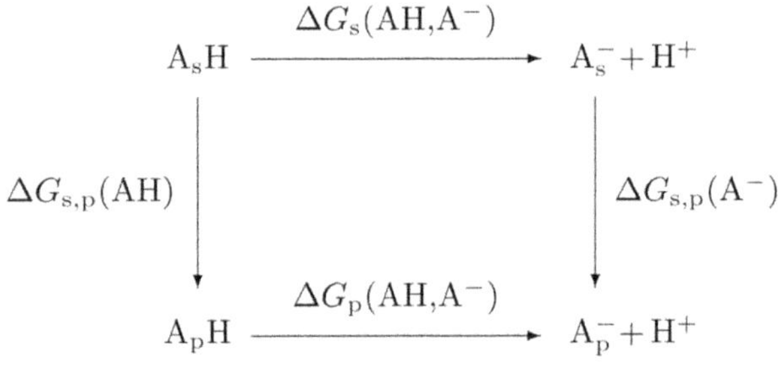

Fig. 4. Thermodynamic cycle for acid dissociation of a titrating residue in a protein, AH/A^- = protonated/deprotonated residue, subscript s/p = solvent/protein environment.

computed from free energy shifts due to protonation/deprotonation.[5] Figure 4 shows the thermodynamic cycle for acid dissociation of a titrating residue in a protein, where AH and A^- stand for the protonated and deprotonated states of the residue, respectively, and subscripts s and p stand for the solvent and protein environments, respectively.[47] The cycle expresses the free energy shifts on transferring the residue from the solvent environment to the protein environment, in its protonated and deprotonated states.

The top arrow in the cycle corresponds to deprotonating the isolated amino acid in the solvent environment and yields the model pKa,

$$\mathrm{pKa}^0 = \frac{1}{RT \ln 10} \Delta G_s(\mathrm{AH}, A^-), \tag{15}$$

which can be determined experimentally. Table 2 records the model pKa values for titrating amino acids.[48] The bottom arrow in the cycle corresponds to deprotonating the amino acid in the protein environment, which can be expressed as

$$\Delta G_p(AH, A^-) = \Delta G_s(AH, A^-) + \Delta G_{s,p}(A^-) - \Delta G_{s,p}(AH), \tag{16}$$

where the first term on the right is the free energy for deprotonating the amino acid in the solvent environment, and the last two terms give the free energy shift for moving the indicated group from the solvent to the protein environment. It is assumed here that the protein conformation remains fixed during acid dissociation, and that

Table 2. Titrating amino acids, residue ID, residue type, model pKa (pKa^0) determined experimentally,[48] residue ID for protonated/deprotonated states with total charge in parentheses.

Residue ID	Residue type	pKa^0	Protonated	Deprotonated
ASP	Acidic	4.0	ASH (0)	ASP (−1)
GLU	Acidic	4.4	GLH (0)	GLU (−1)
CYS	Polar uncharged	9.5	CYX (0)	CYM (−1)
TYR	Polar uncharged	9.6	TYR (0)	TYM (−1)
HIS	Basic	6.3	HIP (+1)	HIE (0)
LYS	Basic	10.4	LYS (+1)	LYN (0)

nonpolar effects are negligible, and hence, the ΔG values in this work refer to the electrostatic free energy in Eq. (9). Starting from the model pKa for a given ionizable group on an amino acid side chain, the procedure computes the intrinsic pKa of the site and then takes site–site interactions into account to determine the site pKa by computational titration.[5,14,49]

5. Procedure for computing protein pKas

The computational procedure is composed of four steps described in the following.

Step 1. Prepare PQR files

This step prepares the PQR files for the protein in various protonation states. The protein structure is obtained from the Protein Data Bank (www.pdb.org), and charges are assigned using PDB2PQR[44] with protonation states specified using the residue IDs in Table 2. For a protein with N_t titrating sites, we need a total of $1 + 2N_t + \frac{1}{2}(N_t^2 - N_t)$ PQR files with charge distributions specified as follows: (a) one file with all titrating sites deprotonated, keeping the background charges of nontitrating sites on, (b) $2N_t$ files having all titrating sites but one deprotonated, once with background charges on and once with background charges off, (c) $\frac{1}{2}(N_t^2 - N_t)$ files with pairs of titrating sites protonated and all other titrating sites deprotonated, and with background charges off.

Step 2. Call TABI solver for electrostatics

This step calls the TABI solver to compute the electrostatic free energy ΔG in Eq. (9) for each protonation state represented by a PQR file, where the structure and charge distribution in the file gives the source term on the right side of the PB equation in Eq. (3). The TABI solver is called with input physical parameters (dielectric constants, ion concentration) and numerical parameters (MSMS density, treecode MAC, order of Taylor approximation). After completing this step, ΔG is available for each protonation state, and these are used to calculate the intrinsic pKas in the next step.

Step 3. Compute intrinsic pKas

The intrinsic pKa for the ith titrating site of the protein accounts for single-site effects, i.e. when the ith site is protonated, the other titrating sites are deprotonated, and the background charges are on; the expression is

$$\begin{aligned}\mathrm{pKa}^{\mathrm{int}}(i) = \mathrm{pKa}^{0}(i) + \frac{1}{RT\ln 10}(\Delta G_p(AH, A^-) \\ - \Delta G_s(AH, A^-)).\end{aligned} \tag{17}$$

The first term on the right, $\mathrm{pKa}^0(i)$, is the model pKa for the residue at the ith titrating site as determined experimentally and recorded in Table 2.[48] The next two terms are electrostatic free energy shifts; $\Delta G_p(AH, A^-)$ is the shift between the protein with only the ith titrating site protonated and the protein with all titrating sites deprotonated, while $\Delta G_s(AH, A^-)$ is the same shift for the residue in the solvent environment. These shifts are computed using the ΔG values obtained in Step 2; note that TABI uses units of kcal/mol for free energy, so the computed ΔG values are scaled by $RT\ln 10$ to obtain $\mathrm{pKa}^{\mathrm{int}}(i)$.

Step 4. Titration with site–site interactions

The pKa of a titrating site is the pH at which the site is equally likely to be protonated or deprotonated. In a protein with N_t titrating sites, let $\theta \in \{0, 1\}^{N_t}$ define a protonation state, where θ_i is the ith entry of θ, and $\theta_i = 1$ means that the ith site is protonated, while $\theta_i = 0$ means it is deprotonated. The pKa of the ith site is the pH satisfying

$<\theta_i, \text{pH}>= 0.5$, where the probability that the ith site is protonated at a particular pH is given by the thermodynamic average,[5,13,14]

$$<\theta_i, \text{pH}>= \frac{\sum_\theta \theta_i \exp(-\Delta G(\theta, \text{pH})/RT)}{\sum_\theta \exp(-\Delta G(\theta, \text{pH})/RT)}. \tag{18}$$

In this expression, the pH-dependent protonation state energy is defined by

$$\Delta G(\theta, \text{pH}) = RT \ln 10 \sum_i \theta_i(\text{pH} - \text{pKa}^{\text{int}}(i)) + \frac{1}{2} \sum_{i \neq j} \theta_i \theta_j \Delta G_{ij}, \tag{19}$$

where $\text{pKa}^{\text{int}}(i)$ is the intrinsic pKa of the ith site computed in Step 3. Equation (19) also requires the site–site interaction energy defined by

$$\begin{aligned} \Delta G_{ij} &= \frac{1}{2}(t_i^T W t_j + t_j^T W t_i) \\ &= \frac{1}{2}(t_i + t_j)^T W (t_i + t_j) - \frac{1}{2} t_i^T W t_i - \frac{1}{2} t_j^T W t_j, \end{aligned} \tag{20}$$

where t_i is a vector of partial charges where the charges of the ith titrating site are on, and the charges of the other titrating sites and background charges are off, while W is the mapping from partial charges to potential values at the charge sites as computed by the TABI solver. The first equality in Eq. (20) expresses ΔG_{ij} as the symmetric interaction of site i and site j, while the second equality in Eq. (20) reduces the cost of computing the ΔG_{ij} values from $O(N_t^2)$ to $O(N_t)$.

Since there are 2^{N_t} protonation states, computing the thermodynamic average in Eq. (18) explicitly is only feasible for relatively small proteins. The proteins considered in this work (BPTI, OMTKY3) were chosen for two reasons: (1) they have few enough titrating sites so that the thermodynamic average can be computed explicitly, enabling us to avoid numerical errors introduced by approximating $<\theta_i, \text{pH}>$, and (2) experimental pKas are available for most of the titrating sites.[12] Future work will consider combining the TABI solver for electrostatics with more efficient procedures for computing the thermodynamic average such as the reduced-site approximation,[5,50] Monte Carlo methods,[13,14,51] and clustering algorithms.[52]

5.1. *Examples of titration curves*

For each titrating site, the protonation probability $< \theta_i, \mathrm{pH} >$ is sampled from pH $= 1$ to pH $= 14$ with step size of 0.2, and cubic spline interpolation is used to find the pH at probability 1/2 giving the predicted site pKa. In this way, each active site has a titration curve and Fig. 5 shows two examples for which the titration curve has the conventional sigmoidal shape. Figure 5(a) is for the basic residue

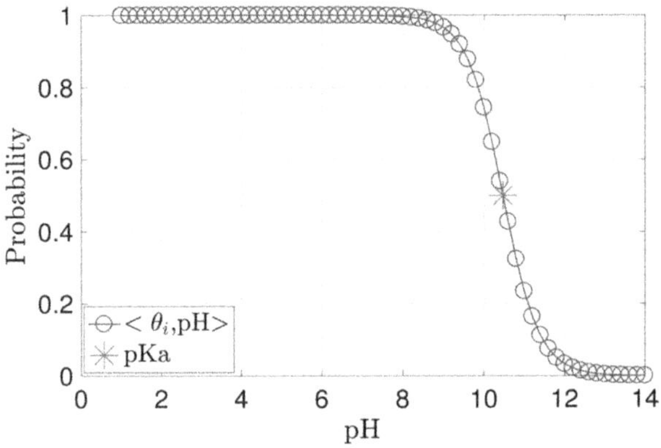

(a)

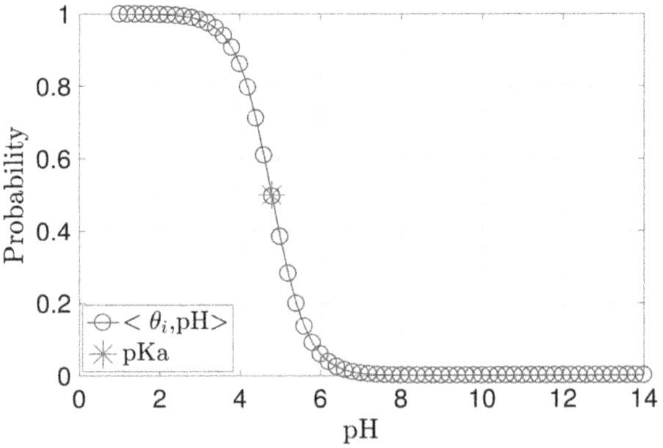

(b)

Fig. 5. Computed titration curves, protonation probability of ionizable residue is plotted versus pH, blue circles ($< \theta_i, \mathrm{pH} >$), red stars (predicted pKa): (a) LYS15 on protein BPTI, predicted pKa $= 10.47$ ($\epsilon_p = 20$), experimental pKa $= 10.4$; (b) GLU43 on protein OMTKY3, predicted pKa $= 4.79$ ($\epsilon_p = 8$), experimental pKa $= 4.8$.

LYS15 on protein BPTI at $\epsilon_p = 20$, which predicts pKa = 10.47 versus the experimental value 10.4, while Fig. 5(b) is for the acidic residue GLU43 on protein OMTKY3 at $\epsilon_p = 8$, which predicts pKa = 4.79 versus the experimental value 4.8. The results indicate that our computational procedure is capable of achieving good agreement with experiment, although already there is a suggestion that the choice of protein dielectric constant ϵ_p plays an important role.

5.2. *Python wrapper*

A convenient Python wrapper was written to pipeline the entire pKa calculating procedure and is publicly available for download on the corresponding author's website (`faculty.smu.edu/wgeng/research/pka_tabi.html`) at Southern Methodist University; also available there is the TABI source code for electrostatics and multi-platform binary versions of MSMS[18] for molecular surface triangulation. The user specifies parameters for the PB equation (dielectric constants, ion concentration) and the TABI solver (MSMS density, treecode MAC, order of Taylor approximation) in the user data file. On a computer with Python and Fortran compilers installed, and the protein specified by its four-digit PDB ID, the user runs the wrapper by typing the following command:

python wrapper_pka.py PDBID.

The wrapper downloads the protein structure from the Protein Data Bank, identifies the titrating sites, calls the TABI solver for electrostatics, computes the intrinsic pKa and protonation probability of each site including site–site interactions as described above, and returns the predicted pKas.

6. Numerical results

Table 3 presents results for and OMTKY3 with two values of the protein dielectric constant, $\epsilon_p = 4, 20$. Successive columns give the residue ID, model pKa (pKa^0), experimental pKa (pKa^{exp}), pKa shift with respect to experiment ($\Delta\text{pKa}^{0,\text{exp}}$), intrinsic pKa ($\text{pKa}^{\text{int}}$), and predicted pKa ($\text{pKa}^{\text{pre}}$); both proteins have 12 titrating residues for

Table 3. Numerical results for BPTI, OMTKY3, protein dielectric constant $\epsilon_p = 4, 20$, residue ID, model pKa (pKa^0), experimental pKa (pKa^{exp}), pKa shift (ΔpKa$^{0,\text{exp}}$), intrinsic pKa (pKa^{int}), predicted pKa (pKa^{pre}), bold value is closest to pKa^{exp}, RMSD is with respect to pKa^{exp}.[12]

BPTI				$\epsilon_p = 4$	$\epsilon_p = 20$	$\epsilon_p = 4$	$\epsilon_p = 20$
ID	pKa^0	pKa^{exp}	ΔpKa$^{0,\text{exp}}$	pKa^{int}		pKa^{pre}	
ASP3	4.0	3.6	−0.4	**3.61**	3.89	3.88	3.96
ASP50	4.0	3.2	−0.8	0.22	**3.23**	3.08	3.87
GLU7	**4.4**	3.9	−0.5	5.38	4.66	5.84	4.69
GLU49	4.4	4.0	−0.4	**3.87**	4.37	5.05	4.70
TYR10	**9.6**	9.4	−0.2	12.01	10.38	12.23	10.46
TYR21	9.6	10.0	+0.4	11.23	**10.15**	13.04	10.51
TYR23	9.6	11.0	+1.4	16.64	**11.24**	17.46	11.42
TYR35	9.6	10.6	+1.0	10.11	9.84	**10.19**	9.86
LYS15	**10.4**	10.4	+0.0	10.38	10.61	10.32	10.47
LYS26	10.4	10.1	−0.3	10.29	10.54	**10.16**	10.39
LYS41	**10.4**	10.6	+0.2	11.65	11.31	12.88	11.44
LYS46	**10.4**	9.9	−0.5	11.61	11.16	11.15	10.60
RMSD	0.63			2.15	0.63	2.43	0.65
OMTKY3							
ASP7	4.0	2.4	−1.6	1.31	3.55	**2.26**	3.82
ASP27	**4.0**	2.2	−1.8	10.20	5.70	6.26	4.90
GLU10	**4.4**	4.1	−0.3	4.89	4.91	7.09	5.36
GLU19	**4.4**	3.2	−1.2	5.76	4.80	8.25	5.26
GLU43	4.4	4.8	+0.4	4.99	**4.72**	4.95	4.69
HIS52	6.3	7.5	+1.2	**7.54**	7.27	7.94	7.30
TYR11	9.6	10.2	+0.6	13.30	**10.68**	13.31	10.76
TYR20	9.6	11.1	+1.5	11.49	10.29	12.79	**10.78**
LYS13	**10.4**	9.9	−0.5	12.16	11.63	12.20	11.48
LYS29	10.4	11.1	+0.7	11.38	**11.24**	11.80	11.30
LYS34	**10.4**	10.1	−0.3	12.53	11.66	12.32	11.45
LYS55	10.4	11.1	+0.7	11.77	11.36	12.10	**11.28**
RMSD	1.04			2.79	1.39	2.50	1.29

which experimental pKas are available.[12] The bold values in each row are the closest to experiment, and the RMSD is with respect to the experimental pKa.

For BPTI, the model pKa and intrinsic pKa are each closest to experiment for five sites, while the predicted pKa is closest for two sites. The RMSD for the model pKa is 0.63 pK units. At the lower dielectric constant $\epsilon_p = 4$, the RMSD for the intrinsic pKa and predicted pKa is more than 2 pK units, while at the higher value $\epsilon_p = 20$ they are essentially the same as the model pKa RMSD. For three of the twelve residues (GLU7, TYR10, LYS46), the calculations gave positive pKa shift whereas the actual shift is negative. However, the calculations do fairly well in capturing the three largest pKa shifts including the negative shift $\Delta pKa^{0,\text{exp}} = -0.8$ for ASP50, and two positive shifts $\Delta pKa^{0,\text{exp}} = +1.4$ for TYR23, $\Delta pKa^{0,\text{exp}} = +1.0$ for TYR35.

For OMTKY3, the model pKa is closest to experiment for five sites, while the intrinsic pKa and predicted pKa are closest for four sites and three sites, respectively. The RMSDs are higher than for BPTI, but the trends are similar. The model pKa has the smallest RMSD at slightly more than 1 pK unit. At the lower dielectric constant $\epsilon_p = 4$, the RMSD for the intrinsic pKa and predicted pKa is 2.5 pK units or more, while at the higher value $\epsilon_p = 20$ they improve to less than 1.4 pK units, but remain higher than the model pKa RMSD. In this case, the predicted pKa has slightly smaller RMSD than the intrinsic pKa. For five of the twelve residues (ASP27, GLU10, GLU19, LYS13, LYS34), the calculations gave positive pKa shifts whereas the actual shifts are negative. Among the five residues with pKa shifts greater than 1 pK unit, the calculations fail to capture two negative shifts $\Delta pKa^{0,\text{exp}} = -1.8$ for ASP27, $\Delta pKa^{0,\text{exp}} = -1.2$ for GLU19; however, the calculations do fairly well in capturing the negative shift $\Delta pKa^{0,\text{exp}} = -1.6$ for ASP7, and two positive shifts $\Delta pKa^{0,\text{exp}} = +1.2$ for HIS52, $\Delta pKa^{0,\text{exp}} = +1.5$ for TYR20.

It is noteworthy that in terms of RMSD, the model pKa does as well or better than the predicted pKa, emphasizing the limitations of continuum dielectric models for predicting protein pKas.[53,54] This could possibly be remedied by including nonpolar effects or spatial variation of the protein dielectric constant. Other improvements suggested in the literature call for including the effect of protein conformational flexibility,[55,56] replacing the discontinuous dielectric interface with a smooth Gaussian function to partially represent solvent penetration[57,58] and using microscopic models to account for atomic polarization and dipole reorientation.[59,60] In Table 3 the

Table 4. Results for five proteins selected from a pKa database,[64] protein dielectric constant $\epsilon_p = 20$, column 1 (PDB ID), column 2 (number of atoms), column 3 (number of residues for which experimental pKas are known), columns 4, 5, 6 (RMSD for intrinsic pKa^{int}, predicted pKa^{pre}, model pKa^0, compared with experimental pKa^{exp}), last row gives results for all five proteins.

PDB ID	N_{atom}	N_{res}	$pKa^{int,exp}$	$pKa^{pre,exp}$	$pKa^{0,exp}$
1PGA	436	13	0.85	0.65	0.51
2QMT	438	13	1.30	1.39	0.51
2CI2	521	10	1.18	1.52	0.84
1IGD	468	10	1.08	1.19	0.46
1BPI	460	8	0.94	1.03	0.56
Total	—	54	1.09	1.19	0.59

higher protein dielectric constant $\epsilon_p = 20$ yields smaller RMSD than the lower value $\epsilon_p = 4$, as found in several previous studies.[13,61,62] Nonetheless, for some residues, $\epsilon_p = 4$ yields better agreement with experiment than $\epsilon_p = 20$ (ASP3 in BPTI, HIS52 in OMTKY3); this is consistent with previous results showing that the best choice of protein dielectric constant for a given residue can depend on the residue's exposure to solvent.[63] Overall, the present results indicate the complexity of these pH-dependent processes and they support the need for further investigations to develop a more physically realistic yet computationally tractable model.

To further test our pKa computing procedure, we chose five other proteins from the pKa database collected by Pahari *et al.*[64] The database contains more than 200 proteins with experimental pKas and our selection is due to the efficiency limitations of our current procedure which solves $O(N_t^2)$ PB equations and involves $O(2^{N_t})$ titrating site–site interactions; a significantly faster version of TABI and a more efficient strategy to accelerate the site–site interactions are under development. Table 4 reports the RMSD of the intrinsic pKa, predicted pKa, and model pKa compared with experimental pKas[64] for each protein and for the total set of residues. From this table, we observe similar trends as seen above with BPTI and OMTKY3, namely that including site–site interactions in the predicted pKa do not significantly improve upon the intrinsic pKa, and the computed pKa (intrinsic or predicted) on average is not as good as the model pKa in approaching the experimental pKa. This reinforces the need to improve the current continuum dielectric model along the lines mentioned above.

7. Conclusion

A common approach to computing protein pKas uses a continuum dielectric model in which the protein is a low dielectric medium with embedded atomic point charges, the solvent is a high dielectric medium with a Boltzmann distribution of ionic charges, and the pKa is related to the electrostatic free energy which is obtained by solving the PB equation. Starting from the model pKa for a titrating residue, the method obtains the intrinsic pKa and then computes the protonation probability for a given pH including site–site interactions.[5,13] This approach assumes that acid dissociation does not affect protein conformation aside from adding or deleting charges at titratable sites. In this work, we demonstrated our TABI solver for the relevant electrostatic calculations.[35] The TABI code accurately resolves the complex charge distributions of protonated and deprotonated states using the appropriate analytic Green's functions. The pKa computing procedure is enclosed in a convenient Python wrapper which is publicly available at the corresponding author's website.

Results were presented for several proteins compared with experimental data. In terms of RMSD with respect to experimental values for the entire set of titratable sites on each protein, the model pKa does as well or better than the intrinsic pKa and predicted pKa. Nonetheless, the calculations did succeed for several residues with the largest pKa shifts from the model pKa. Among ongoing efforts to improve protein pKa calculations,[6] the advantage of TABI is that it reduces the numerical errors in the electrostatic calculations so that attention can be focused on modeling assumptions. Further improvements to TABI are expected in future work using our recently developed GPU-accelerated barycentric treecode.[65,66]

Acknowledgments

This work was supported by the National Science Foundation grants DMS-1819094 and DMS-1819193. J. Hu and Y. Xu were partially supported by the Hamilton Scholarship from Southern Methodist University.

References

1. Ullmann, G. M.; Knapp, E. W. Electrostatic Models for Computing Protonation and Redox Equilibria in Proteins. *Eur. Biophys. J.* **1999**, *28*, 533–551.
2. Onufriev, A. V.; Alexov, E. Protonation and pK Changes in Protein-Ligand Binding. *Q. Rev. Biophys.* **2013**, *46*, 181–209.
3. Nielsen, J. E. Analyzing Protein NMR pH-Titration Curves, in *Annual Reports in Computational Chemistry*, eds., Wheeler, R. A.; Spellmeyer, D. C. Elsevier 2008, pp. 89–106.
4. Lee, A. C.; Crippen, G. M. Predicting pKa. *J. Chem. Inf. Model.* **2009**, *49*, 2013–2033.
5. Bashford, D.; Karplus, M. pK_a's of Ionizable Groups in Proteins: Atomic Detail from a Continuum Electrostatic Model. *Biochemistry* **1990**, *29*, 10219–10225.
6. Alexov, E.; Mehler, E. L.; Baker, N.; Baptista, A. M.; Huang, Y.; Milletti, F.; Nielsen, J. E.; Farrell, D.; Carstensen, T.; Olsson, M. H. M.; Shen, J. K.; Warwicker, J.; Williams, S.; Word, J. M. Progress in the Prediction of pKa Values in Proteins. *Proteins* **2011**, *79*, 3260–3275.
7. Antosiewicz, J. M.; Shugar, D. Poisson-Boltzmann Continuum-Solvation Models: Applications to pH-Dependent Properties of Biomolecules. *Mol. BioSyst.* **2011**, *7*, 2923–2949.
8. Kilambi, K. P.; Gray, J. J. Rapid Calculation of Protein pKa Values Using Rosetta. *Biophys. J.* **2012**, *103*, 587–595.
9. Meyer, T.; Knapp, E. W. pKa Values in Proteins Determined by Electrostatics Applied to Molecular Dynamics Trajectories. *J. Chem. Theory Comput.* **2015**, *11*, 2827–2840.
10. Riccardi, D.; Schaefer, P.; Cui, Q. pKa Calculations in Solution and Proteins with QM/MM Free Energy Perturbation Simulations: A Quantitative Test of QM/MM Protocols. *J. Phys. Chem. B* **2005**, *109*, 17715–17733.
11. Gosink, L. J.; Hogan, E. A.; Pulsipher, T. C.; Baker, N. A. Bayesian Model Aggregation for Ensemble-Based Estimates of Protein pKa Values. *Proteins* **2014**, *82*, 354–363.
12. Li, H.; Robertson, A. D.; Jensen, J. H. Very Fast Empirical Prediction and Rationalization of Protein pKa Values. *Proteins* **2005**, *61*, 704–721.
13. Juffer, A. H.; Argos, P.; Vogel, H. J. Calculating Acid-Dissociation Constants of Proteins using the Boundary Element Method. *J. Phys. Chem. B* **1997**, *101*, 7664–7673.
14. Ho, K. L. *Fast Direct Methods for Molecular Electrostatics.* PhD Thesis, New York University, **2012**.

15. Fogolari, F.; Brigo, A.; Molinari, H. The Poisson-Boltzmann Equation for Biomolecular Electrostatics: A Tool for Structural Biology. *J. Mol. Recognit.* **2002**, *15*(6), 377–92.
16. Baker, N. A. Poisson-Boltzmann Methods for Biomolecular Electrostatics. *Methods Enzymol.* **2004**, *383*, 94–118.
17. Connolly, M. L. Depth Buffer Algorithms for Molecular Modelling. *J. Mol. Graphics* **1985**, *3*, 19–24.
18. Sanner, M. F.; Olson, A. J.; Spehner, J. C. Reduced Surface: An Efficient Way to Compute Molecular Surfaces. *Biopolymers* **1996**, *38*, 305–320.
19. Geng, W. A Boundary Integral Poisson–Boltzmann Solvers Package for Solvated Bimolecular Simulations. *Comput. Math. Biophys.* **2015**, *3*, 43–58.
20. Holst, M. J. *The Poisson-Boltzmann Equation: Analysis and Multilevel Numerical Solution.* PhD Thesis, UIUC, **1994**.
21. Lu, B. Z.; Zhou, Y. C.; Holst, M. J.; McCammon, J. A. Recent Progress in Numerical Methods for the Poisson-Boltzmann Equation in Biophysical Applications. *Comm. Comput. Phys.* **2008**, *3*(5), 973–1009.
22. Baker, N. A. Improving Implicit Solvent Simulations: A Poisson-Centric View. *Curr. Opin. Struct. Biol.* **2005**, *15*(2), 137–143.
23. Yang, A. S.; Gunner, M. R.; Sampogna, R.; Sharp, K.; Honig, B. On the Calculation of pKas in Proteins. *Proteins Struct. Funct. Genet.* **1993**, *15*, 252–265.
24. Holst, M. J.; Saied, F. Numerical Solution of the Nonlinear Poisson-Boltzmann Equation: Developing more Robust and Efficient Methods. *J. Comput. Chem.* **1995**, *16*, 337–364.
25. Honig, B.; Nicholls, A. Classical Electrostatics in Biology and Chemistry. *Science* **1995**, *268*, 1144–1149.
26. Im, W.; Beglov, D.; Roux, B. Continuum Solvation Model: Electrostatic Forces from Numerical Solutions to the Poisson-Boltzmann Equation. *Comput. Phys. Commun.* **1998**, *111*, 59–75.
27. Baker, N. A.; Sept, D.; Joseph, S.; Holst, M. J.; McCammonJ. A. Electrostatics of Nanosystems: Application to Microtubules and the Ribosome. *Proc. Natl. Acad. Sci.* **2001**, *98*, 10037–10041.
28. Luo, R.; David, L.; Gilson, M. K. Accelerated Poisson-Boltzmann Calculations for Static and Dynamic Systems. *J. Comput. Chem.* **2002**, *23*, 1244–1253.
29. Chen, D.; Chen, Z.; Chen, C.; Geng, W.; Wei, G. W. MIBPB: A Software Package for Electrostatic Analysis. *J. Comput. Chem.* **2011**, *32*, 657–670.
30. Deng, W.; Zhufu, X.; Xu, J.; Zhao, S. A New Discontinuous Galerkin Method for the Nonlinear Poisson-Boltzmann Equation. *Appl. Math. Lett.* **2015**, *257*, 1000–1021.
31. Zauhar, R. J.; Morgan, R. S. A New Method for Computing the Macromolecular Electric Potential. *J. Mol. Biol.* **1985**, *186*, 815–820.
32. Yoon, B. J.; Lenhoff, A. M. A Boundary Element Method for Molecular Electrostatics with Electrolyte Effects. *J. Comput. Chem.* **1990**, *11*, 1080–1086.
33. Juffer, A.; Botta, E. F. F.; van Keulen, B.; van der Ploeg, A.; Berendsen, H. The Electric Potential of a Macromolecule in a Solvent: A Fundamental Approach. *J. Comput. Phys.* **1991**, *97*, 144–171.

34. Boschitsch, A. H.; Fenley, M. O.; Zhou, H. X. Fast Boundary Element Method for the Linear Poisson-Boltzmann Equation. *J. Phys. Chem. B* **2002**, *106*(10), 2741–2754.
35. Geng, W.; Krasny, R. A Treecode-Accelerated Boundary Integral Poisson-Boltzmann Solver for Electrostatics of Solvated Biomolecules. *J. Comput. Phys.* **2013**, *247*, 62–78.
36. Zhang, B.; Lu, B.; Cheng, X.; Huang, J.; Pitsianis, N. P.; Sun, X.; McCammon, J. A. Mathematical and Numerical Aspects of the Adaptive Fast Multipole Poisson-Boltzmann Solver. *Commun. Comput. Phys.* **2013**, *13*, 107–128.
37. Zhong, Y.; Ren, K.; Tsai, R. An Implicit Boundary Integral Method for Computing Electric Potential of Macromolecules in Solvent. *J. Comput. Phys.* **2018**, *359*, 199–215.
38. Cancès, E., Maday, Y.; Stamm, B. Domain Decomposition for Implicit Solvation Models. *J. Chem. Phys.* **2013**, *139*(5), 054111.
39. Saad, Y.; Schultz, M. GMRES: A Generalized Minimal Residual Algorithm for Solving Nonsymmetric Linear Systems. *SIAM J. Sci. Stat. Comput.* **1986**, *7*, 856–869.
40. Barnes, J.; Hut, P. A Hierarchical O(NlogN) Force-Calculation Algorithm. *Nature* **1986**, *324*, 446–449.
41. Li, P.; Johnston, H.; Krasny, R. A Cartesian Treecode for Screened Coulomb Interactions. *J. Comput. Phys.* **2009**, *228*, 3858–3868.
42. Marquart, M.; Walter, J.; Deisenhofer, J.; Bode, W.; Huber, R. The Geometry of the Reactive Site and of the Peptide Groups in Trypsin, Trypsinogen and its Complexes with Inhibitors. *Acta Cryst. Sec. B* **1983**, *39*(4), 480–490.
43. Bode, W.; Epp, O.; Huber, R.; Laskowski, M.; Ardelt, W. The Crystal and Molecular Structure of the Third Domain of Silver Pheasant Ovomucoid (omsvp3). *Eur. J. Biochem.* **1985**, *147*(2), 387–395.
44. Dolinsky, T. J.; Czodrowski, P.; Li, H.; Nielsen, J. E.; Jensen, J. H.; Klebe, G.; Baker, N. A. PDB2PQR: Expanding and Upgrading Automated Preparation of Biomolecular Structures for Molecular Simulations. *Nucleic Acids Res.* **2007**, *35*, W522–W525.
45. Sitkoff, D.; Sharp, K. A.; Honig, B. Accurate Calculation of Hydration Free Energies Using Macroscopic Solvent Models. *J. Phys. Chem.* **1994**, *98*, 1978–1988.
46. Chen, J.; Geng, W. On Preconditioning the Treecode-Accelerated Boundary Integral (TABI) Poisson-Boltzmann Solver. *J. Comput. Phys.* **2018**, *373*, 750–762.
47. Warshel, A. Calculations of Enzymatic Reactions: Calculations of pKa, Proton Transfer Reactions, and General Acid Catalysis Reactions in Enzymes. *Biochemistry* **1981**, *20*, 3167–3177.
48. Nozaki, Y.; Tanford, C. Examination of Titration Behaviour. *Methods Enzymol.* **1967**, *11*, 715–734.
49. Hu, J.; Zhao, S.; Geng, W. Accurate pKa Computation using Matched Interface and Boundary (MIB) Method Based Poisson-Boltzmann Solver. *Commun. Comput. Phys.* **2018**, *23*, 520–539.

50. Bashford, D.; Karplus, M. Multiple-Site Titration Curves of Proteins: An Analysis of Exact and Approximate Methods for their Calculation. *J. Phys. Chem.* **1991**, *95*, 9556–9561.
51. Beroza, P.; Fredkin, D. R.; Okamura, M. Y.; Feher, G. Protonation of Interacting Residues in a Protein by a Monte Carlo Method: Application to Lysozyme and the Photosynthetic Reaction Center of *Rhodobacter sphaeroides*. *Proc. Natl. Acad. Sci.* **1991**, *88*, 5804–5808.
52. Myers, J.; Grothaus, G.; Narayanan, S.; Onufriev, A. A Simple Clustering Algorithm can be Accurate Enough for Use in Calculations of pKs in Macromolecules. *Proteins* **2006**, *63*, 928–938.
53. Anandakrishnan, R.; Aguilar, B.; Onufriev, A. V. H++3.0: Automating pK Prediction and the Preparation of Biomolecular Structures for Atomistic Molecular Modeling and Simulations. *Nucleic Acids Res.* **2012**, *40*, W537–W541.
54. Gunner, M. R.; Baker, N. A. Continuum Electrostatics Approaches to Calculating pKas and Ems in Proteins, in *Computational Approaches for Studying Enzyme Mechanism Part B*, ed., Voth, G. A. Academic Press, 2016, pp. 1–20.
55. Alexov, E. G.; Gunner, M. R. Incorporating Protein Conformational Flexibility into the Calculation of pH-Dependent Protein Properties. *Biophys. J.* **1997**, *74*, 2075–2093.
56. Georgescu, R. E.; Alexov, E. G.; Gunner, M. R. Combining Conformational Flexibility and Continuum Electrostatics for Calculating pKas in Proteins. *Biophys. J.* **2002**, *83*, 1731–1748.
57. Word, J. M.; Nicholls, A. Application of the Gaussian Dielectric Boundary in Zap to the Prediction of Protein pKa Values. *Proteins* **2011**, *79*, 3400–3409.
58. Wang, L.; Li, L.; Alexov, E. pKa Predictions for Proteins, RNAs, and DNAs with the Gaussian Dielectric Function using DelPhiPKa. *Proteins* **2015**, *83*, 2186–2197.
59. Warshel, A.; Levitt, M. Theoretical Studies of Enzymic Reactions: Electrostatic and Steric Stabilization of the Carbonium Ion in the Reaction of Lysozyme. *J. Mol. Biol.* **1976**, *103*, 227–249.
60. Sham, Y. Y.; Chu, Z. T.; Warshel, A. Consistent Calculations of pKa's of Ionizable Residues in Proteins: Semi-Microscopic and Microscopic Approaches. *J. Phys. Chem. B* **1997**, *101*, 4458–4472.
61. Antosiewicz, J.; McCammon, J. A.; Gilson, M. K. Prediction of pH-Dependent Properties of Proteins. *J. Mol. Biol.* **1994**, *238*, 415–436.
62. Antosiewicz, J.; McCammon, J. A.; Gilson, M. K. The Determinants of pKas in Proteins. *Biochemistry* **1996**, *35*, 7819–7833.
63. Demchuk, E.; Wade, R. Improving the Continuum Dielectric Approach to Calculating pKas of Ionizable Groups in Proteins. *J. Phys. Chem.* **1996**, *100*, 17373–17387.
64. Pahari, S.; Sun, L.; Alexov, E. PKAD: A Database of Experimentally Measured pKa Values of Ionizable Groups in Proteins. *Database J. Biol. Database Curation* **2019**, *2019*, article ID baz024.

65. Wang, L.; Krasny, R.; Tlupova, S. A Kernel-Independent Treecode Based on Barycentric Lagrange Interpolation. *Commun. Comput. Phys.* **2020**, *28*, 1415–1436.
66. Vaughn, N.; Wilson, L.; Krasny, R. A. GPU-Accelerated Barycentric Lagrange Treecode, in *2020 IEEE Int. Parallel and Distributed Processing Symp. Workshop (IPDPSW)* IEEE, 2020, pp. 701–710.

https://doi.org/10.1142/9789819808199_0008

Chapter 8

Characterizing the Water Wire in the Gramicidin Channel Found by Monte Carlo Sampling Using Continuum Electrostatics and in Molecular Dynamics Trajectories with Conventional or Polarizable Force Fields*

Yingying Zhang[†,‡], Kamran Haider[†], Divya Kaur[†,§], Van A. Ngo[¶], Xiuhong Cai[†,‡], Junjun Mao[||], Umesh Khaniya[†,‡], Xuyu Zhu[†,‡], Sergei Noskov[**], Themis Lazaridis[§,††] and M. R. Gunner[†,‡,‡‡]

[†]*Department of Physics, City College of New York, City University of New York, New York, NY 10031, USA*
[‡]*Department of Physics, The Graduate Center, City University of New York, New York, NY 10016, USA*
[§]*Department of Chemistry, The Graduate Center, City University of New York, New York, NY 10016, USA*
[¶]*Center for Nonlinear Studies, Los Alamos National Laboratory, Los Alamos, NM 87544, USA*
[||]*Levich Institute, School of Engineering, City College of New York, City University of New York, New York, NY 10031, USA*
[**]*Department of Biological Sciences, Centre for Molecular Simulation, University of Calgary, Calgary, AB, Canada*
[††]*Department of Chemistry, City College of New York, City University of New York, New York, NY 10031, USA*
[‡‡]*mgunner@ccny.cuny.edu*

*To cite this article, please refer to its earlier version published in *J. Comput. Biophys. Chem.* **2021**, *20* (2), 111–130, DOI: 10.1142/S2737416520420016

‡‡Corresponding author.

Water molecules play a key role in all biochemical processes. They help define the shape of proteins, and they are reactant or product in many reactions and are released as ligands are bound. They facilitate the transfer of protons through transmembrane proton channel, pump and transporter proteins. Continuum electrostatics (CE) force fields used by program Multiconformation CE (MCCE) capture electrostatic interactions in biomolecules with an implicit solvent, which captures the averaged solvent water equilibrium properties. Hybrid CE methods can use explicit water molecules within the protein surrounded by implicit solvent. These hybrid methods permit the study of explicit hydrogen bond networks within the protein and allow analysis of processes such as proton transfer reactions. Yet hybrid CE methods have not been rigorously tested. Here, we present an explicit treatment of water molecules in the Gramicidin A (gA) channel using MCCE and compare the resulting distributions of water molecules and key hydration features against those obtained with explicit solvent Molecular Dynamics (MD) simulations with the nonpolarizable CHARMM36 and polarizable Drude force fields. CHARMM36 leads to an aligned water wire in the channel characterized by a large absolute net water dipole moment; the MCCE and Drude analysis lead to a small net dipole moment as the water molecules change orientation within the channel. The correct orientation is not as yet known, so these calculations identify an open question.

Keywords: Water simulation; molecular dynamics; Monte Carlo; polarized force field.

1. Introduction

Water is a unique solvent, with a high dielectric constant and a facile ability to make hydrogen bonds. It plays an important role in determining the structure and cellular localization as well as the dynamics, and function of proteins and other biomolecules.[1] It is the solvent for biochemical reactions as well as the substrate or product in many reactions. Water drives the formation of membrane bilayers and of properly folded proteins by solvating charged groups on the surface, while favorable water–water interactions lead to the burial of hydrophobic moieties.[2] Stabilization of charged groups by water modulates the proton affinity of acidic and basic side chains and ligands. Its high dielectric constant screens long-range electrostatic interactions.[3] The release of water molecules from apo-protein cavities can have significant effects on ligand binding selectivity and affinity.[4] Water is a weak acid and base, so leads to a low concentrations of solvated hydronium and hydroxyl ions in neutral, aqueous solutions.

Protons play key roles in cellular reactions. H^+ is often a biochemical reactant. Protons transfer into buried active sites or through membrane embedded proton pumps, such as bacteriorhodopsin[5] and the heme–copper oxidases.[6,7] The transmembrane pH gradient, is a vital store of cellular energy. Protons are transferred via water molecules or polar side chains.[8,9] While we will refer to the mobile proton, a proton is always associated with one or more water molecules. Hydronium is a minimum unit, but the proton is more likely to be associated with two water molecules as a Zundel cation or as a larger, Eigen complex. Water associated protons move through proton specific channels[10–13] and play a role in proton dependent ion and substrate transporters.[14,15] Transiently hydrated cavities help control the connectivity of proton pathways between the two sides of the membrane.[5–7] Naturally occurring short peptide antibiotics, such as gramicidin work by forming open channels allowing protons and other cations to pass through the targeted cell membrane, dissipating essential gradients.[16,17]

Water molecules in oriented water wires with aligned dipoles facilitate proton transfer via the Grotthuss mechanism.[18–20] The water wire starts with dipoles oriented in one direction and ends facing the opposite direction. The water wire then must flip back to the original dipole direction to prepare for the next proton transfer.[21–23]

The stability of an aligned water wire and the barrier to reorientation is likely to depend crucially on the local dielectric environment of the narrow channel. Classical MD simulations with additive force fields have been used to study the dynamics of water molecules confined in ion channels.[21,24,25] The classical force fields such as AMBER or CHARMM36 assign fixed partial charges to all atoms, fixing the water molecule dipole moment. However, the complex chemistry of permeation pathways presented by narrow pores such as the gramicidin A (gA) channel should lead to a significant, dynamic, induced component of the water molecular dipole moment.[26–30] The importance of induced polarizability for water molecules has been shown in various systems such as the stability and dynamics of individual hydrogen-bonds[31–33] as well as of water molecule binding to the surface of nanomolecule or proteins.[34–36] Polarizability may thus affect potential barriers and ion transport properties across the narrow channels.[30,37] Therefore, an account of the induced dipole component of water molecule interactions are critical to assess the accuracy of models as they try to account for the thermodynamics of proton-induced wire reorganization in proteins. A detailed understanding of the role of water molecules confined in the narrow pores and the interplay between the solvent–protein and solvent–ion interactions can help provide more clues for accurate descriptions of proton transport in gA and other channels.

gA is a model system that has long been used to study water molecules, protons and ions in channels.[12,30,37,38] gA consists of 15 alternating l- and d-amino acid residues in two right-handed β-helices, with the N-terminus of both helices meeting in the center of the membrane. The N-terminus is capped with a formyl group and the C-terminus with an ethanolamine. The β-helical structure has side chains pointing away from the channel pore, leaving sufficient space for water molecules to form a single-file water wire within the channel.[39] gA is a simple model system, with few conformational degrees of freedom and no protonatable residues. Thus, the balance of water:water, water:protein and water:ion interactions can be dissected (relatively) cleanly. Various computational methods have been used to study the protein stability,[40,41] as well as the water, proton and ion conductions and selectivity.[38,42] Implicit and explicit water models have been used[43] with nonpolarizable[40] and polarizable[30] MD force fields. MD analysis has been compared with experimentally determined barriers and rates for conductance of ions and protons.[17]

gA provides an ideal, simple model system to directly compare the properties of water molecules confined in a protein found by different simulation techniques and levels of theory. The presence of a single-file water wire in gA is well established and its dynamical reorganization is known to play a significant role in monovalent cation and proton permeation. The work presented here briefly reviews methods to treat water molecules in molecular simulations then compares the behavior of water wires in the gA channel obtained with classical MD trajectories, trajectories using the Drude polarizable force field and the Multiconformation Continuum Electrostatics (MCCE) program which is a Continuum Electrostatics/Monte Carlo (CE/MC) method. MCCE simulations are run on snapshots from the classical MD trajectory. The MD simulations use explicit water molecules inside and outside of the channel. The MCCE calculations monitors explicit individual water molecules within the channel with an implicit, continuum solvent outside of the membrane embedded protein. The resulting average number of water molecules in the channel, and the water dipole orientation are compared for different methods. The different methods give surprisingly different results. The dipole moment of water confined in gA can be used as a reporter of the net alignment of the water molecules. In the classical MD trajectories obtained with the CHARMM36 force field, the water molecules are highly aligned through the whole permeation pathway. They change direction quickly with rare turns. In contrast, the dipole moment for water modeled with the Drude polarizable force field has the peak in the average dipole moment near zero showing they are not aligned through the channel and the orientation changes often through the trajectory. In MC simulations with TIP3P water parameters and CE energies, water molecules in the channel also have a small summed dipole moment. Increasing the size of the water atomic charges in MCCE results in a dipole distribution closer to that found in the classical MD force field with an aligned water wire. MCCE is used to calculate the barrier for proton transfer through the gA channel, which is found to be in good agreement with the results of earlier empirical valence bond (EVB) calculations.[44]

1.1. *Explicit water molecules*

Water molecules are challenging to include in molecular simulations. They are present at a concentration of $\approx$55.5 M so many individual molecules are required to reproduce the solvent properties. The charge distribution is only approximated by standard fixed charge force fields since the electron distribution is polarized by the local environment.[45]

The parameters for water molecules in MD simulations can use fixed charge force fields[46] or polarizable descriptions of the charge.[27,47,48] The dielectric constant of water is time-dependent, starting with the optical dielectric constant of 2 on the picosecond time scale, then taking picoseconds to nanoseconds to relax to the its static value near 80.[49,50] The time for dielectric relaxation provides an estimation of how long calculations with explicit water molecules need to bring the water to equilibrium in an MD simulation. Water molecule force fields remain a continuing focus of innovation and modification.[51–59]

1.2. *Properties of water modeled with implicit solvent*

Water molecules can also be included in simulations by averaging their behavior with implicit models of solvent water.[3] This method aims to reproduce the equilibrated electrostatic dielectric properties of water, including ionic strength dependent solvation and screening of ions, as well as the contribution of surface area dependent hydrophobic energy and attractive Lennard–Jones interactions.[60] CE provides the foundation for implicit solvent analysis. For MC calculations, the electrostatic potential is often determined by the Poisson–Boltzmann equation of CE.[3] In MD simulations, the Generalized Born implementation is often used.[61] Implicit solvent methods have strengths and weaknesses when compared to those that use explicit water molecules.[62] Implicit solvent is always at equilibrium with the charges in the system, so greatly reduces the burden of sampling water molecule configurations.[62] However, implicit water loses all information about individual water molecule dynamics, localization and orientation. In contrast, explicit water captures the motion of individual water molecule at the expense of requiring sampling and equilibrating the water molecules around the molecule of interest for tens of nanoseconds.

1.3. *Molecular dynamics or Monte Carlo sampling*

In addition to the choice of force fields to describe the system, molecular modeling relies on the method of sampling the available degrees of freedom. Molecular dynamics (MD) moves atoms following

Newton's laws of motion given the forces generated by the force fields at the time-dependent atomic positions. Monte Carlo (MC) sampling is an alternative approach to generate a Boltzmann ensemble of microstates of molecules.[63] MC has advantages and disadvantages when compared to MD methods. While MD generates moves from knowledge of local forces, MC moves are random, with their acceptance controlled by the system free energy. The goal is to achieve an equilibrium ensemble of individual states of the molecule of interest, losing information about the pathway or rates of processes. If equilibrium is achieved thermodynamic analysis can be straightforward in MC simulations as the energies of all microstates in the equilibrated ensemble are known.[64] MC moves can change atomic positions of the macromolecule, the distribution of bound and free ligand. In addition, MC can sample different chemical states of the system.

A group of MC methods have been developed for accurate calculation of the pK_a and redox midpoints, E_ms, of residues and ligands in proteins.[3,65] Changing protonation states represents a chemical reaction so cannot be treated by standard MD analysis, although constant pH[66–68] or E_h.[69] MD techniques are being developed to handle these processes. MC methods is often combined with a CE force field to calculate electrostatic energies.[70,71] These CE/MC methods have the advantage of providing accurate electrostatic energies, taking into account electronic polarization with a dielectric constant of 2 that is missing in classical nonpolarizable MD, which uses a dielectric constant of 1.[72] Higher dielectric constants for the protein are used to model protein flexibility.[73,74] These CE/MC methods generate ensembles of protonation or redox states following changes in proton chemical potential (i.e. the pH).[3] Ligand binding has been treated by Grand canonical MC (GCMC), measuring the affinity of explicit ions[75,76] or hydronium[77] binding to proteins in implicit solvent water. MD simulations follow the position changes of atoms over time in an unchanging chemical state and thus fix the protonation states of all residues. GCMC sampling thus allows ionization state changes which conventional MD simulations do not.[78]

1.3.1. *Degrees of freedom in the MCCE program*

The MCCE is the CE/MC program that will be used here. It retains a rigid protein backbone, but allows full rotamer side chain search.[79] It also allows GCMC sampling of binding of explicit water, ions[80] and other small ligands to access relative binding affinities.[81] One microstate is made up of one choice of position and protonation states for each side chain, ligand and associated explicit water. The choice for each group is called a conformer. The energy of each state is the sum of interactions with the continuum solvent (solvation energy) and pairwise interactions calculated with the rigid backbone and amongst all side chain and ligand conformers in that microstate. The final result is the probability of choosing each conformer of each residue in the Boltzmann distribution. Changing the ligand chemical potential or pH changes the equilibrium distributions so gives the binding affinity of ligands and the pK_a of acids and bases.

1.4. *Use of Monte Carlo sampling with explicit water*

In most CE/MC applications, water molecules are removed from the simulation and replaced by an implicit solvent with a dielectric constant of 80. Hybrid CE/MC methods use MC sampling, with a CE force field, with implicit solvent outside of the macromolecule, while explicit water molecules with many conformational choices are available for sampling within cavities. Calculations comparing the pK_as of residues deeply buried within bacteriorhodopsin using only implicit solvent with those using explicit water molecules near key residues showed remarkably similar values.[79] Explicit water molecules have been used to follow proton pathways in proteins via MC sampling within CE/MC simulations.[7]

1.5. *Modeling proton transfer in molecular dynamics simulations*

Because proton transfer involves making and breaking of covalent bonds, it cannot be treated with classical MD simulations. Quantum mechanical methods, such as Car-Parrinello *ab initio* MD,[82] have offered important insights but at a substantial computational cost.

Semiempirical methods based on valence bond theory[83] are much more efficient and have been used widely on a variety of problems.[84,85] Dissociative models that view water as a collection of free H and O particles have also been proposed,[86–91] but have so far been applied mostly to pure water or material science problems. The constant pH approach for modeling proton transfer combines classical MD simulation with periodic MC steps that move a proton to an eligible neighboring site.[12,92] However, the existing approaches violate detailed balance.[93] A method based on λ-dynamics[94] has also been proposed[95,93] but so far has been applied only to proton transfer between water molecules.

1.6. *The barrier for hydronium transfer in Gramicidin channel*

The proton gradient across energized bacterial, mitochondrial and chloroplast membranes provides the energy source for ATP synthesis and other processes. It is therefore necessary to control proton conduction without introduction of leaks.[20] gA collapses the proton gradient by poking a hole in the membrane. Protons are conducted through gA by the Grotthuss mechanism.[20] MD simulation,[96,97] umbrella sampling,[96] and multistate EVB (MS-EVB)[98] have been used to analyze the thermodynamic properties of the proton in channels leading to conduction. In this paper, MC sampling is used to calculate the energy of a hydronium molecule moving through the gA channel. The energy difference for a hydronium in water or in the gA channel is compared with that calculated using the EVB method.[44]

2. Methods

2.1. *gA protein model*

The starting structure of the gA dimer was derived from the X-ray structure (PDB ID: 1JNO). Each of the gA monomers has the sequence: formyl-Val_1-Gly_2-Ala_3-$DLeu_4$-Ala_5-$DVal_6$-Val_7-$DVal_8$-Trp_9-$DLeu_{10}$-Trp_{11}-$DLeu_{12}$-Trp_{13}-$DLeu_{14}$-Trp_{15}-ethanolamine. The two monomers are each oriented with their N-terminus in the middle of the membrane.

2.2. *Classical MD simulations*

The simulation protocol was generated within the CHARMM-GUI, based on optimal settings for simulations of protein-membrane systems in OpenMM.[99] The parameters for the N- and C-termini formyl and ethanolamine caps for both classical MD and MCCE are taken from Ref. 12. For Classical MD simulations, the protein was embedded in a 1,2-dimyristoyl-sn-glycero-3-phosphocholine (DMPC) membrane using the CHARMM-GUI Web server.[99] The CHARMM36 force field is used for the protein and lipids and TIP3P model is used for water molecules. This puts a charge of −0.834 on each water oxygen.

An initial energy minimization is carried out for 500 steps with heavy atoms of the protein and lipid molecules restrained with a force constant 4000, 2000 and 1000 $\mathrm{kJ/mol/nm^2}$, for protein backbone, side chain and lipid heavy atoms respectively. A step-wise equilibration was performed in the NPT ensemble, where restraints on the heavy atoms are gradually released. The pressure was regulated by MC membrane barostat in OpenMM at a pressure of 1 bar and a pressure coupling frequency of 100 simulation steps. The temperature was maintained at 303.15 K using Langevin thermostat. The simulations were carried out using OpenMM (GPU version).[100] Snapshots were saved every 1 ps for 200 ns. The restrained MD trajectories were carried out with the channel restrained with a positional restraint force 50 $\mathrm{kJ/mol/nm^2}$ for the gA backbone.

2.3. *MD simulations with Drude force field*

The 2013b version of CHARMM Drude force field for proteins and lipids was used.[101–104] The parameters for ethanolamine and formyl groups present in gA were taken from Ref. 105. The SWM4-NDP model is used for water molecules in the channel and in bulk.[101] It consists of two hydrogens (with charge 0.557330), one oxygen (with charge −1.11466). Charged auxiliary particles are attached to the oxygen with springs.

The equilibrated system from Refs. 28 and 101 was used to generate the starting configurations for MD simulations with the Drude polarizable force field. The resulting simulation systems contained the gA channel embedded into a lipid bilayer comprised of 96 DMPC molecules solvated by 4003 SWM4-NDP water.[28,101] A hexagonal periodic cell with a side of 61 Å and length of 77 Å was used for all production simulations. Langevin dynamics with a dual-thermostat scheme was used to propagate the atoms and auxiliary Drude particles with an extended Lagrangian formalism implemented in the NAMD package.[102,106] The thermostat acting on heavy (non-Drude) particles was set to $T_{\text{heavy}} = 330\,\text{K}$ to enhance sampling of lipid dynamics as suggested previously.[107] The Langevin damping coefficient was set to 1.0 ps$^-$1. The weak damping coefficient enables stable Drude production runs with an integration time step of 1.0 fs.[108] For all production runs, $T_{\text{Drude}} = 0.1\,\text{K}$ was used with a spring constant for the atom–Drude bond of 1000 kcal/mol/Å^2.[109,110] A damping constant of 20.0 ps^{-1} was applied to Drude particles.

2.4. *MCCE simulations*

MCCE is used to carry out the CE/MC simulations of the water molecules in the gA channel.[79] In MCCE a conformer is a specified set of positions and charges that can be chosen in MC sampling for the residue or ligand. These provide the degrees of freedom for a protein with a fixed backbone. Conformers are made prior to MC sampling and all energy terms are pre-calculated. Microstates, which specify the conformer for each residue and ligand, are subjected to GCMC sampling.[78]

The energy of a microstate includes self-energy electrostatic solvation energy and torsion energy for each conformer, which are independent of the choices for other residues. To these are added the pair-wise Lennard–Jones and electrostatic interactions between all conformers in the microstate. For GCMC binding calculations as carried out here, the water molecules chemical potential is considered.[81] For pH titrations, the free energy of group ionization at the pH of the calculation is included.[111] Thus, the free energy cost of producing a hydronium in solution is a component of the calculation of the free energy of hydronium transfer through the channel.

Electrostatic interactions are calculated with DelPhi[112] using PARSE charges and radii[113] with a dielectric constant of 4 for protein and featureless slab which mimics the membrane and 80 for water with 150 mM implicit ions.[78] Amber non-electrostatic parameters are used.[114] As with the classical MD simulations the N- and C-termini formyl and ethanolamine caps are taken from earlier calculations[12] and the default water model is TIP3P with the partial charges of O: −0.834 and H: 0.417. The standard MCCE calculations were carried out at 298 K. Addition simulations are presented where the temperature we raised to 330 K for Metropolis sampling and the dielectric constant for water was reduced to 70. The internal dielectric constant remained at 4.

2.5. *Placing explicit water molecules in the channel in MCCE*

Each MCCE calculation starts with a snapshot from the backbone restrained classical MD trajectory with all water molecules and lipids removed. As used for the MCCE simulations of other membrane proteins the protein is embedded in a 35 Å tall low dielectric slab using the Implement Protein Environment for CE (IPECE) utility.[79] This is a minimal, implicit membrane model, whose sole purpose is to form a region with a low dielectric constant around a protein.

Protein coordinates are taken from nine snapshots of the MD restrained trajectory. These are chosen based on the summed dipole moment of the water molecules in that snapshot. Thus, the water molecules in three of the snapshots have a net positive dipole moment, three have a negative dipole moment and three have summed dipole moments near zero. The results presented for the MCCE calculation are the sum of the results starting with each snapshot, weighted by probability of each dipole moment in the trajectory.

Water molecules were added to the channel on a 7 Å × 7 Å × 22 Å rectangular grid with 0.5 Å spacing oriented along the channel z-axis. Grid points that clash with the protein atoms were removed. An oxygen atom was placed at the approximately 700 remaining grid points. For each oxygen, a fixed number of water molecule conformers were generated by applying 10 random rotations to place protons on the fixed oxygen position (Fig. SI.1). Thus, ≈ 7000 water molecule conformers are sampled. A comparison analysis considering 5, 10, 15, or 20 different proton positions (water conformers) is shown in Fig. SI.2. The default calculations included 10 water conformers/oxygen.

2.6. *Tests for convergence of the water molecules in the gA channel in MC sampling*

In MCCE we evaluated convergence in two ways. One is the stability of measurable outcomes, these being the distribution of the number of water molecules in the channel, the number of hydrogen bonds and the summed dipole moment of these water molecules. The more rigorous test is the difference in the Boltzmann averaged energy for 100 restarts of MC sampling. With the TIP3P water charges the full width at half maximum (FWHM) of the average energy distribution is generally ≈ 0.2–0.3 kcal/mol and all MC restarts have an average total energy range within ≈ 0.4 kcal/mol (Fig. SI.3). Microstates created in different input snapshots do have different energy resulting from the differences in the input protein backbone and side chain conformation.

Three methods are used to enhance convergence. (1) The oxygens are grouped so that all water conformers, where the oxygen has the same z-coordinate, are treated as one residue, so that only one conformer from the slice can contribute to a given microstate. All water molecules in one layer are too close to each other so they can never be present in the same accepted microstate. Thus, grouping them in one residue eliminates the chance that any microstate with two water molecules in the same layer would be evaluated as a proposed microstate and then rejected. There are 44 layers, separated by 0.5 Å. (2) The ratio of conformers in the channel and outside in the bulk is varied. There are ≈ 7000 conformers to choose from but only approximately 7 will be in the channel at one time. Each layer has ≈ 10 oxygens and thus 100 different choices for water molecule conformers in the gA channel. Six to ten chances to move out of the protein are included for each layer. Water molecules transferred to the bulk have an electrostatic solvation (reaction field) energy of −2.6 kcal/mol, which is the value calculated for a TIP3P water molecule in solvent with a dielectric constant of 4 inside and 80 outside. (3) The number of new conformer choices made for each microstate helps to reduce the memory of the system. Standard MCCE makes from 1 to 3 changes before evaluating the energy. Here 10 to 20 changes are applied before calculating the energy of the new microstate and applying Metropolis criteria for acceptance.

The calculations with TIP3P water atomic charges are well converged, giving both consistent averaged microstate energy and physical properties. We will present some information about MCCE calculations with a larger water partial charge on the water molecule that can reproduce elements of the behavior of the classical MD trajectories. For larger charge of water model, the measurable properties we present are robust, but the energy difference between multiple restarts of MC sampling have significantly different energies. Thus, while the large charge calculations provide useful information, they are very difficult to converge.

2.7. *Water chemical potential*

We have not set out to independently calculate the absolute binding affinity of the ligand water molecule in the MCCE simulations. Rather energy is added to the ligand solution chemical potential in a binding titration to match experimental or other computational results. In previous studies of ion binding, an experimental K_d is used to define the concentration at a given MCCE energy.[80] Here, we will compare the distribution of the number of water molecules in the channel in MCCE and in the MD calculations.

In MCCE, it is recognized that the ratio of available reactant to product conformers changes the outcome. A correction for this imbalance, which excludes the conformers that are at high energy and so are never in accepted microstates need not be considered, is include in the standard MCCE program.[78]

2.8. *Analysis of the water molecules in the gA channel*

When carrying out calculations using different force fields it is difficult to directly compare energy terms. In addition, as the different methods have different degrees of freedom, contribution of entropy to the results will be different. A Solvation Structure and Thermodynamic Mapping (SSTmap)[115] analysis of the classical MD trajectories and an analogous analysis of energy terms in MCCE for water molecules are found in Table SI.2. In the main text, we will focus on observable water properties in the channels where the outcome of the different calculations can be directly compared.

2.8.1. *Counting water molecules*

The number of water molecules in the channel is a function of the relative energy of the water in bulk solvent and in the channel. For the MD simulation, the total number of water molecules inside the gA channel is determined using VMD.[116] Frames separated by 20 ps are aligned with each other along the z-axis, the root-mean-square deviation (RMSD) of atomic positions of 200 ns restrained and unrestrained structures are shown in Fig. SI.4.

The water molecules where the oxygen coordinates are within the limits of the gA channel protein are considered inside the channel. In the MCCE simulation, explicit water molecule conformers are only placed within the channel. The water molecules count is the average number of explicit water molecules bound in the GCMC accepted microstates.

2.8.2. *Dipole orientation*

The dipole moment describes the overall orientation of water molecules in the gA channel in a given MD snapshot or MC microstate. The channel is aligned along the z-axis, through the membrane. The z-component of the dipole moment is given by

$$\mu_z = q_0 * r_z, \tag{1}$$

where q_0 is the charge on the oxygen and r_z is the vector sum of the z-components for each O–H bond. The dipole moment is determined individually for each MD snapshot or MC microstate as the sum of μ_z for each water molecule in the channel.

The Dipole Moment Watcher tool in VMD is used to calculate the dipole moment for every frame in the MD trajectory. For the Drude MD trajectory, only the real atoms are considered, the oscillator positions are not included. For MCCE simulation the dipole moment is calculated independently for the water molecules in each accepted microstate. A histogram of the dipole moments for each set of conditions used a standard bin of 0.1 D. The z-component of the dipole moment of the gA protein backbone in each frame of the classical MD trajectory is also calculated with the VMD tool (Fig. SI.5).

2.8.3. *Count of hydrogen bonds between water molecules and from water molecules to protein backbone*

Hydrogen bonds made between neighboring water molecules N_{ww}^{HB} or between a water molecule and the protein backbone N_{pw}^{HB} are counted if the distance between heavy atoms is ≤ 3.5 Å and hydrogen–donor–acceptor angle is $\leq 30°$.

3. Results and discussion

3.1. *Number of water molecules within the gA channel*

The number of water molecules in the gA channel at equilibrium is a function of both the free energy of the water molecules in bulk solvent and their energy in the channel. Explicit water MD simulations are carried out using the standard CHARMM36 force field as well as with the polarizable Drude oscillators. Table 1 shows the average number of water molecules in the channel and Fig. 1 shows the distribution of the count of the number of water molecules in individual MD

Table 1. Average number of water molecules in gA channel.

Simulation methods	Water charges	Water count
CHARMM36 (restrained)	O: −0.834, H: 0.417	7.84 ± 0.93
CHARMM36 (unrestrained)	O: −0.834, H: 0.417	7.38 ± 0.95
Drude	O: −1.114, H: 0. 557	5.93 ± 1.09
MCCE_303K	O: −0.834, H: 0.417	7.37 ± 0.19
MCCE_330K	O: −0.834, H: 0.417	7.35 ± 0.18

Note: The classical MD trajectories were carried out at 303.15 K.

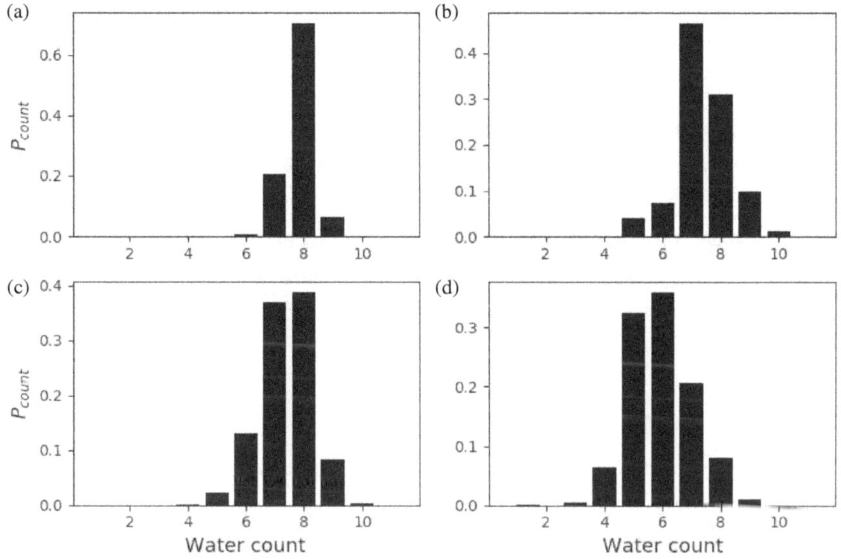

Fig. 1. Distribution of the number of water molecules inside the gA channel found in: (a) restrained MD trajectory with CHARMM36 force field; (b) unrestrained MD trajectory with CHARMM36 force field; (c) MC sampling using MCCE with TIP3P charges; and (d) MD trajectory with Drude force field.

snapshots or MC microstates. The 200 ns restrained MD trajectory has a narrow distribution, strongly favoring eight water molecules. The 200 ns unrestrained classical MD trajectory prefers seven water molecules in the channel. Previous experiment results also showed around seven water molecules inside the gA channel.[117] The MD simulation with the Drude force fields has on average only six water molecules and has a wider range of water molecule numbers in the channel.

With the standard MCCE force field and added chemical potential of zero and no correction for the implicit water van der Waals interaction, the channel occupancy in MCCE is close to that seen with the unrestrained MD simulations. This may reflect the PARSE charge and radii used for the solution of the Poisson–Boltzmann equation in the calculation were optimized to replicate solute transfer data so these parameters model the energy of water molecule transfer without significant correction needed.[118] There are only modest differences in the distribution of water molecules in the gA channel for MCCE calculations initiated with different MD snapshots (Fig. SI.6).

The water chemical potential was modified for a single snapshot in MCCE to examine the binding isotherm Fig. 2. The channel saturates at approximately 14 water molecules, however after there are ≈ 8 water molecules the binding becomes somewhat anti-cooperative. Convergence of the MC sampling is evaluated by the distribution of the Boltzmann averaged energy for 100 independent MC runs. As the channel becomes more packed it becomes more difficult for the MC

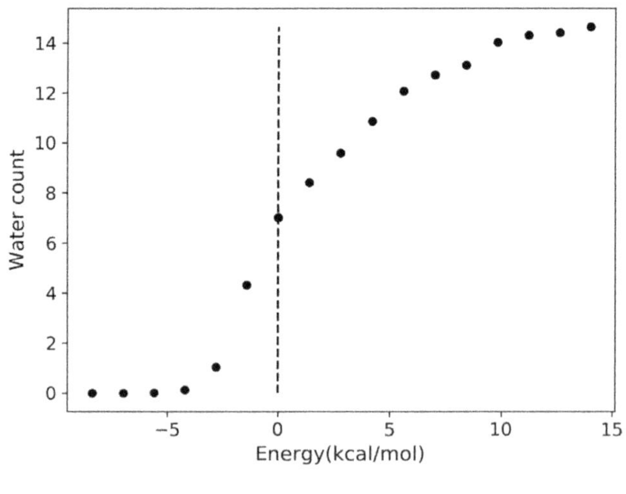

Fig. 2. Average number of water molecules in the gA channel as a function of the extra, unfavorable chemical potential added to water molecules in bulk solvent in the GCMC simulation in MCCE. All other calculations reported here have zero energy added to bulk water chemical potential.

sampling to converge as the behavior of all water molecules become increasingly interdependent Fig. SI.7.

3.2. *Dipole orientation of water molecules inside gA channel*

3.2.1. *MD simulation*

The gA protein is made up of 2 identical β-helices each 15 amino acids long,[16] oriented so the two N-termini meet in the middle of the membrane. The N- and C-termini are capped with the neutral formyl and ethanolamine groups, respectively, so there is not charge on the termini. The β-helix alternates the orientation of adjacent amides (as in a β-strand) so it does not build up a helix macro-dipole as would be found in an α-helix. The summed amide dipole moment of the backbone along the membrane z-axis, $\mu_{z,\mathrm{prot}}$ is close to 0 (Fig. SI.5). The question then is whether the water molecules inside the gA channel align with the helix, flipping in the middle, or if the water: water interactions ensure that they keep one orientation through the channel. The three computational methods find different equilibrium water orientations.

The dipole moment of the water molecules in the channel was calculated along a 200 ns restrained, classical MD trajectory. (Fig. 3(a)) A histogram of the summed water dipole moments ($\mu_{z,\mathrm{wat}}$) show two clear peaks at ± 14 Debye (Fig. 3(a)) with a FWHM of $\approx$ 3D as has been found previously.[119] The sign of the orientation is arbitrary but is used consistently for a given trajectory. Typical water molecule positions in states with positive, negative and near 0 Debye dipole moment are shown in Fig. 4. After 200 ns, the probability 45.5% of the snapshots have water molecules oriented in the positive direction, 1.1% have a dipole moment near zero (+5 Debye to −5 Debye), while 53.4% are oriented in the negative direction. The flips between positive and negative orientation are rare. Thus, these water molecules maintain an ordered water wire that persists through the channel, rather than changing orientation in the middle of the channel. It should be noted there is no transmembrane potential nor

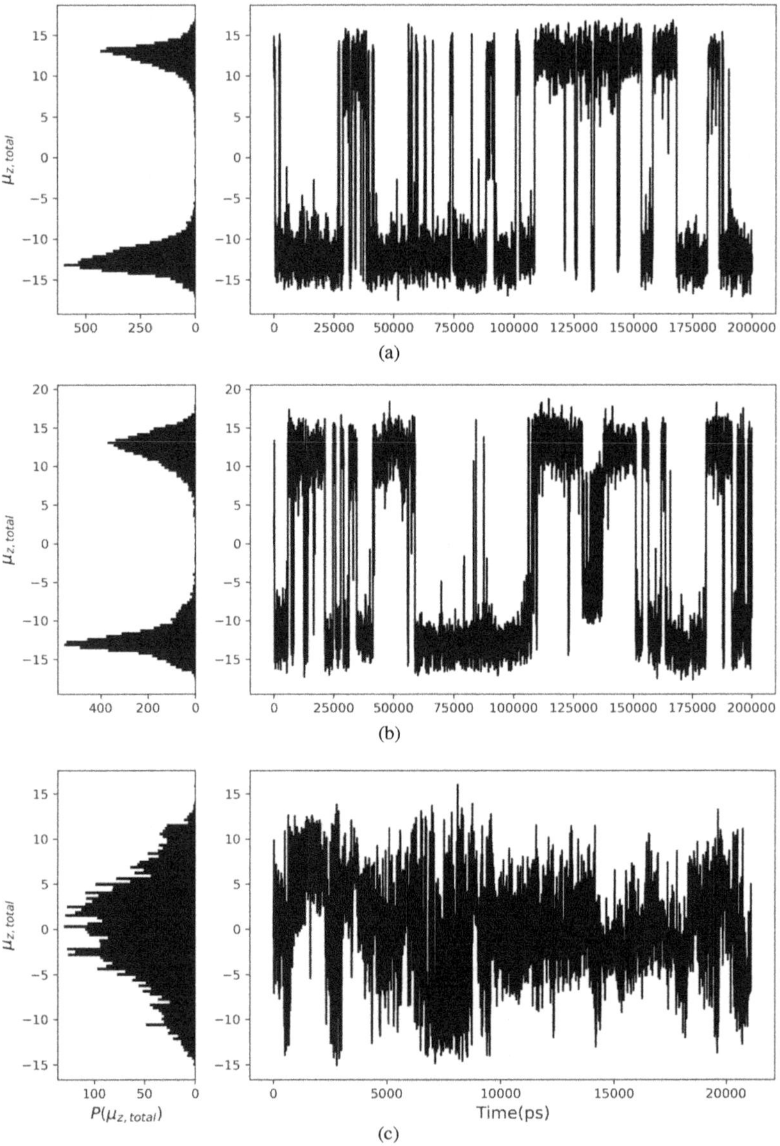

Fig. 3. Summed dipole moment for the water molecules in the gA channel in each snapshot in the MD trajectories. Left: The distribution of the water dipole moments. Right: The dipole moment of the water molecules inside the gA channel through each trajectory as a function of time. (a) Restrained MD trajectory with CHARMM36 force field; (b) unrestrained MD trajectory with CHARMM36 force field and (c) MD trajectory with Drude force field.

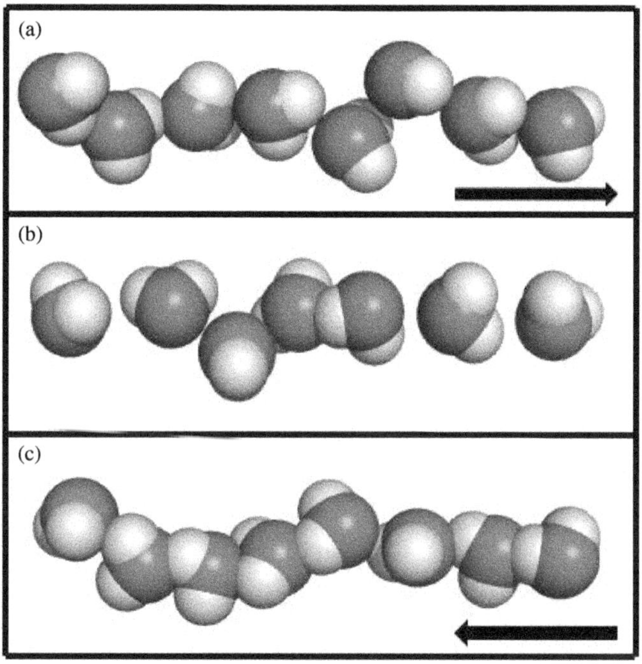

Fig. 4. Example of water orientation for the microstates with: (a) positive dipole moment; (b) near zero dipole moment and (c) negative dipole moment. Each figure represents one snapshot from the restrained MD trajectory.

ions in any of these simulations to orient the water molecules in the channel.

The transition between states with positive and negative dipole moments states of the water wire takes only ≈ 6 ps, starting at one end and zipping along to the other side (Fig. 5). This is the orientation change required for the Grotthuss mechanism.[20,38] With 97 flips/200 ns the flip rate is $\approx 4.9 \times 10^8$ flips/s. The rate (k) is $k = k_0 \times 10^{-\mathrm{Ea/kT}}$. Using the standard value of k_0 of $\approx 10^{13}$/s, the flip barrier is ≈ 5.8 kcal/mol. This estimate of the barrier obtained from the measured frequency of transition is higher than the 2–3 kcal/mol estimated previously from PMF simulations at 330 K. The trajectory here is run at 303.15 K, which may contribute to there being fewer flips.[20,119]

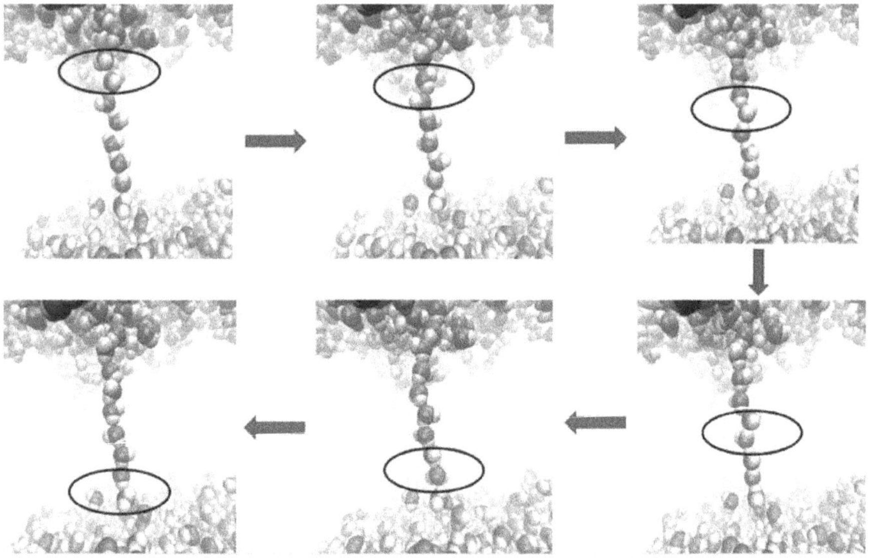

Fig. 5. Series of snapshots separated by 1 ps from in the restrained classical force field MD trajectory which shows the process by which the water wire in the gA channel changes direction. The reorientation takes approximately 6 ps.

When the restraints on the amide backbone are released in the classical CHARMM36 MD trajectory the water wire remains oriented through the channel, with two peaks in the $\mu_{z,\mathrm{wat}}$ histogram near ± 14 Debye (Fig. 3(b)). The flip rate is $\approx 3.25 \times 10^8$ flips/s. This suggests the barrier is ≈ 6.0 kcal/mol, similar to the value found with the restrained simulation. A running sum of the averaged dipole moment with time shows that the dipole moment requires at least 150 ns to converge to an approximate 50:50 mix of positive and negative facing water wires (Fig. SI.8).[20] In contrast, the water wire orientation in the restrained MD trajectory converges in less than 50 ns. However, in all trajectories, the probability that the sum of the water dipole is near zero remains very low at $\approx 1\%$.

The behavior of the water molecules in the gA channel was explored with the Drude polarizable force field and SWM4-NDP water model. The results are quite different than found in the classical CHARMM36 MD simulations. The cumulative water dipole now has a broad distribution peaked between 0 D and ± 4 D

(Fig. 3(c)). The water molecules are rarely fully oriented in either direction. In addition, the dipole orientation is very dynamic, changing often through the trajectory.

3.2.2. *MC simulation*

The total dipole moment of the water molecules inside the channel is determined for each microstate, where a microstate is analogous to a frame in an MD trajectory. Here we have saved ≈ 50 million microstates for each independent MC sampling with the results from 100 independent MC runs added together for each snapshot.

The distribution of $\mu_{z,\mathrm{wat}}$ was determined. The single input MD snapshot has one water orientation with a unique dipole moment. However, MC sampling from any snapshot yields a broad distribution of dipole moments, which usually has a single peak (Fig. SI.9). Thus, while MCCE analysis of an individual MD snapshot cannot capture the full spread of the full MD trajectory, it does show a broad distribution of water orientations in the Boltzmann distribution of microstates that are acceptable in this protein.

The results were obtained using protein structures from the restrained MD snapshots with positive, negative or near zero dipole moment. The MCCE results from each snapshot were summed, weighted by the probability of these positions in the MD trajectory to generate the properly weighted dipole moment distribution (Fig. 6). The overall dipole moment distribution shows one broad peak centered near 0 Debye, more similar to that found with the Drude, polarizable force field than with the classical MD analysis. The dipole distribution is little changed when the temperature for the Metropolis acceptance is carried out at 298 K or 330 K.

3.2.3. *The orientation of the states with a net dipole moment near zero*

In the dipole moment distribution from the MD trajectory with the Drude force field and in the MCCE simulation with TIP3P water model, there are many states which have a dipole moment near 0

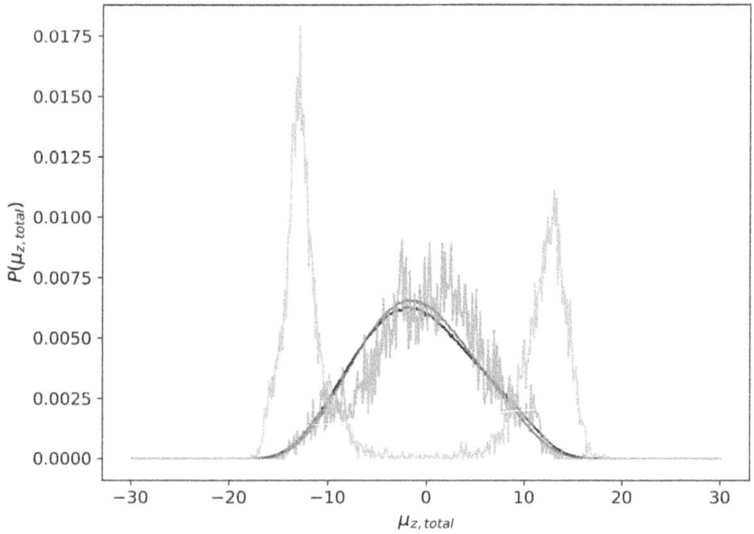

Fig. 6. (Color online) Comparison of the distribution of water molecule dipole moments in each MD snapshot or MC microstate. Water molecules in MCCE: Black line: 303.15 K; Purple line: 330 K. The dipole moment peak is slightly higher at the higher temperature. The results of MCCE simulations with TIP3P water, initiated with nine snapshots were, with three each for positive, negative and near zero net dipole moment and weighted by the probability of each dipole orientation in the MD trajectory (Fig. SI.9). Distribution of summed water molecule dipole moment in each frame of restrained CHARMM36 MD trajectory (yellow line) or Drude force field (cyan line) are reproduced from Figs. 3(a) and 3(c).

Debye. There are two ways to get a summed dipole moment near zero. One is that the water molecules flip in the middle of the gA channel pointing their protons towards the NTR (or CTR). The other way is that they are disordered so are not orientated strongly to the protein or to each other (Fig. 7).

The dipole moment is calculated separately for water molecules inside the two helices. For the states with a net dipole between −1 and 1 Debye, 30% of the snapshots in the Drude trajectory are disordered, while 20% of the microstates are disordered in the MCCE MC sampling (Fig. 7(a)). The rest are oriented, but reverse direction in the middle when the two peptides come together. In the Drude trajectory, approximately 35% the water molecules have their oxygens pointing towards the center and the same number keep the oxygens pointing

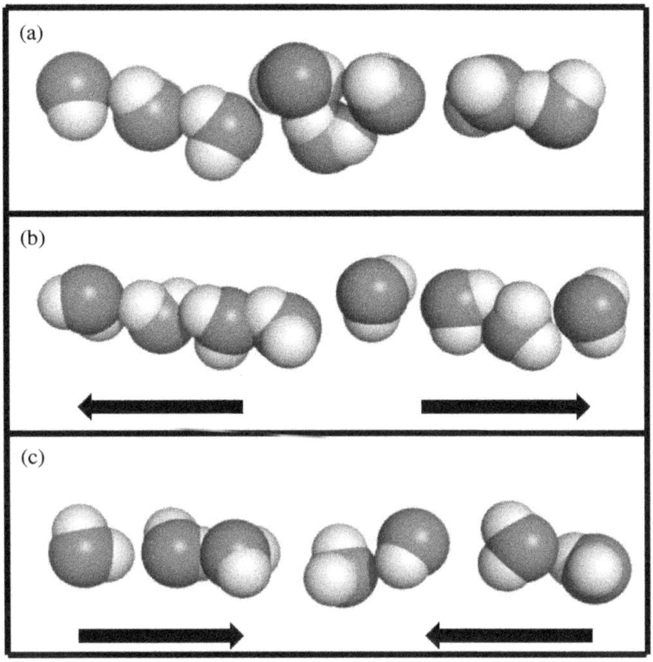

Fig. 7. Examples of water molecules in gA channel whose summed dipole moment is between ±1 Debye: (a) water molecules are disordered; water molecules flip orientation in the middle; with protons oriented towards: (b) the outside of the channel and (c) the center of the channel.

toward the outside of the channel (Fig. 7(b)). In the MCCE analysis the orientation with the oxygens pointing to the inside is preferred (Fig. 7(c)).

3.3. *Hydrogen bond analysis of MD and MC simulations*

Hydrogen bonds can be formed with neighboring water molecules or with the nearby protein backbone carbonyls (Fig. 8). The resultant number of hydrogen bonds have been compared for the restrained Classical MD simulation and MCCE simulation (Table 2). SSTMap was used to determine the average hydrogen bond number for each cluster.[115] This program divides all oxygen positions along the whole trajectory into eight clusters in a given MD frame or MC microstate

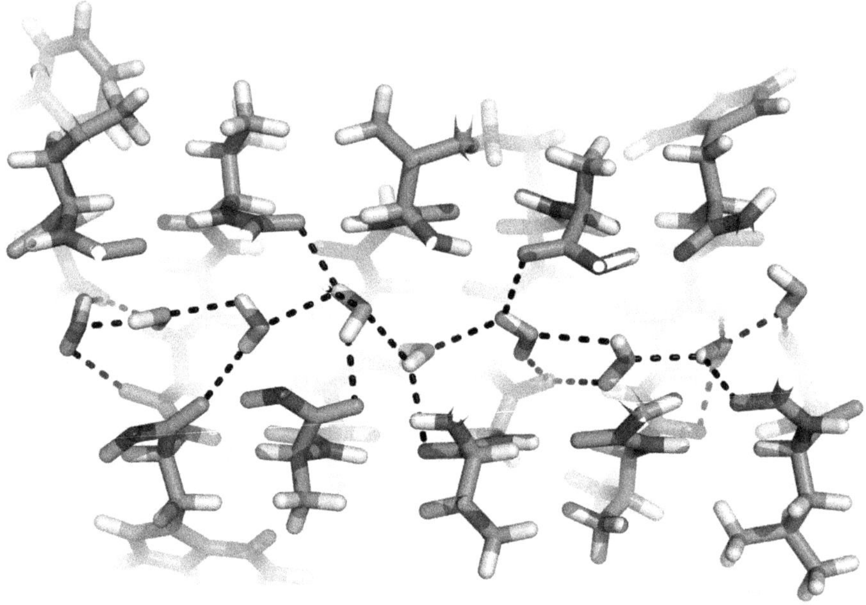

Fig. 8. Water molecules inside the gA channel taken from a single MD snapshot. The black dashed lines show hydrogen bonds between water molecules or between water molecules and the protein backbone carbonyls. This example has eight water molecules with seven water unique hydrogen bonds and 11 water:protein hydrogen bonds. The dipole moment of the eight water molecules in the gA channel is −15 Debye.

(Fig. SI.10). The hydrogen bond connections between the specific water molecules found between water molecules assigned to a specific cluster in that frame or microstate are evaluated.

In the classical MD simulation, each water molecule has on average 1.60 hydrogen bonds with neighboring water molecules and 0.99 hydrogen bond with protein. These values are consistent with those found previously.[20] For the MCCE calculations there are only 0.61 hydrogen bonds with neighboring water molecules, fewer than MD. However, there are on average 1.22 hydrogen bonds to the protein which is more than the MD simulation.

The decrease in water:water hydrogen bonds in the MCCE calculations is a reflection of the less organized water wires (Fig. 9). All MCCE microstates were divided by their summed dipole moments

Table 2. Average properties of water molecules in gA channel obtained from MCCE GCMC simulations and restrained MD trajectory with classical CHARMM36 force field.

Cluster	MC_f_{wat}	MC_N_{ww}^{HB}	MC_N_{pw}^{HB}	MD_f_{wat}	MD_N_{ww}^{HB}	MD_N_{pw}^{HB}
1	0.96	0.29	1.06	0.88	1.32	1.30
2	0.82	0.69	1.28	0.91	1.79	0.85
3	0.92	0.72	1.31	0.93	1.84	0.86
4	0.97	0.69	1.08	0.93	1.84	0.85
5	0.57	0.76	1.10	0.92	1.80	0.85
6	0.89	0.72	1.55	0.87	1.54	1.10
7	0.64	0.64	1.41	0.74	1.24	1.14
8	0.84	0.34	0.95	0.79	1.46	0.98
Mean	0.83	0.61	1.22	0.87	1.60	0.99
Std. Dev.	0.15	0.18	0.20	0.07	0.24	0.17

Notes: TIP3P water molecules are grouped into eight clusters as described in Section 8 (Supporting Information). SSTMap is used for the MD trajectory and the algorithm was repurposed to analyze the water molecules in each accepted MC microstate. The analysis of a MCCE simulation initiated with a single MD snapshot is shown. f_{wat}: Fraction of MD frames or MCCE snapshots that have a water in this cluster; N_{ww}^{HB}: Average number of hydrogen bonds between water molecules in this cluster with water molecules in adjacent clusters; N_{pw}^{HB}: Average number of hydrogen bonds between water molecules in the given cluster and the protein backbone. Mean and standard deviation average the properties of the eight clusters.

and the number of hydrogen bonds was determined for each microstate. The number of hydrogen bonds between water molecules increases as the water molecules become more aligned. When MD snapshots and MC microstates have the same degree of alignment they have essentially the same number of water:water hydrogen bonds.

3.4. *Increasing the water molecule atomic partial charges in MCCE*

The MCCE parameters were modified, increasing the charge on the oxygen to be −1.2, to see if we could match the enhanced alignment of the water wires in the classical MD trajectory. In MCCE, the dielectric constant is 4, while it is 1 in the classical MD trajectory,

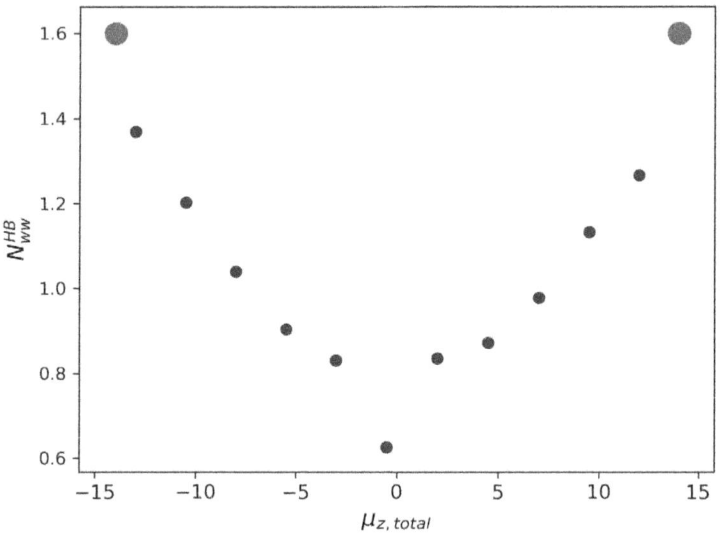

Fig. 9. (Color online) The number of hydrogen bonds between water molecules in MCCE microstates as a function of dipole moment of the microstate. MCCE microstates are binned into groups with the same dipole moment ±3 D. Water molecules are divided into eight clusters as described in Table 2; Blue points: the average number of water:water hydrogen bonds between adjacent water clusters as a function of the microstate dipole moment; Purple points: average number of water: water hydrogen bonds between water clusters in the restrained MD trajectory.

thus the electrostatic screening reduces the interactions between the water molecules. Increasing the water partial charges increases the alignment of the MC water wire. The model with larger atomic partial charges has much larger total energies and proved to be much more difficult to come to convergence. A number of modifications were made for the MCCE MC sampling methods, which are described in Fig. SI.11. However, despite not being fully converged we present data on the system with large water partial charges because the properties of interest are found to be quite reproducible. The key result is with the larger charges the orientation of the water wire becomes significantly closer to that found in the classical MD trajectory (Fig. 10). As suggested in Fig. 9, the water wire becomes more oriented and more water:water hydrogen bonds are made (Table 3). Thus, with larger partial charges, the water wire

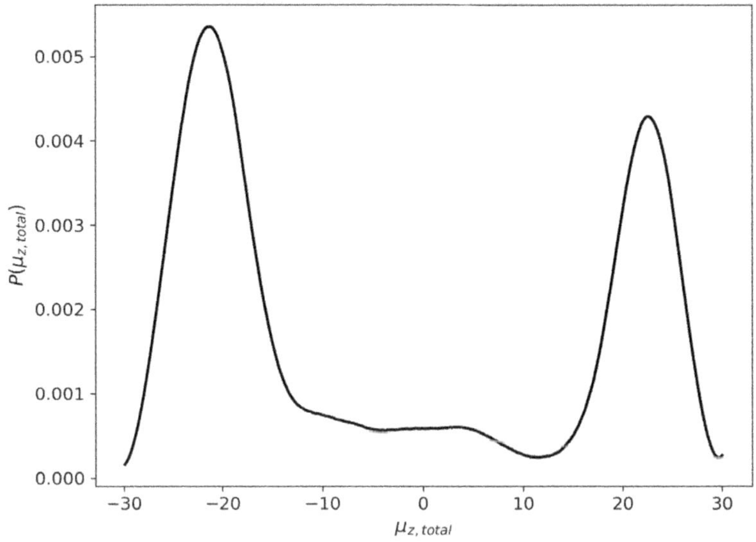

Fig. 10. The summed dipole moment distribution for enhanced water charge with water oxygen partial charge of −1.2. The MCCE analysis was initiated with the same MD snapshots and water positions as used in Fig. 6.

Table 3. Average properties of water molecules in gA channel obtained from MCCE GCMC simulations with O partial charge of −1.2.

Cluster	f_{wat}	N_{ww}^{HB}	N_{pw}^{HB}
1	0.84	0.66	1.00
2	0.92	1.39	1.32
3	0.98	1.58	0.94
4	0.99	1.86	0.35
5	0.97	1.91	1.64
6	0.99	1.60	1.61
7	0.92	0.92	1.26
8	0.37	0.58	0.74
Mean	0.87	1.31	1.11
Std. Dev.	0.21	0.53	0.44

Note: See Table 2 for a description of the table entries.

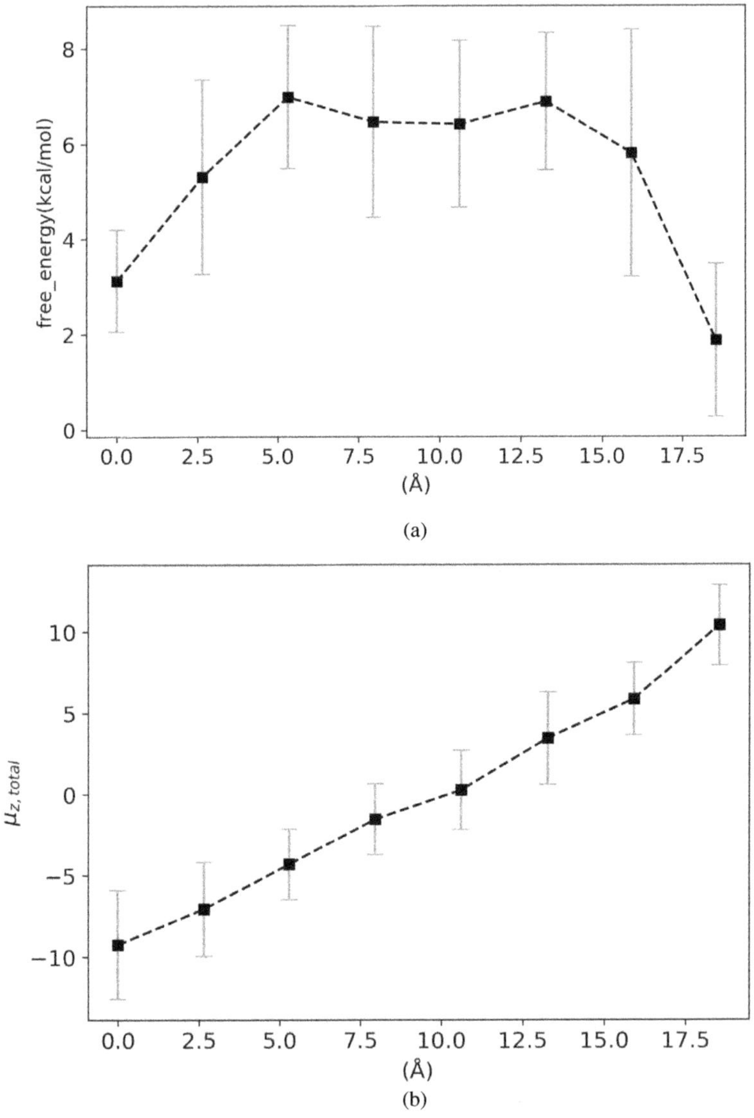

Fig. 11. (a) The free energy found for the gA channel with seven water molecules and one hydronium as a function of the hydronium position. The reference state has one hydronium in solution. (b) The dipole moment of the water wire as the hydronium is moved through the gA channel. The water wire is fully oriented with protons pointing away from the hydronium.

orientation and hydrogen bond count approach that of the classical MD simulation.

3.5. *Barrier to hydronium transfer*

The gA channel has been used as a model system for the study of ion and hydronium transfer through channels.[98,119–121] An excess proton is assumed to go through the channel via the Grotthuss mechanism.[20,98,120] To determine the energy of moving a hydronium through the channel in the MCCE simulation, we started with a microstate with eight water molecules within the 25 Å channel. One water at a time was replaced with a hydronium (H_3O^+), which had five possible proton orientations. The protein structure and oxygen positions of water and hydronium are fixed. The proton positions on the hydronium and surrounding water molecules were allowed to come to equilibrium and the total system free energy was obtained. Thus, each calculation has seven water molecules and one hydronium. The results initiated with 10 different snapshots were averaged.

Generating hydronium from water in solution requires 12.36 kcal/mol at pH 7.4. The average barrier for moving a hydronium going through the channel is 7.0 ± 1.1 kcal/mol (Fig. 11(a)). The value of the barrier for hydronium in the gA channel can be compared with earlier calculations which provided an estimate of 6 kcal/mol with an uncertainty of 1–2 kcal/mol[44] using the EVB model and Langevin dynamics to simulate the proton transportation in gA channel. The EVB simulations highlighted the importance of electrostatic energy in determining the proton transport rate. As CE/ME methods are optimized for the calculation of processes that change the charge state of the system, this may allow them to estimate the hydronium energy using a classical model. In the presence of the hydronium all water molecules orient towards the positive charge (Fig. 11(b)).

4. Conclusion

Gramicidin is a simple water and cation conducting channel and so has provided a model system to explore the properties of water and ions in a confined space. We have investigated the ability of MCCE, which uses MC sampling with CE energies to recover the properties of water found with MD simulations. MD simulations without a polarizable force field yield a water wire with a single orientation through the channel. In contrast, with a polarizable Drude force field the water molecules have a summed dipole moment of the water molecules near zero. Thus, it is not certain what the water orientation in the gA channel should be.

MCCE with standard TIP3P water parameters yields water dipole orientations much closer to that found with the polarizable Drude MD simulation. This may reflect CE including polarization in the dielectric response. Increasing the charge on the water molecules in MCCE yields water orientations closer to that found in the nonpolarizable simulation. The barrier for hydronium in the channel of 7 ± 1 kcal/mol calculated with MCCE, which agrees remarkably well with earlier calculations that used a more sophisticated EVB model.[44]

CE with implicit solvent is generally used to remove explicit water molecules, which have many degrees of freedom and can be difficult to come to equilibrium. However, using gA as a model system, we see that explicit water molecules provide appropriate numbers of water molecules in the channel, degree of dipole orientation and barrier to hydronium found with other simulation techniques that use quite different methods of sampling and different force fields. This supports the use of hybrid continuum models that include explicit water in regions where they play a functional role, with implicit water elsewhere.

Acknowledgments

MG and YZ are supported by NSF MCB 1519640; TL by NSF MCB 1855942; SN and VN by the Natural Sciences and Engineering Research Council of Canada (NSERC) (Discovery Grant RGPIN-315019 to SYN) and the Alberta Innovates Technology Futures (AITF) Strategic Chair in BioMolecular Simulations through funding for Centre for Molecular Simulation. All calculations with polarizable force fields were performed on the CFI-supported TNK cluster at the University of Calgary and on the West-Grid/Compute Canada clusters under Research Allocation Award to SYN. The classical MD calculations used the resources of the OLCF at the Oak Ridge National Laboratory, which is supported by the Office of Science at DOE under Contract No. DE-AC05-00OR22725, made available via the INCITE program.

References

1. Bellissent-Funel, M.-C.; Hassanali, A.; Havenith, M.; Henchman, R.; Pohl, P.; Sterpone, F.; van der Spoel, D.; Xu, Y.; Garcia, A. E. Water determines the structure and dynamics of proteins. *Chem. Rev.* **2016**, *116*, 7673–7697.
2. Ball, P. More than a bystander. *Nature* **2011**, *478*, 467–468.
3. Gunner, M.; Baker, N. Continuum electrostatics approaches to calculating pKas and Ems in proteins. *Meth. Enzymol.* **2016**, *578*, 1–20.
4. Mobley, D. L.; Gilson, M. K. Predicting binding free energies: Frontiers and benchmarks. *Annu. Rev. Biophys.* **2017**, *46*, 531–558.
5. Freier, E.; Wolf, S.; Gerwert, K. Proton transfer via a transient linear water-molecule chain in a membrane protein. *Proc. Natl. Acad. Sci.* **2011**, *108*, 11435–11439.
6. Goyal, P.; Lu, J.; Yang, S.; Gunner, M. R.; Cui, Q. Changing hydration level in an internal cavity modulates the proton affnity of a key glutamate in cytochrome c oxidase. *Proc. Natl. Acad. Sci. USA* **2013**, *110*, 18886–18891.
7. Cai, X.; Haider, K.; Lu, J.; Radic, S.; Son, C. Y.; Cui, Q.; Gunner, M. Network analysis of a proposed exit pathway for protons to the P-side of cytochrome c oxidase. *Biochim. Biophys. Acta. Bioenerg.* **2018**, *1859*, 997–1005.
8. Gunner, M. R.; Madeo, J.; Zhu, Z. Modification of quinone electrochemistry by the proteins in the biological electron transfer chains: Examples from photosynthetic reaction centers. *J. Bioenerg. Biomembr.* **2008**, *40*, 509–519.
9. Maupin, C. M.; Voth, G. A. Proton transport in carbonic anhydrase: Insights from molecular simulation. *Biochim. Biophys. Acta. Proteins Proteomics* **2010**, *1804*, 332–341.
10. Chamberlin, A.; Qiu, F.; Rebolledo, S.; Wang, Y.; Noskov, S. Y.; Larsson, H. P. Hydrophobic plug functions as a gate in voltage-gated proton channels. *Proc. Natl. Acad. Sci. USA* **2014**, *111*, E273–E282.
11. DeCoursey, T. E.; Hosler, J. Philosophy of voltage-gated proton channels. *J. R. Soc. Interface* **2014**, *11*, 20130799.
12. Lazaridis, T.; Hummer, G. Classical molecular dynamics with mobile protons. *J. Chem. Inf. Model* **2017**, *57*, 2833–2845.
13. van Keulen, S. C.; Gianti, E.; Carnevale, V.; Klein, M. L.; Rothlisberger, U.; Delemotte, L. Does proton conduction in the voltage-gated H^+ channel hHv1

involve grotthuss-like hopping via acidic residues? *J. Phys. Chem. B.* **2017**, *121*, 3340–3351.
14. Chenal, C.; Gunner, M. Two Cl Ions and a Glu Compete for a Helix Cage in the CLC Proton/Cl-Antiporter. *Biophys. J.* **2017**, *113*, 1025–1036.
15. Lee, S.; Mayes, H. B.; Swanson, J. M. J.; Voth, G. A. The origin of coupled chloride and proton transport in a Cl^-/H^+ antiporter. *J. Am. Chem. Soc.* **2016**, *138*, 14923–14930.
16. Kelkar, D. A.; Chattopadhyay, A. The gramicidin ion channel: A model membrane protein. *Biochim. Biophys. Acta. Biomembr.* **2007**, *1768*, 2011–2025.
17. Li, H.; Anuwongcharoen, N.; Malik, A.; Prachayasittikul, V.; Wikberg, J.; Nantasenamat, C. Roles of D-amino acids on the bioactivity of host defense peptides. *Int. J. Mol. Sci.* **2016**, *17*, 1023.
18. Agmon, N.; Bakker, H. J.; Campen, R. K.; Henchman, R. H.; Pohl, P.; Roke, S.; Thämer, M.; Hassanali, A. Protons and hydroxide ions in aqueous systems. *Chem. Rev.* **2016**, *116*, 7642–7672.
19. Bozdaganyan, M. E.; Lokhmatikov, A. V.; Voskoboynikova, N.; Cherepanov, D. A.; Steinhoff, H.-J.; Shaitan, K. V.; Mulkidjanian, A. Y. Proton leakage across lipid bilayers: Oxygen atoms of phospholipid ester linkers align water molecules into transmembrane water wires. *Biochim. Biophys. Acta. Bioenerg.* **2019**, *1860*, 439–451.
20. Pomés, R.; Roux, B. Molecular mechanism of H^+ conduction in the single-file water chain of the Gramicidin channel. *Biophys. J.* **2002**, *82*, 2304–2316.
21. Rasaiah, J. C.; Garde, S.; Hummer, G. Water in nonpolar confinement: From nanotubes to proteins and beyond. *Annu. Rev. Phys. Chem.* **2008**, *59*, 713–740.
22. Kopfer, D. A.; Song, C.; Gruene, T.; Sheldrick, G. M.; Zachariae, U.; de Groot, B. L. Ion perme-ation in K^+ channels occurs by direct Coulomb knock-on. *Science* **2014**, *346*, 352–355.
23. Kratochvil, H. T.; Carr, J. K.; Matulef, K.; Annen, A. W.; Li, H.; Maj, M.; Ostmeyer, J.; Serrano, A. L.; Raghuraman, H.; Moran, S. D.; Skinner, J. L.; Perozo, E.; Roux, B.; Valiyaveetil, F. I.; Zanni, M. T. Instantaneous ion configurations in the K^+ ion channel selectivity filter revealed by 2D IR spectroscopy. *Science* **2016**, *353*, 1040–1044.
24. Ricci, M. A.; Bruni, F.; Gallo, P.; Rovere, M.; Soper, A. K. Water in confined geometries: Experiments and simulations. *J. Phys. Condens. Matter* **2000**, *12*, A345–A350.
25. Mann, D. J.; Halls, M. D. Water alignment and proton conduction inside carbon nanotubes. *Phys. Rev. Lett.* **2003**, *90*, 195503.
26. Jing, Z.; Liu, C.; Cheng, S. Y.; Qi, R.; Walker, B. D.; Piquemal, J.-P.; Ren, P. Polarizable force fields for biomolecular simulations: Recent advances and applications. *Annu. Rev. Biophys.* **2019**, *48*, 371–394.
27. Lemkul, J. A.; Huang, J.; Roux, B.; MacKerell, A. D. An empirical polarizable force field based on the classical Drude oscillator model: Development history and recent applications. *Chem. Rev.* **2016**, *116*, 4983–5013.

28. Lamoureux, G.; MacKerell, A. D.; Roux, B. A simple polarizable model of water based on classical Drude oscillators. *J. Chem. Phys.* **2003**, *119*, 5185–5197.
29. Ren, P.; Ponder, J. W. Polarizable atomic multipole water model for molecular mechanics simulation. *J. Phys. Chem. B* **2003**, *107*, 5933–5947.
30. Peng, X.; Zhang, Y.; Chu, H.; Li, Y.; Zhang, D.; Cao, L.; Li, G. Accurate Evaluation of ion conductivity of the Gramicidin A channel using a polarizable force field without any corrections. *J. Chem. Theory Comput.* **2016**, *12*, 2973–2982.
31. Xu, H.; Stern, H. A.; Berne, B. J. Can water polarizability be ignored in hydrogen bond kinetics? *J. Phys. Chem. B* **2002**, *106*, 2054–2060.
32. Leontyev, I. V.; Stuchebrukhov, A. A. Electronic polarizability and the effective pair potentials of water. *J. Chem. Theory Comput.* **2010**, *6*, 3153–3161.
33. Ngo, V. A.; Fanning, J. K.; Noskov, S. Y. Comparative analysis of protein hydration from MD simulations with additive and polarizable force fields. *Adv. Theory Simul.* **2019**, *2*, 1800106.
34. Huang, J.; Lopes, P. E. M.; Roux, B.; MacKerell, A. D. Recent advances in polarizable force fields for macromolecules: Microsecond simulations of proteins using the classical Drude oscillator model. *J. Phys. Chem. Lett.* **2014**, *5*, 3144–3150.
35. Takei, K.; Fang, H.; Kumar, S. B.; Kapadia, R.; Gao, Q.; Madsen, M.; Kim, H. S.; Liu, C.-H.; Chueh, Y.-L.; Plis, E.; Krishna, S.; Bechtel, H. A.; Guo, J.; Javey, A. Quantum confinement effects in nanoscale-thickness InAs membranes. *Nano Lett.* **2011**, *11*, 5008–5012.
36. Mosaddeghi, H.; Alavi, S.; Kowsari, M. H.; Najafi, B. Simulations of structural and dynamic anisotropy in nano-confined water between parallel graphite plates. *J. Chem. Phys.* **2012**, *137*, 184703.
37. Allen, T. W.; Andersen, O. S.; Roux, B. Ion permeation through a narrow channel: Using Gramicidin to ascertain all-atom molecular dynamics potential of mean force methodology and biomolecular force fields. *Biophys. J.* **2006**, *90*, 3447–3468.
38. Roux, B. Computational studies of the Gramicidin channel. *Acc. Chem. Res.* **2002**, *35*, 366–375.
39. Andersen, O. S.; Gramicidin channels. *Annu. Rev. Physiol.* **1984**, *46*, 531–548.
40. Allen, T. W.; Baštug, T.; Kuyucak, S.; Chung, S.-H. Gramicidin a channel as a test ground for molecular dynamics force fields. *Biophys. J.* **2003**, *84*, 2159–2168.
41. Basu, I.; Chattopadhyay, A.; Mukhopadhyay, C. Ion channel stability of Gramicidin A in lipid bilayers: Effect of hydrophobic mismatch. *Biochim. Biophys. Acta. Biomembr.* **2014**, *1838*, 328–338.
42. Allen, T. W.; Andersen, O. S.; Roux, B. Molecular dynamics potential of mean force calculations as a tool for understanding ion permeation and selectivity in narrow channels. *Biophys. Chem.* **2006**, *124*, 251–267.

43. Bransburg-Zabary, S.; Kessel, A.; Gutman, M.; Ben-Tal, N. Stability of an ion channel in lipid bilayers: Implicit solvent model calculations with Gramicidin. *Biochemistry* **2002**, *41*, 6946–6954.
44. Braun-Sand, S.; Burykin, A.; Chu, Z. T.; Warshel, A. Realistic simulations of proton transport along the Gramicidin channel: Demonstrating the importance of solvation effects. *J. Phys. Chem. B* **2005**, *109*, 583–592.
45. Gregory, J. K. The water dipole moment in water clusters. *Science* **1997**, *275*, 814–817.
46. Riniker, S. Fixed-charge atomistic force fields for molecular dynamics simulations in the condensed phase: An overview. *J. Chem. Inf. Model* **2018**, *58*, 565–578.
47. Baker, C. M. Polarizable force fields for molecular dynamics simulations of biomolecules. *WIREs Comput. Mol. Sci.* **2015**, *5*, 241–254.
48. Laury, M. L.; Wang, L.-P.; Pande, V. S.; Head-Gordon, T.; Ponder, J. W. Revised parameters for the AMOEBA polarizable atomic multipole water model. *J. Phys. Chem. B* **2015**, *119*, 9423–9437.
49. Fernández, D. P.; Mulev, Y.; Goodwin, A. R. H.; Sengers, J. M. H. L. A database for the static dielectric constant of water and steam. *J. Phys. Chem. Ref. Data* **1995**, *24*, 33–70.
50. Zasetsky, A. Y. Dielectric relaxation in liquid water: Two fractions or two dynamics? *Phys. Rev. Lett.* **2011**, *107*, 117601.
51. Head-Gordon, T.; Paesani, F. Water is not a dynamic polydisperse branched polymer. *Proc. Natl. Acad. Sci. USA* **2019**, *116*, 13169–13170.
52. Jorgensen, W. L.; Chandrasekhar, J.; Madura, J. D.; Impey, R. W.; Klein, M. L. Comparison of simple potential functions for simulating liquid water. *J. Chem. Phys.* **1983**, *79*, 926–935.
53. Levy, R. M.; Gallicchio, E. Computer simulations with explicit solvent: Recent progress in the thermodynamic decomposition of free energies and in modeling electrostatic effects. *Annu. Rev. Phys. Chem.* **1998**, *49*, 531–567.
54. Morawietz, T.; Singraber, A.; Dellago, C.; Behler, J. How van der Waals interactions determine the unique properties of water. *Proc. Natl. Acad. Sci. USA* **2016**, *113*, 8368–8373.
55. Qiu, Y.; Nerenberg, P. S.; Head-Gordon, T.; Wang, L.-P. Systematic optimization of water models using liquid/vapor surface tension data. *J. Phys. Chem. B* **2019**, *123*, 7061–7073.
56. Zhang, H.; Yin, C.; Jiang, Y.; van der Spoel, D. Force field benchmark of amino acids: hydration and diffusion in different water models. *J. Chem. Inf. Model* **2018**, *58*, 1037–1052.
57. Best, R. B.; Mittal, J. Protein simulations with an optimized water model: Cooperative helix formation and temperature-induced unfolded state collapse. *J. Phys. Chem. B* **2010**, *114*, 14916–14923.
58. Nerenberg, P. S.; Jo, B.; So, C.; Tripathy, A.; Head-Gordon, T. Optimizing solute–water van der Waals interactions to reproduce solvation free energies. *J. Phys. Chem. B* **2012**, *116*, 4524–4534.

59. Ashbaugh, H. S.; Collett, N. J.; Hatch, H. W.; Staton, J. A. Assessing the thermodynamic signatures of hydrophobic hydration for several common water models. *J. Chem. Phys.* **2010**, *132*, 124504.
60. Levy, R. M.; Zhang, L. Y.; Gallicchio, E.; Felts, A. K. On the Nonpolar hydration free energy of proteins: Surface area and continuum solvent models for the solute–solvent interaction energy. *J. Am. Chem. Soc.* **2003**, *125*, 9523–9530.
61. Feig, M.; Brooks, C. L. Recent advances in the development and application of implicit solvent models in biomolecule simulations. *Curr. Opin. Struct. Biol.* **2004**, *14*, 217–224.
62. Onufriev, A. V.; Case, D. A. Generalized born implicit solvent models for biomolecules. *Annu. Rev. Biophys.* **2019**, *48*, 275–296.
63. Ross, G. A.; Bodnarchuk, M. S.; Essex, J. W. Water sites, networks, and free energies with grand canonical Monte Carlo. *J. Am. Chem. Soc.* **2015**, *137*, 14930–14943.
64. Barillari, C.; Taylor, J.; Viner, R.; Essex, J. W. Classification of water molecules in protein binding sites. *J. Am. Chem. Soc.* **2007**, *129*, 2577–2587.
65. Nielsen, J. E.; Gunner, M. R.; Garca-Moreno, E. B. The pKa cooperative: A collaborative effort to advance structure-based calculations of pKa values and electrostatic effects in proteins. *Proteins* **2011**, *79*, 3249–3259.
66. Machuqueiro, M.; Baptista, A. M. Is the prediction of pKa values by constant-pH molecular dynamics being hindered by inherited problems? *Proteins* **2011**, *79*, 3437–3447.
67. Radak, B. K.; Chipot, C.; Suh, D.; Jo, S.; Jiang, W.; Phillips, J. C.; Schulten, K.; Roux, B. Constant-pH molecular dynamics simulations for large biomolecular systems. *J. Chem. Theory Comput.* **2017**, *13*, 5933–5944.
68. Swails, J. M.; York, D. M.; Roitberg, A. E. Constant pH replica exchange molecular dynamics in explicit solvent using discrete protonation states: Implementation, testing, and validation. *J. Chem. Theory Comput.* **2014**, *10*, 1341–1352.
69. Cruzeiro, V. W. D.; Amaral, M. S.; Roitberg, A. E. Redox potential replica exchange molecular dynamics at constant pH in AMBER: Implementation and validation. *J. Chem. Phys.* **2018**, *149*, 072338.
70. Baker, N. A. Improving implicit solvent simulations: A Poisson-centric view. *Curr. Opin. Struct Biol.* **2005**, *15*, 137–143.
71. Gilson, M. K.; Rashin, A.; Fine, R.; Honig, B. On the calculation of electrostatic interactions in proteins. *J. Mol. Biol.* **1985**, *184*, 503–516.
72. Leontyev, I.; Stuchebrukhov, A. Accounting for electronic polarization in non-polarizable force fields. *Phys. Chem. Chem. Phys.* **2011**, *13*, 2613.
73. Gunner, M. R.; Zhu, X.; Klein, M. C. MCCE analysis of the pKas of introduced buried acids and bases in staphylococcal nuclease. *Proteins* **2011**, *79*, 3306–3319.
74. Sham, Y. Y.; Muegge, I.; Warshel, A. The effect of protein relaxation on charge-charge interactions and dielectric constants of proteins. *Biophys. J.* **1998**, *74*, 1744–1753.

75. Csnyi, V.; Boda, D.; Gillespie, D.; Kristf, T. Current and selectivity in a model sodium channel under physiological conditions: Dynamic Monte Carlo simulations. *Biochim Biophys Acta Biomembr* **2012**, *1818*, 592–600.
76. Song, Y.; Gunner, M. Using multiconformation continuum electrostatics to compare chloride binding motifs in a-amylase, human serum albumin, and Omp32. *J. Mol. Biol.* **2009**, *387*, 840–856.
77. Till, M. S.; Essigke, T.; Becker, T.; Ullmann, G. M. Simulating the proton transfer in Gramicidin A by a sequential dynamical Monte Carlo Method. *J. Phys. Chem. B* **2008**, *112*, 13401–13410.
78. Song, Y.; Mao, J.; Gunner, M. R. MCCE2: Improving protein *pKa* calculations with extensive side chain rotamer sampling. *J. Comput. Chem.* **2009**, 2231–2247.
79. Song, Y.; Mao, J.; Gunner, M. R. Calculation of proton transfers in bacteriorhodopsin bR and Mintermediates. *Biochemistry* **2003**, *42*, 9875–9888.
80. Song, Y.; Gunner, M. R. Halorhodopsin pumps Cl^- and bacteriorhodopsin pumps protons by a common mechanism that uses conserved electrostatic interactions. *Proc. Natl. Acad. Sci. USA* **2014**, *111*, 16377–16382.
81. Zheng, Z.; Dutton, P. L.; Gunner, M. R.; The measured and calculated affnity of methyl- and methoxy-substituted benzoquinones for the Q(A) site of bacterial reaction centers. *Proteins* **2010**, *78*, 2638–2654.
82. Hassanali, A. A.; Cuny, J.; Verdolino, V.; Parrinello, M. Aqueous solutions: State of the art in *ab initio* molecular dynamics. *Philos. Trans. R. Soc. A* **2014**, *372*, 20120482.
83. Warshel, A. *Computer modeling of chemical reactions in enzymes and solutions*, Wiley, New York, 1991.
84. Voth, G. A. Computer simulation of proton solvation and transport in aqueous and biomolecular systems. *Acc. Chem. Res.* **2006**, *39*, 143–150.
85. Knight, C.; Voth, G. A. The curious case of the hydrated proton. *Acc. Chem. Res.* **2012**, *45*, 101–109.
86. Stillinger, F. H.; David, C. W. Polarization model for water and its ionic dissociation products. *J. Chem. Phys.* **1978**, *69*, 1473–1484.
87. Weber, T. A.; Stillinger, F. H. Dynamical study of the $H_5O_2^+ + H_3O_2^-$ neutralization reaction using the polarization model. *J. Chem. Phys.* **1982**, *77*, 4150–4155.
88. Ojamae, L.; Shavitt, I.; Singer, S. J. Potential models for simulations of the solvated proton in water. *J. Chem. Phys.* **1998**, *109*, 5547–5564.
89. Lussetti, E.; Pastore, G.; Smargiassi, E. A fully polarizable and dissociable potential for water. *Chem. Phys. Lett.* **2003**, *381*, 287–291.
90. Mahadevan, T. S.; Garofalini, S. H. Dissociative water potential for molecular dynamics simulations. *J. Phys. Chem. B* **2007**, *111*, 8919–8927.
91. van Duin, A. C. T.; Zou, C.; Joshi, K.; Bryantsev, V.; Goddard, W. A.; in Asthagiri, A.; Janik, M. J. (eds.), A reaxff reactive force-field for proton transfer reactions in bulk water and its applications to heterogeneous catalysis, *Catalysis Series*; Royal Society of Chemistry, Cambridge, 223–243, **2013**.

92. Lill, M. A.; Helms, V. Molecular dynamics simulation of proton transport with quantum mechanically derived proton hopping rates (Q-HOP MD). *J. Chem. Phys.* **2001**, *115*, 7993–8005.
93. Wolf, M.; Grubmüller, H.; Groenhof, G. Anomalous surface diffusion of protons on lipid membranes. *Biophys. J.* **2014**, *107*, 76–87.
94. Kong, X.; Brooks, C. L. λ-dynamics: A new approach to free energy calculations. *J. Chem. Phys.* **1996**, *105*, 2414–2423.
95. Wolf, M. G.; Groenhof, G. Explicit proton transfer in classical molecular dynamics simulations. *J. Comput. Chem.* **2014**, *35*, 657–671.
96. Pomés, R.; Roux, B. Free energy profiles for H^+ conduction along hydrogen-bonded chains of water molecules. *Biophys. J.* **1998**, *75*, 33–40.
97. Marrink, S.; Jhnig, F.; Berendsen, H. Proton transport across transient single-file water pores in a lipid membrane studied by molecular dynamics simulations. *Biophys. J.* **1996**, *71*, 632–647.
98. Swanson, J. M. J.; Maupin, C. M.; Chen, H.; Petersen, M. K.; Xu, J.; Wu, Y.; Voth, G. A. Proton solvation and transport in aqueous and biomolecular systems: Insights from computer simulations. *J. Phys. Chem. B* **2007**, *111*, 4300–4314.
99. Jo, S.; Kim, T.; Iyer, V. G.; Im, W. CHARMM-GUI: A web-based graphical user interface for CHARMM. *J. Comput. Chem.* **2008**, *29*, 1859–1865.
100. Bowers, K. J.; Chow, D. E.; Xu, H.; Dror, R. O.; Eastwood, M. P.; Gregersen, B. A.; Klepeis, J. L.; Kolossvary, I.; Moraes, M. A.; Sacerdoti, F. D.; Salmon, J. K.; Shan, Y.; Shaw, D. E.Scalable algorithms for molecular dynamics simulations on commodity clusters, *ACM/IEEE SC 2006 Conference (SC'06)*, Tampa, FL.; pp. 43–43, 2006.
101. Lamoureux, G.; Harder, E.; Vorobyov, I. V.; Roux, B.; MacKerell, A. D. A polarizable model of water for molecular dynamics simulations of biomolecules. *Chem. Phys. Lett.* **2006**, *418*, 245–249.
102. Lopes, P. E. M.; Huang, J.; Shim, J.; Luo, Y.; Li, H.; Roux, B.; MacKerell, A. D. Polarizable force field for peptides and proteins based on the classical Drude oscillator. *J. Chem. Theory Comput.* **2013**, *9*, 5430–5449.
103. Luo, Y.; Jiang, W.; Yu, H.; MacKerell, A. D.; Roux, B. Simulation study of ion pairing in concentrated aqueous salt solutions with a polarizable force field. *Faraday Discuss* **2013**, *160*, 135–149.
104. Yu, H.; Whitfield, T. W.; Harder, E.; Lamoureux, G.; Vorobyov, I.; Anisimov, V. M.; MacKerell, A. D.; Roux, B. Simulating monovalent and divalent ions in aqueous solution using a Drude polarizable force field. *J. Chem. Theory Comput.* **2010**, *6*, 774–786.
105. Ngo, V.; Li, H.; MacKerell Jr, A. D.; Allen, T. W.; Roux, B.; Noskov, S. Polarization effects in water-mediated selective cation transport across a narrow transmembrane channel, *Submitted.*
106. Phillips, J. C.; Braun, R.; Wang, W.; Gumbart, J.; Tajkhorshid, E.; Villa, E.; Chipot, C.; Skeel, R. D.; Kal, L.; Schulten, K. Scalable molecular dynamics with NAMD. *J. Comput. Chem.* **2005**, *26*, 1781–1802.

107. Allen, T. W.; Andersen, O. S.; Roux, B. Structure of Gramicidin A in a lipid bilayer environment determined using molecular dynamics simulations and solid-state NMR data. *J. Am. Chem. Soc.* **2003**, *125*, 9868–9877.
108. Jiang, W.; Hardy, D. J.; Phillips, J. C.; MacKerell, A. D.; Schulten, K.; Roux, B. High-performance scalable molecular dynamics simulations of a polarizable force field based on classical Drude oscillators in NAMD. *J. Phys. Chem. Lett.* **2011**, *2*, 87–92.
109. Lamoureux, G.; Roux, B. Modeling induced polarization with classical Drude oscillators: Theory and molecular dynamics simulation algorithm. *J. Chem. Phys.* **2003**, *119*, 3025–3039.
110. Sprik, M. Computer simulation of the dynamics of induced polarization fluctuations in water. *J. Phys. Chem.* **1991**, *95*, 2283–2291.
111. Gunner, M. R.; Murakami, T.; Rustenburg, A. S.; Işik, M.; Chodera, J. D. Standard state free energies, not p*Ka*s, are ideal for describing small molecule protonation and tautomeric states. *J. Comput-Aided Mol. Des.* **2020**, *34*, 561–573.
112. Honig, B.; Sharp, K.; Yang, A. S. Macroscopic models of aqueous solutions: Biological and chemical applications. *J. Phys. Chem.* **1993**, *97*, 1101–1109.
113. Sitkoff, D.; Sharp, K. A.; Honig, B. Accurate calculation of hydration free energies using macroscopic solvent models. *J. Phys. Chem.* **1994**, *98*, 1978–1988.
114. Pearlman, D. A.; Case, D. A.; Caldwell, J. W.; Ross, W. S.; Cheatham, T. E.; DeBolt, S.; Ferguson, D.; Seibel, G.; Kollman, P. AMBER, a package of computer programs for applying molecular mechanics, normal mode analysis, molecular dynamics and free energy calculations to simulate the structural and energetic properties of molecules. *Comput. Phys. Commun.* **1995**, *91*, 1–41.
115. Haider, K.; Cruz, A.; Ramsey, S.; Gilson, M. K.; Kurtzman, T. Solvation structure and thermodynamic mapping (SSTMap): An open-source, flexible package for the analysis of water in molecular dynamics trajectories. *J. Chem. Theory Comput.* **2018**, *14*, 418–425.
116. Humphrey, W.; Dalke, A.; Schulten, K. VMD: Visual molecular dynamics. *J. Mol. Graphics* **1996**, *14*, 33–38.
117. Horner, A.; Pohl, P. Single-file transport of water through membrane channels. *Faraday Discuss* **2018**, *209*, 9–33.
118. Sitkoff, D.; Sharp, K. A.; Honig, B. Correlating solvation free energies and surface tensions of hydrocarbon solutes. *Biophys. Chem.* **1994**, *51*, 397–409.
119. Allen, T. W.; Andersen, O. S.; Roux, B. Energetics of ion conduction through the gramicidin channel. *Proc. Natl. Acad. Sci.* **2004**, *101*, 117–122.
120. Lee, M.; Bai, C.; Feliks, M.; Alhadeff, R.; Warshel, A. On the control of the proton current in the voltage-gated proton channel Hv1. *Proc. Natl. Acad. Sci. USA* **2018**, *115*, 10321–10326.
121. Setiadi, J.; Kuyucak, S. Computational investigation of the effect of lipid membranes on ion permeation in Gramicidin A. *Membranes* **2016**, *6*, 20.

https://doi.org/10.1142/9789819808199_0009

Chapter 9

pH-Dependent Interactions of Apolipophorin-III with a Lipid Disk*

Yunhui Peng[†,¶], Rudolfs Kelle[†,‡], Chandler Little[†,‡], Ekaterina Michonova[‡], Kostantin G. Kornev[§] and Emil Alexov[†,§,||]

[†]*Department of Physics and Astronomy, College of Sciences, Clemson University, Clemson, SC 29634, USA*
[‡]*Department of Chemistry, Erskine College, Due West, SC 29639, USA*
[§]*Department of Material Sciences and Engineering, Clemson University, Clemson, SC 29634, USA*
[¶]*yunhuip@g.clemson.edu*
[||]*ealexov@clemson.edu*

Apolipophorin-III (ApoLp-III) is required for stabilization of molecular shuttles of lipid fuels in insects and is found to contribute to the insect immune reaction. Rearrangement of its five α-helices enables ApoLp-III to reversibly associate with lipids. We investigate computationally the conformational changes of ApoLp-III and the pH-dependence of the binding free energy of ApoLp-III association with a lipid disk. A dominant binding mode along with several minor, low population, modes of the ApoLp-III binding to a lipid disk was identified. The pH-dependence of the binding energy for ApoLp-III with the lipid disk is predicted to be significant, with the pH-optimum at pH= 8.0. The calculations suggest that there are no direct interactions between the lipid head groups and titratable residues of ApoLp-III. In the physiological pH range from 6.0 to 9.0, the binding free energy of ApoLp-III with the lipid disk decreases significantly with respect to its optimal value at pH 8.0 (at pH= 6.0, it is 1.02 kcal/mol and at

*To cite this article, please refer to its earlier version published in *J. Comput. Biophys. Chem.* **2021**, *20* (2), 153–164, DOI: 10.1142/S2737416520420041
||Corresponding author.

pH= 9.0 it is 0.23 kcal/mol less favorable than at the optimal pH= 8.0), indicating that the pH is an important regulator of ApoLp-III lipid disk association.

Keywords: pH-dependence; apolipophorin-III; lipids; molecular dynamics; pKa's calculations; free energy.

1. Introduction

A living organism produces hydrophobic lipids, such as triacylglycerols and cholesterols, which are not solvable in aqueous biofluids. To move lipids from one site of the organism to the other, animals develop special shuttles-lipoproteins that can be loaded with lipids and are able to circulate through the animal blood vessels.[1–7] Lipoprotein is a special assembly of lipids and proteins representing a single noncovalently bound particle. In mammals and insects, lipoproteins have similar structures and multiple functions. Some lipoprotein particles form very complex structures consisting of other lipoproteins, which are bound together and cannot be broken; these are nonexchangeable lipoproteins.[8,9] There are exchangeable lipoproteins which, upon a physiological command of the organism, are able to temporally and reversibly bind the nonexchangeable proteins and then leave them to move to another nonexchangeable lipoprotein.[10,11] While in mammals, the lipid transport involves different types of lipoproteins, the number of lipoproteins in insects is much smaller.

Insect lipoproteins are called lipophorins.[12–15] The nonexchangeable lipoprotein in insects is called High Density Lipophorin particle (HDLp). It is assembled from two proteins that bind lipids, Apolipophorin-I (ApoLp-I) and Apolipophorin-II (ApoLp-II) and phospholipids, sterols, and hydrocarbons. In insects, the lipids are stored in the fat body in the form of triglycerides (TAG).[4,15] The insects use a set of complex biochemical reactions to convert TAG into diglycerides (DAG). The latter is loaded to HDLp to be transported to the flight muscles where the DAG is used to generate metabolic energy required for sustained flight. As more and more DAGs are collected by HDLp, the particle increases in size and needs some molecular belts to keep it intact. Fortunately, at the surface of the DAG-loaded HDLps, the phospholipid packing is disrupted by the hydrophobic fatty acid end groups. These groups attract a surfactant-like.[16,17] exchangeable protein called Apolipophorin-III (ApoLp-III), which is freely available in insect blood. In the blood, free ApoLp-III resides as a compact globular protein composed of a bundle of five amphiphilic α-helices.[18–20] Interacting with the surface of HDLp, ApoLp-III undergoes large conformational changes opening up its structure to form molecular belts interacting with the hydrophobic fatty acid end groups of DAG.[21–23] Up to 16 ApoLp–III, belts can be attached to a single HDLp preventing this particle against breakup.[24,25] This new structure — the HDLp covered with DAGs and ApoLp-III has a lower density and is consequently named Low Density Lipophorin particle (LDLp).

The mechanism of interactions of ApoLp-III with HDLp generated a considerable interest not only in the community of entomologists, but in much broader community, because of its structural and functional similarity with mammalian proteins ApoE[26] and TIP47/perilipin 3.[27–29] To model interactions of ApoLp-III with HDLp, different phospholipid model vesicles have been proposed. For example, dimyristoylphosphatidylcholine (DMPC) vesicles mimic the HDLp surface fairy well and hence become quite popular.[23] These experiments led to the development of the so-called disc assay where association of a lipoprotein in question with DMPC vesicles results in the formation of stable nanodiscs.[26,30–35]

The experiments showed that ApoLp-III unfolds by preserving the α-helical secondary structure and exposing its hydrophobic residues to the hydrophobic *acyl groups* of lipids.[23,36] Thus, the conformational change involves helices rearrangement via connecting loops.[36–41]

Here, we will investigate the conformational change of ApoLp-III via molecular dynamics (MD) simulations to provide mechanistic interpretation of experimental data. In our modeling protocol, the size of the lipid disk is 20 nm and its thickness is 5 nm.[23] Such a large macromolecular assemblage along with several copies of ApoLp-III is a challenge for MD simulations. However, we apply nano-scale molecular dynamics simulations with NAMD[42,43] of free and the disk-bound-ApoLp-III in open conformation to investigate the following questions: (i) is the lipid disk coated with the ApoLp-III belt a stable structure; (ii) which conformation is more stable, when an open ApoLp-III belt fastens the lipid disk over the disk border or when an open ApoLp-III belt resides free in water; (iii) which helical topology of ApoLp-III results in the most stable association with the disk?

We study these questions by investigating the free energy differences between a compact soluble ApoLp-III and the-disk-bound ApoLp-III. Both the lipids and ApoLp-III have titratable groups in which ionization state depends on pH of the environment. However, the lipid headgroups pKa's are much lower than neutral pH, while the ApoLp-III titratable groups may titrate at such a pH. In addition, the binding may induce pKa's changes, i.e. it may be pH-dependent. In this work, we investigate how the solution pH affects ApoLp-III binding to a lipid disk. Addressing these questions, we expect to get a deeper insight into the pH dependence of the transition from HDLP to LDLP.

2. Methods

2.1. *Compact structure of Apolipophorin-III*

The NMR structure of Manduca Sexta ApoLp-III (PDB:1EQ1) was used for the calculation of compact (a lipid free) state of ApoLp-III. The PDB file contains 25 models, which are quite similar except for the conformations for the N-terminal short loop. The first model was used in our MD simulations. However, for the pKa calculations and pH-dependence of the binding, all 25 models were investigated, and the results were averaged.

2.2. *Construction of Apolipophorin-III open conformation models*

In order to bind to the lipid disk, the ApoLp-III must open its compact structure. According to numerous experimental works,[23,44–47] the opening involves rearrangement of α-helices without structural changes of α-helices themselves. Thus, we used the NMR structure mentioned above as a starting structure to build models of open conformations of ApoLp-III. In the compact state, the ApoLp-III

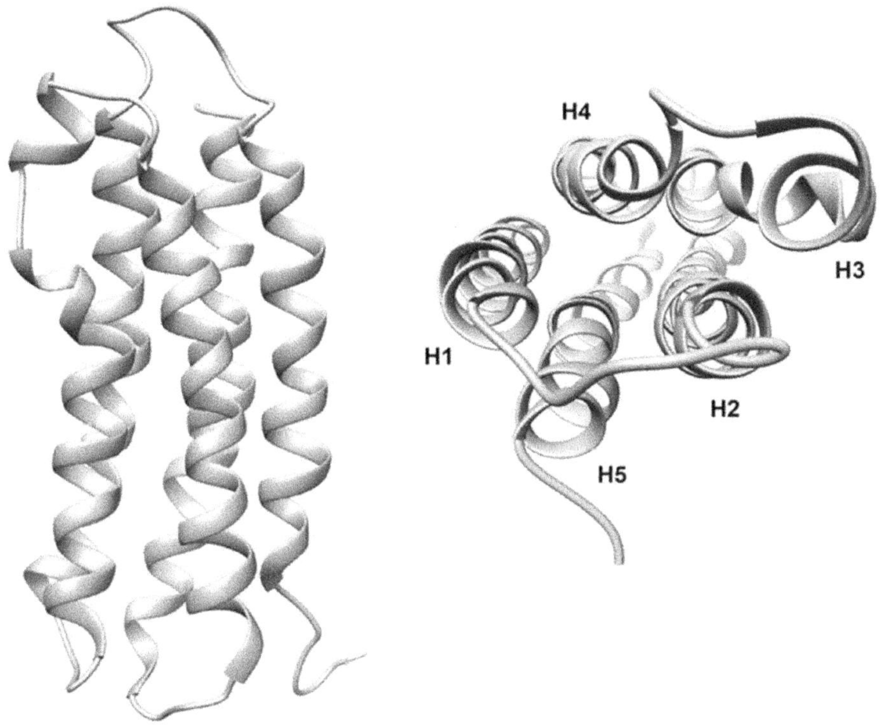

Fig. 1. Structure of ApoLp-III of *Manduca Sexta* in compact conformation ((PDB ID:1EQ1). The α-helices are counted starting from the N-terminal.

adopts a bundle of five α-helices connected by short loops (following the convention discussed in Ref. 45, the helices are labeled as 1,2, ..., 5 beginning from the N-terminal, Fig. 1). The previous studies have pointed out that the opening should involve two α-helices resulting in formation of two sub-bundles,[19,23] which allow the hydrophobic interior of the protein to interact directly with lipids. Based on the topology of ApoLp-III, there are four plausible open conformations (Fig. 2). Here, starting from the original NMR structure (PDB ID:1EQ1), we manipulated torsion angles of the loops connecting the α-helices with Chimera[48] and rearranged five α-helices bundles in two sub-bundles.[48] For each model, we rotated the two α-helices in a way to make their hydrophobic interior facing outward. In total, we generated four open conformation models (Fig. 2).

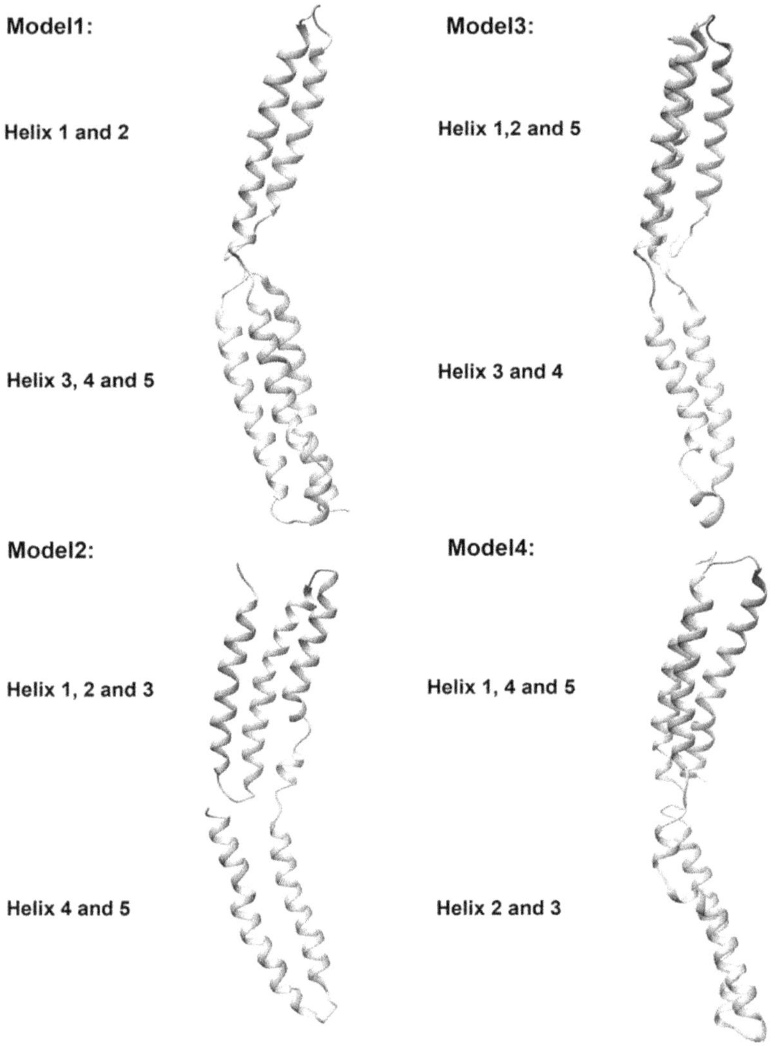

Fig. 2. Four plausible structural model of unfolded conformation of ApoLp-III.

2.3. *Model for Apolipophorin-III bound to a lipid disk*

The lipid disk was constructed with VMD membrane plugin.[49] Association of ApoLp-III with DMPC vesicles results in the formation of uniform discs with an average diameter of 18.5 ± 2.0 nm.[23] Therefore, the disk radius was selected to be 100 Å and the disk is

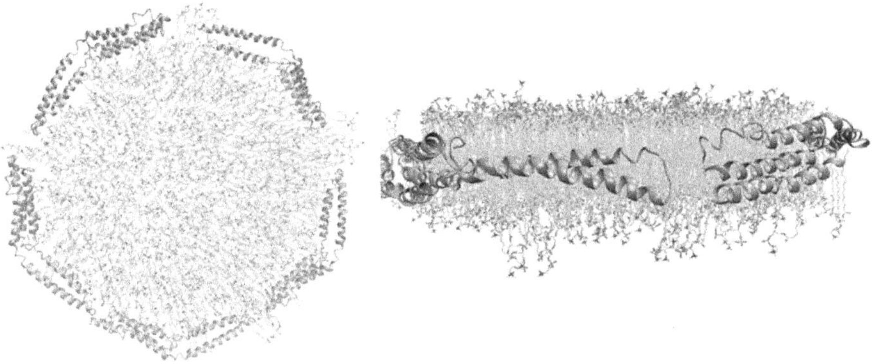

Fig. 3. The lipid disk — ApoLp-III complex. The left panel is the top view, the right panel is the side view.

made of POPC lipids. As a result, the disk had a total of 683 POPC lipids. Then using Chimera,[48] five copies of the open ApoLp-III were manually wrapped over the lipid disk border, fitting the entire circumference of the lipid disk (Fig. 3, and the movie apolipophorin-disk.mp4). Thus, four lipid-disk — ApoLp-III models were created, using each of the four open conformation modes of ApoLp-III described above. The resulting 3D structures can be downloaded via http://compbio.clemson.edu/downloads.

2.4. *Electrostatics potential calculations*

The electrostatic potential calculations were performed with the Delphi program.[50,51] The dielectric constants for protein and solvent were set to 2 and 80, respectively, and the salt concentration was set to be 0.15 M. The percentage filling of the box was 70 with the scale of 1 grid/Å. The surface potential of the protein was calculated with DelPhi, the phimap output file and the electric field lines were drawn with VMD.

2.5. *Calculations of the pH-dependence of the Apolipophorin-III net charge*

The net charge of ApoLp-III was calculated with DelPhiPKa.[52,53] The ion concentration was set at 0.15 M and the dielectric constants for protein and water phase were 8 and 80, respectively. The pH range was set from 0 to 14 with an interval of 1. Based on the modeling of pKa's of ionizable groups, the DelPhiPKa program calculates the net charge of the protein at each pH. Two sets of calculations were done. A calculation of the pH-dependence of the net charge of compact ApoLp-III and of open ApoLp-III bound to a lipid disk. In the last case, the calculations were done by keeping the lipid charges constant, while allowing ApoLp-III ionizable groups to titrate. As mentioned above, in the pKa's calculations of compact ApoLp-III, all NMR models (25 models) were used and then the results were averaged. To have multiple conformations for the bound state, for each Model, five conformations were extracted from the MD simulations at $t = 20$, 40, 60, 80, and 100 ns, resulting in five snapshots per model. Due to the large size of the ApoLp-III-lipid-disk complex, adding more snapshots was computationally prohibited. These snapshots were used for the pKa's calculations, and the results were averaged.

2.6. *pH-dependence of binding free energy of Apolipophorin-III to a lipid disk*

The change of the net charge of a system from one state to another as a function of pH can be used to compute the pH-dependence of the free energy between the states. In the previous paragraph, we outlined the modeling steps for calculations of the net charge of compact and bound-to-the-lipid-disk ApoLp-III. The net charge difference Δq(pH) has been identified from these calculations and analyzed as a function of pH. Then the pH-dependence of the free energy difference between compact ApoLp-III in water and open ApoLp-III bound to the lipid disk is calculated as follows:

$$\Delta\Delta G(\mathrm{pH}) = \int_{\mathrm{pH}_0}^{\mathrm{pH}} \Delta q d\mathrm{pH} + \Delta G(\mathrm{pH}_0), \quad (1)$$

where $\Delta\Delta G(\text{pH})$ is the pH-dependence of the free energy, $\Delta G(\text{pH}_0)$ is the free energy at pH_0, and the integral goes from pH_0 to the desired pH.

2.7. *Molecular dynamics simulations*

We performed the all-atom MD simulations for each of the four open models of ApoLp-III. To compare the structural dynamics of a free and the disk bound ApoLps, the MD simulations were done on ApoLp-III alone and five ApoLp-III-lipid disk complexes. All MD simulations were performed by NAMD 2.11[43] with CHARMM36m force field[54] and TIP3P water model.[16] The input files were prepared with VMD.[49] The systems were solvated with 0.15 M NaCl in a cubic water box with no less than 10 Å spacing from the protein to the edge of the box. The Langevin dynamics with periodic boundary conditions were applied in the simulations. The van der Waals and electrostatic interactions were truncated at 12 Å distance with a switching function from 10 Å. The particle mesh ewald was applied for calculations of long-range electrostatic interactions.

In simulations of the open ApoLp-III conformations alone, the system first underwent a 5000-step minimization with the backbone fixed and then followed by a subsequent 5000-step minimization without any constraints. Furthermore, all atoms of the protein were fixed for 100 ps equilibration of the water molecules and ions and then the system was gradually heated from 0 K to 310 K with 1000-step/K in NVT. Next, the system was subjected to the 1 ns equilibration with the alpha carbon atoms (CA) constrained and another 2 ns equilibration without any constraints at 310 K. Lastly, the system was switched to NPT simulations and all constraints were removed for the 100 ns production run.

In the simulation of ApoLp-III–disk complexes, the system first underwent a 5000-step minimization by fixing all the atoms except of the lipid tails, and then a subsequent 5000-step minimization without constraints. Next, the systems were subjected to 500 ps equilibration at 310 K in NVT simulations and a harmonic constraint of 1 kcal mol was applied to the CAs. The system was then equilibrated without any constrains for another 500 ps in NPT simulation and after that, we performed the 100 ns production run.

2.8. *Analysis of trajectories*

The analysis of trajectories were performed using TCL scripts in VMD.[49] We first calculated the Root Mean Square Deviation (RMSD) and Radius of Gyration (R_g) for the simulations of ApoLp-III alone, where the frames are superimposed onto the initial structures and then the RMSD of C^α atoms were calculated. The R_g was calculated using the following equation:

$$R_g^2 = \sum m_i (r_i - R_c)^2 \Big/ \sum m_i, \tag{2}$$

where m_i is the mass of the ith nonhydrogen atoms and r_i is its coordinates. R_c is the center of mass of the protein.

In the simulations of ApoLp-III -disk complexes, we calculated the RMSD and R_g for each of ApoLp-III separately and then took the mean values.

2.9. *Surface area calculations*

The solvent accessible surface area (SASA) was calculated using VMD[49] with a probe distance of 1.4 Å using the last frames from each simulation runs. The binding interfacial area (measured in squared Å) of ApoLp-III to lipid disk was calculated as the subtraction between the SASA of free ApoLp-III and SASA of bound ApoLp-III.

3. Results

We question (a) what is the most favorable conformational changes induced by the lipid disk binging and (b) what is the pH dependence of the binding free energy?

3.1. *Investigating the most favorable open conformation of ApoLp-III*

To bind a lipid disk, ApoLp-III has to undergo large conformational change to partially unfold its bundle. To test the ability of MD simulations to sample such a change, we simulated ApoLp-III in water and in a medium mimicking a lipid bilayer. It is known that in water the ApoLp-III bundle is not very stable since the folding free energy is only 1.24 kcal/mol.[19,55] We carried MD simulations of the compact structure of ApoLp-III in water, modeled as a high dielectric constant medium ($\epsilon_{\rm out} = 80$), and in lipid environment modeled as a low dielectric constant medium ($\epsilon_{\rm out} = 4$). This way, one can evaluate whether the ApoLp-III stability depends on the presence of a lipid phase. We did not observe any significant difference between the RMSD of ApoLp-III in high and in low dielectric media. In both cases, the protein demonstrated instability manifested by significant conformational deviations of its initial structure after 100 ns simulation time. This indicates that the protein is not very stable, indeed. Despite these observations, we were unable to model a conformational change expected for the lipid-bound protein. Perhaps much longer simulations are needed along with enhanced sampling techniques. To avoid this, the following strategy was adopted: instead of modeling the process of the protein opening, the open state was generated manually (see below).

As mentioned in Sec. 2, four plausible open conformations of ApoLp- III were generated and docked to a lipid disk. The open conformations and those bound to the disk were subjected to MD simulations and the outcome was investigated in terms of RMSD, R_g and the interfacial area of ApoLp- III bound to the lipid disk (Figs 4 and 5 and Table 1).

It is expected that ApoLp-III should unfold its bundle in the presence of a lipid bilayer exposing its hydrophobic core to the disk border be able to fasten it up. However, during this unfolding process prior to binding to the lipid disk, the open ApoLp-III will still be in water, suggesting that the open ApoLp-III should be stable in water as well.

Thus, it can be anticipated that the optimal configuration of the unfolded ApoLp-III bundle should be the one with the smallest RMSD and the smallest RMSD fluctuations. This conformation would provide the most probable configuration of the ApoLp-III belt which is about to land on the disk border. We also assume that the

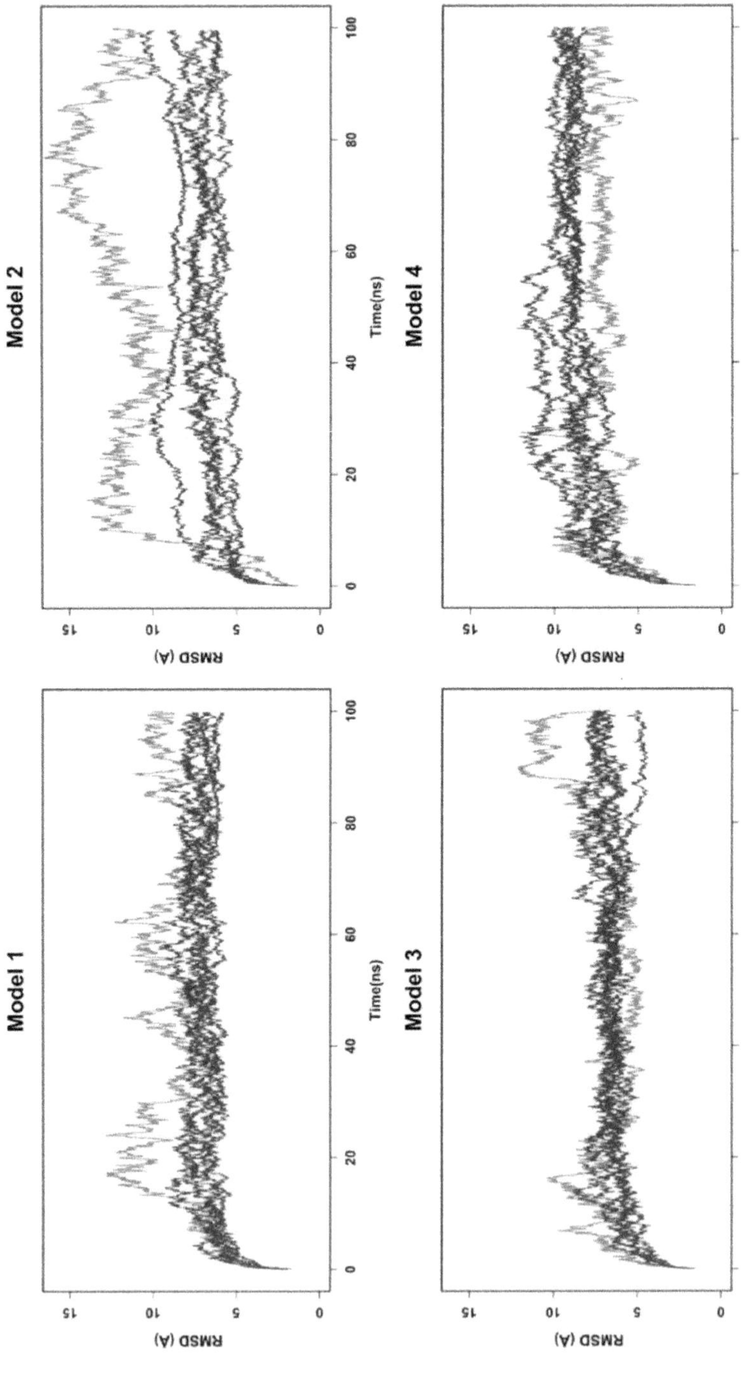

Fig. 4. The RMSDs for *Manduca Sexta* ApoLp-III in unfolded conformations. The RMSDs were calculated for ApoLp-III alone (red color) and for the disk-bound states for all four models (blue color).

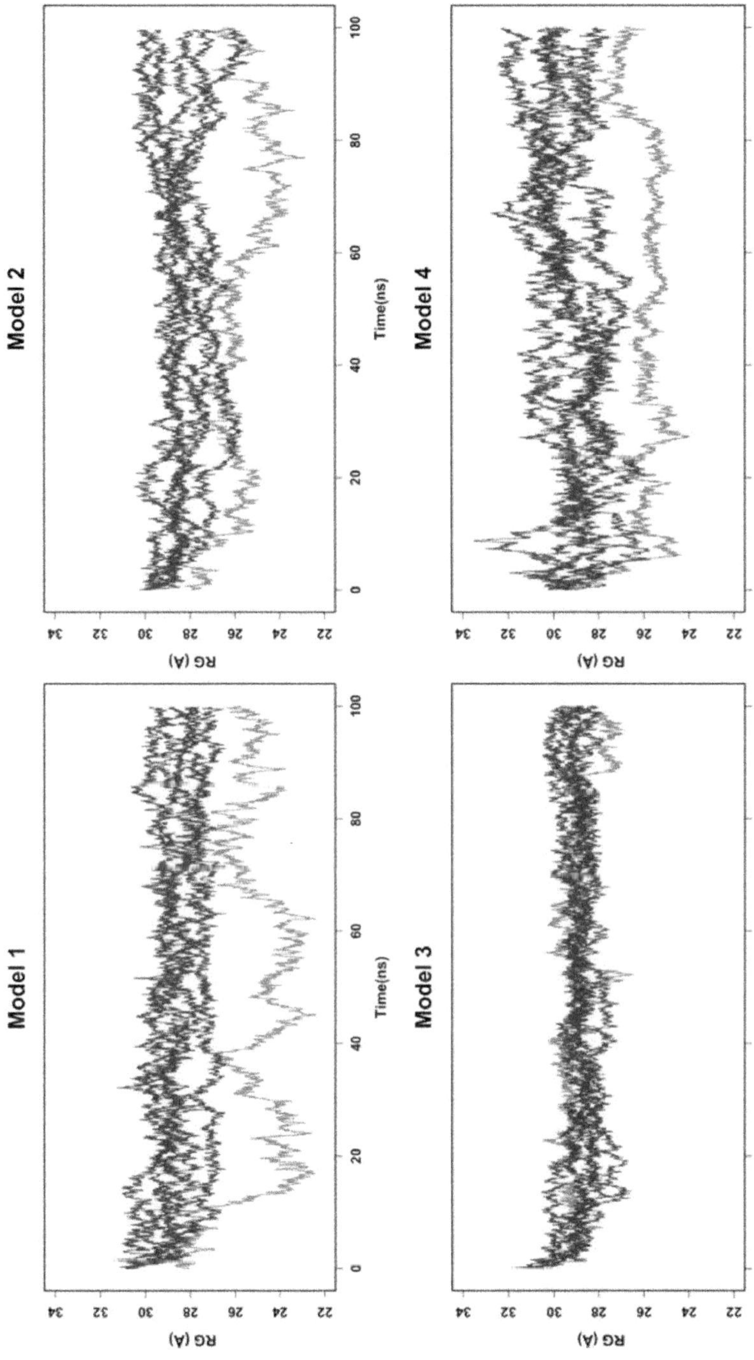

Fig. 5. The R_g for *Manduca Sexta* ApoLp-III in unfolded conformations. The $R_g s$' were calculated for ApoLp-III alone (red color) and for the disk-bound states for all four models (blue color).

Table 1. The interfacial area (measured in squared Å) of ApoLp-III bound to the lipid disk. For each model, the interfacial area of ApoLp-III was calculated using the last frame of simulations. Since the lipid disk was bound with five copies of ApoLp-III, we calculated the binding interface area for each of them and calculated the mean value.

	1	2	3	4	5	Average	Stdev
Mode 1	4249.83	2374.90	3687.11	4534.00	3000.17	3569.20	888.56
Model 2	4403.92	3601.80	4645.46	4635.66	4446.37	4346.64	430.36
Model 3	4193.38	4670.48	5116.60	4605.20	4472.36	4611.60	336.46
Model 4	4281.19	2897.81	3877.76	4064.11	3912.68	3806.71	532.39

best model should be such that the RMSD of the disk-bound ApoLp-III is the smallest, and the R_g remains almost a constant while radius of gyration of unfolded ApoLp-III in water should have a tendency to decrease (indicating that ApoLp-III would tend to fold back into a soluble compact configuration). Lastly, the best configuration should result in a relatively tight binding to the lipid disk, which should result in the largest interfacial area covered (Table 1).

An analysis of the results suggests that the Model 3 is the best candidate following experimental observations.[56] However, the other modes cannot be ruled out completely. Indeed, the Model 2 has the second largest interfacial area and also small RMSD of the bound state. This indicates that while Model 3 is putatively correct and describes the anticipated mode of ApoLp-III association with the lipid disk, there may be a mixture of binding modes with low populations coming from the Models 1, 2 and 4.

3.2. *Investigating the pH-dependence of binding free energy*

The first step in this investigation is to check the validity of the computational protocol by benchmarking it against the experimental data. We used the PDB structure of compact soluble ApoLp-III and performed pKa's calculations with DelPhiPKa. The results are shown in Table S1. We are not aware of any experiments to compare our calculated pKa's of ionizable residues with. However, there are experimental data on the net charge of ApoLp-III at a particular pH

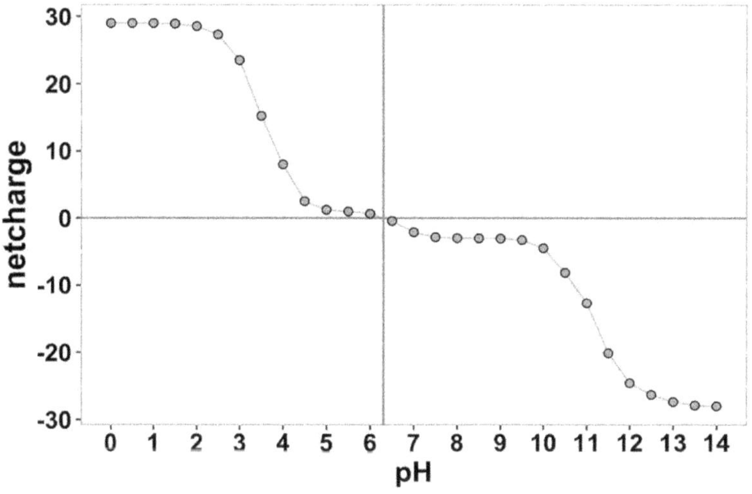

Fig. 6. The net charge of a soluble compact ApoLp-III bundle as a function of pH.

and about the isoelectric point, pI (the pH as which the net charge is zero). Figure 6 shows the calculated net charge of a compact ApoLp-III as a function of pH. One can see that the pI is calculated to be 6.2 which is very close to experimentally measured pI of 5.7. At pH= 8.6, the experiment gives the net charge of −4.7e which is again closely matched by our calculations resulting in the net charge of −3.0e. These results indicate that the computational protocol is adequate.

The pH dependence of the transition from the compact soluble ApoLp-III to the ApoLp-III lipid disk complex was modeled using Eq. (1) and computing the net charge differences as a function of pH. The pH dependence of the net charges of the compact ApoLp-III bundle and four models of to the ApoLp-III lipid disk complexes are provided in Fig. S1. The results are shown in Fig. 7 for each of the models separately (for completeness, Fig. 7(a) shows the pH-dependence in the range of pH from 0 to 14, while Fig. 7(b) shows the results in the range of physiological pH). In most of the cases, the pH-dependence of the free energy is an inverted bell-shaped curve and has a minimum at a particular pH. This pH is termed the pH-optimum[57–59] since the binding affinity is the most favorable at this pH. There is no much difference between pH-dependence of biding

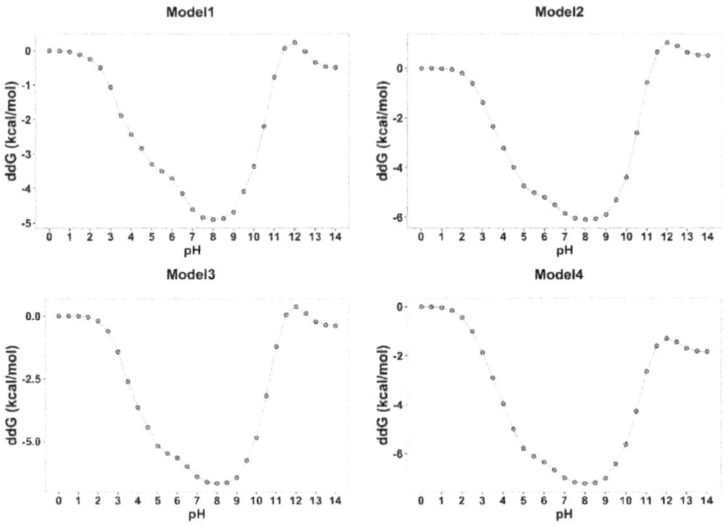

(a)

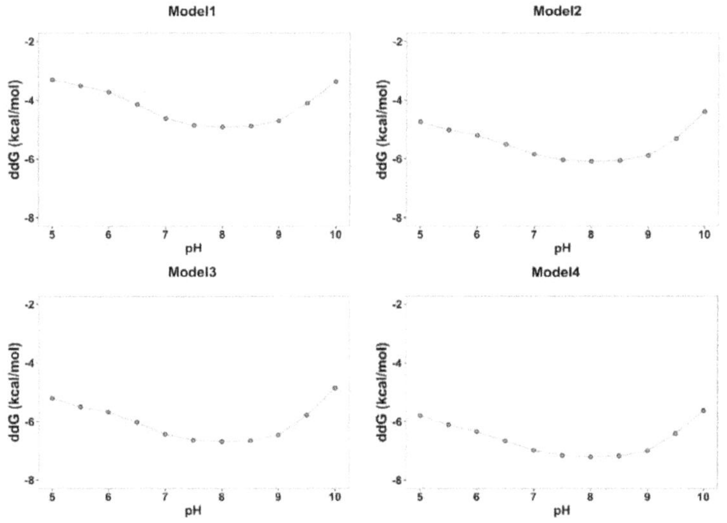

(b)

Fig. 7. The calculated pH-dependence of the free energy difference between the compact soluble ApoLp-III bundle and ApoLp-III bound to a lipid disk. The profiles are adjusted to result in zero energy at pH= 0. (a) pH ranges from 0.0 to 14.0. (b) pH ranges from 6.0 to 10.0.

free energy among the four models, indicating that there are no specific interactions between ApoLp-III titratable residues and the lipid disk that differ among the models (Note that pKa calculations were done at atomistic level of detail, thus including interactions between all atoms of ApoLp-III and lipids).

It is important to notice that we are talking about the relative energy values since Eq. (1) has a constant, the binding free energy at a particular pH, which in this case it taken to be zero at pH= 0.0. In the range of physiological pH, from 6.0 to 9.0, the magnitude of the binding free energy for Model 3 changes from −5.67 kcal/mol at pH= 6.0 to −6.69 kcal/mol at pH= 8.0 to −6.46 kcal/mol at pH= 9.0. In the other models, these changes are very similar. Having in mind that the folding free energy of ApoLp-III is only 1.24 kcal/mol,[55] such a change represents a significant barrier for the transition of ApoLp-III from the free to the lipid-bound state. Thus, the pH is an important regulator of the ApoLp-III association with a lipid disk and perhaps the same is valid for ApoLp-III binding to LDLp.

The pKa's of ionizable residues of the soluble compact ApoLp-III and in the ApoLp-III-disk complexes are provided in Tables S1, S2, S3, S4 and S5. A titratable group that has different pKa in compact versus bound ApoLP-III will have contribution to the pH-dependence, while a titratable group that has the same pKa in both states will not contribute. Thus, one can compare the individual pKa's and seek change in pKa's from soluble to bound state to identify which titratable groups cause the pH-dependence of the binding free energy. While there are many such changes, we are mostly interested in the cases affecting the pH dependence around the pH-optimum = 8.0. There is no pKa= 8.0 in any of the Tables S1, S2, S3, S4 and S5, however there are several Histidine and Lysine residues that undergo pKa shifts that make them not fully ionized at pH= 8.0. Comparing pKa's of such residues across all the models, one cannot spot any specific trend. Thus, Histidine 15, 151 and 160 have mostly unperturbed pKa (with respect to pKa in the unbound state) in all the Models, while Histidine 94 is predicted to have some shifted pKa in Models 2 and 4, but not in Models 1 and 3. Similarly, the behavior of Lysine residues varies across the models. This indicates that the

ApoLp-III binding to lipid disk is not specific, indeed, and it is driven by hydrophobicity which is nonspecific effect. Thus, while almost the same pH-dependence profile is predicted for all the models, the ionizable groups contributing to it are different in different models.

3.3. *Role of electrostatics in Apolipophorin III binging to a lipid disk*

While the hydrophobic effect is the main reason for ApoLp-III physical binding to a lipid disk, the electrostatics provides long-range guiding force to initiate the process. Indeed, the electrostatic potential mapped onto the compact soluble ApoLp-III bundle is shown in Fig. S2 and it can be seen that ApoLp-III surface is mostly negatively charged and in its compact form the protein will be repelled from the negatively charged lipids. The electrostatic field lines indicate the same phenomena — there are electrostatic funnels that will provide guidance for binding to the positively charged partners, but not for the negatively charged ones.

In the case of open ApoLp-III conformation (Fig. S3), the negatively and positively charged surface patches are equally distributed over the protein surface. Especially in the case of Models 3 and 4, the positively charged patches dominate. These patches will provide guidance toward binding to the negatively change partners like lipids.

When ApoLp-III is bound to the lipid disk (Fig. S4), the negatively charged patches face away from the binding interface, while the positively charged patches are mostly facing the lipid disk. This stabilizes the ApoLp-III-lipid-disk complex and makes it soluble in water.

4. Discussion

The work suggests that the ApoLp-III binding to a lipid disk may be multi-modal binding. While we confirm the experimental observation that the dominant binding mode involves an opening of hydrophobic code of ApoLp-III such that helices 3,4 move away their original positions adopted in the 3D structure of soluble ApoLp-III, we suggest that other open conformations may be present as well. This prompted us to consider that ApoLP-III association with a lipid disk is nonspecific and is mostly governed by the hydrophobic effect. Indeed, the binding involves an opening of ApoLp-III such that the hydrophobic core turns into a curved patch of hydrophobic residues and the curvature of the patch matches the curvature of the lipid disk (see the video). Thus, the ApoLp-III hydrophobic patch binds to the hydrophobic lipid tails and completes the physical binding. It is known that hydrophobic effect is nonspecific, thus, any structural rearrangement of ApoLp-III that results in such hydrophobic patch will promote the binding to the lipid disk.

From electrostatics point of view, we predict strong pH-dependence of the binding free energy of ApoLp-III to a lipid disk. The predicted free energy change as pH varies in physiological pH ranges is comparable with the folding free energy of ApoLp-III. This indicates that pH is an important regulator for ApoLp-III binding to lipid disk and perhaps to LDLP. It should be mentioned that all four open ApoLp-III models resulted in very similar pH-dependence of the binding free energy to a lipid disk, which further support the hypothesis of nonspecific binding.

Acknowledgments

This work was supported in part by National Science Foundation SC EPSCoR/IDeA Program under NSF Award No. OIA-1655740. The views, perspectives and content do not necessarily represent the official views of the SC EPSCoR/IDeA Program nor those of the NSF. E. A and Y. P. were supported by a grant from NIH, grant number R01GM093937. We thank Alex Alexov for generating the movie illustrating ApoLp-III binding to a lipid disk.

References

1. Ryan, R. O.; van der Horst, D. J. Lipid transport biochemistry and its role in energy production, *Annu. Rev. Entomol.* **2000**, *45*, 233–260.
2. Van der Horst, D. J.; Rodenburg, K. W. Lipoprotein assembly and function in an evolutionary perspective, *Biomol. Concepts*, **2010**, *1* (2), 165–183.

3. Van der Horst, D. J.; Roosendaal, S. D.; Rodenburg, K. W. Circulatory lipid transport: Lipoprotein assembly and function from an evolutionary perspective, *Mol. Cell Biochem.* **2009**, *326* (1–2), 105–119.
4. Van der Horst, D. J.; Ryan, R. O. Lipid transport, in Gilbert, L. I. (ed.), *Insect Molecular Biology and Biochemistry*, Elsevier Academic Press Inc., San Diego, pp. 317–345, 2012.
5. Van der Horst, D. J.; *et al.*, Alternative lipid mobilization: The insect shuttle system, *Mol. Cell Biochem.* **2002**, *239* (1–2), 113–119.
6. Plump, A. S.; Breslow, J. L. Apolipoprotein E and the apolipoprotein E-deficient mouse, *Annu. Rev. Nutr.* **1995**, *15*, 495–518.
7. Breslow, J. L. Human apolipoprotein molecular biology and genetic variation, *Annu. Rev. Biochem.* **1985**, *54*, 699–727.
8. Soulages, J. L.; Wells, M. A. Lipophorin — the structure of an insect lipoprotein and its role in lipid transport in insects, *Adv. Protein Chem.* **1994**, *45*, 371–415.
9. Zdybicka-Barabas, A.; Cytrynska, M. Apolipophorins and insects immune response, *Isj-Invertebrate Survival J.* **2013**, *10* (1), 58–68.
10. Weers, P. M. M.; Ryan, R. O. Apolipophorin III: Role model apolipoprotein, *Insect. Biochem. Mol. Biol.* **2006**, *36* (4), 231–240.
11. Narayanaswami, V.; Kiss, R. S.; Weers, P. M. M. M.; The helix bundle: A reversible lipid binding motif, *Comparative Biochem Physiol A Molecular Integrative Physiol* **2010**, *155* (2), 123–133.
12. Law, J. H.; Ribeiro, J. M. C.; Wells, M. A.; Biochemical insights derived from insect diversity, *Annual Rev. Biochem.* **1992**, *61*, 87–111.
13. Law, J. H.; Wells, M. A.; Insects as biochemical models, *J. Biol. Chem.* **1989**, *264* (28), 16335–16338.
14. Canavoso, L. E.; *et al.*, Fat metabolism in insects, *Annual Rev. Nutrition* **2001**, *21*, 23–46.
15. Arrese, E. L.; Soulages, J. L. Insect fat body: Energy, metabolism, and regulation, *Annual Rev. Entomol.* **2010**, *55*, 207–225.
16. Mirheydari, M.; Mann, E. K.; Kooijman, E. E. Interaction of a model apolipoprotein, apoLp-III, with an oil-phospholipid interface, *Biochim Et Biophys Acta-Biomembranes*, **2018**, *1860* (2), 396–406.
17. Rathnayake, S. S.; *et al.*, Insertion of apoLp-III into a lipid monolayer is more favorable for saturated, more ordered, acyl-chains, *Biochim Et Biophys Acta-Biomembranes*, **2014**, *1838* (1), 482–492.
18. Breiter, D. R.; *et al.*, Molecular structure of an apolipoprotein determined at 2.5 A° resolution, *Biochemistry*, **1991**, *30* (3), 603–608.
19. Wang, J.; Sykes, B. D.; Ryan, R. O. Structural basis for the conformational adaptability of apolipophorin III, a helix-bundle exchangeable apolipoprotein, *Proc Natl Acad Sci USA*, **2002**, *99* (3), 1188–1193.
20. Fan, D. P.; *et al.*, NMR solution structure and dynamics of an exchangeable apolipoprotein, Locusta migratoria apolipophorin III, *J. Biol. Chem.* **2003**, *278* (23), 21212–21220.
21. Ryan, R. O.; Oikawa, K.; Kay, C. Conformational, thermodynamic, and stability properties of Manduca sexta apolipophorin III, *J. Biol. Chem.* **1993**, *268* (3), 1525–1530.

22. Weers, P. M. M.; Ryan, R. A.; Apolipophorin III: A lipid-triggered molecular switch, *Insect Biochem. Mol. Biol.* **2003**, *33* (12), 1249–1260.
23. Wientzek, M.; *et al.*, Binding of insect apolipophorin III to dimyristoylphosphatidylcholine vesicles. Evidence for a conformational change, *J. Biol. Chem.* **1994**, *269* (6), 4605–4612.
24. Wells, M. A.; *et al.*, The role of apolipophorin III in *in vivo* lepoprotein interconversions in adult Manduca sexta, *J. Biol. Chem.* **1987**, *262* (9), 4172–4176.
25. Kawooya, J. K.; *et al.*, Lipophorin structure analyzed by in vitro treatment with lipases, *J. Lipid. Res.* **1991**, *32* (11), 1781–1788.
26. Saito, H.; *et al.*, Lipid binding-induced conformational change in human apolipoprotein E. Evidence for two lipid-bound states on spherical particles, *J. Biol. Chem.* **2001**, *276* (44), 40949–40954.
27. Hickenbottom, S. J.; *et al.*, Structure of a lipid droplet protein: The PAT family member TIP47, *Structure*, **2004**, *12* (7), 1199–1207.
28. Gimenez-Andres, M.; Copic, A.; Antonny, B. The many faces of amphipathic helices, *Biomolecules*, **2018**, *8* (3), 14.
29. Itabe, H.; *et al.*, Perilipins: A diversity of intracellular lipid droplet proteins, *Lipids Health Dis.* **2017**, *16*, 11.
30. Wan, C. P. L.; *et al.*, Apolipoprotein-induced conversion of phosphatidylcholine bilayer vesicles into nanodisks, *Biochim Et Biophys Acta. Biomembranes*, **2011**, *1808* (3), 606–613.
31. Weers, P. M. M.; *et al.*, A laboratory exercise to illustrate protein-membrane interactions, *Biochem Mol. Biol. Edu.* **2016**, *44* (1), 86–94.
32. Nath, A.; Atkins, W. M.; Sligar, S. G. Applications of phospholipid bilayer nanodiscs in the study of membranes and membrane proteins, *Biochemistry*, **2007**, *46* (8), 2059–2069.
33. Mizuguchi. C.; *et al.*, Fluorescence analysis of the lipid binding-induced conformational change of apolipoprotein E4, *Biochemistry*, **2012**, *51* (28), 5580–5588.
34. Fukuda, M.; *et al.*, Conformational change of apolipoprotein A-I and HDL formation from model membranes under intracellular acidic conditions, *J. Lipid. Res.*, **2008**, *49* (11), 2419–2426.
35. Wang, S. X.; Sun, Y. T.; Sui, S. F. Membrane induced conformational change in human apolipoprotein H, *Biochem. J.* **2000**, *348* Pt 1, 103–106.
36. Sahoo, D.; *et al.*, Lipid-triggered conformational switch of apolipophorin III helix bundle to an extended helix organization, *J. Mol. Biol.* **2002**, *321* (2), 201–214.
37. Chetty, P. S.; *et al.*, Role of helices and loops in the ability of apolipophorin-III to interact with native lipoproteins and form discoidal lipoprotein complexes, *Biochemistry*, **2003**, *42* (51), 15061–15067.
38. Soulages, J. L.; Arrese, E. L. Interaction of the alpha-helices of apolipophorin III with the phospholipid acyl chains in discoidal lipoprotein particles: A fluorescence quenching study, *Biochemistry*, **2001**, *40* (47), 14279–14290.
39. Soulages, J. L.; *et al.*, Essential role of the conformational flexibility of helices 1 and 5 on the lipid binding activity of apolipophorin-III, *J. Biol. Chem.* **2001**, *276* (36), 34162–34166.

40. Soulages, J. L.; Arrese, E. L. Dynamics and hydration of the alpha-helices of apolipophorin III, *J. Biol. Chem.* **2000**, *275* (23), 17501–17509.
41. Raussens, V.; *et al.*, Alignment of the apolipophorin-III alpha-helices in complex with dimyristoylphosphatidylcholine. A unique spatial orientation, *J. Biol. Chem.* **1995**, *270* (21), 12542–12547.
42. Wang, Y.; *et al.*, Implementation of accelerated molecular dynamics in NAMD, *Comput. Sci. Discov.* **2011**, *4* (1), 015002.
43. Phillips, J. C.; *et al.*, Scalable molecular dynamics with NAMD, *J. Comput. Chem.* **2005**, *26* (16), 1781–1802.
44. Wijeratne, T. U.; Weers, P. M. M. M. Lipid-bound apoLp-III is less effective in binding to lipopolysaccharides and phosphatidylglycerol vesicles compared to the lipid-free protein, *Mol. Cell Biochem.* **2019**, *458* (1–2), 61–70.
45. Leon, L. J.; *et al.*, Apolipophorin III: Lipopolysaccharide binding requires helix bundle opening, *Biochem. Biophys. Res. Commun.* **2006**, *348* (4), 1328–1333.
46. Dwivedi, P.; *et al.*, Deletion of the N- or C-terminal helix of apolipophorin III to create a four-helix bundle protein, *Biochemistry*, **2016**, *55* (26), 3607–3615.
47. Narayanaswami, V.; *et al.*, A molecular trigger of lipid binding-induced opening of a helix bundle exchangeable apolipoprotein, *Proc. Natl. Acad. Sci. USA*, **1999**, *96* (8), 4366–4371.
48. Pettersen, E. F.; *et al.*, UCSF Chimera–a visualization system for exploratory research and analysis, *J. Comput. Chem.* **2004**, *25* (13), 1605–1612.
49. Humphrey, W.; Dalke, A.; Schulten, K. VMD: Visual molecular dynamics, *J. Mol. Graph.* **1996**, *14* (1), 27–28, 33–38.
50. Li, C.*et al.*, DelPhi Suite: New developments and review of functionalities, *J. Comput. Chem.* **2019**, *40* (28), 2502–2508.
51. Li, L.; *et al.*, DelPhi: A comprehensive suite for DelPhi software and associated resources, *BMC Biophys.* **2012**, *5*, 9.
52. Wang, L.; Li, L.; Alexov, E. pKa predictions for proteins, RNAs, and DNAs with the Gaussian dielectric function using DelPhi pKa, *Proteins*, **2015**, *83* (12), 2186–2197.
53. Pahari, S.; *et al.*, DelPhiPKa: Including salt in the calculations and enabling polar residues to titrate, *Proteins*, **2018**, *86* (12), 1277–1283.
54. Huang, J.; *et al.*, CHARMM36m: An improved force field for folded and intrinsically disordered proteins, *Nat. Meth.* **2017**, *14* (1), 71–73.
55. Ryan, R. O.; Oikawa, K.; Kay, C. M. Conformational, thermodynamic, and stability properties of Manduca sexta apolipophorin III, *J. Biol. Chem.* **1993**, *268* (3), 1525–1530.
56. Soulages, J. L.; Wells, M. A. Lipophorin: The structure of an insect lipoprotein and its role in lipid transport in insects, *Adv. Protein. Chem.* **1994**, *45*, 371–415.
57. Alexov, E. Numerical calculations of the pH of maximal protein stability. The effect of the sequence composition and three-dimensional structure, *Eur. J. Biochem.* **2004**, *271* (1), 173–185.
58. Mitra, R. C.; Zhang, Z.; Alexov, E. In silico modeling of pH-optimum of protein-protein binding, *Proteins*, **2011**, *79* (3), 925–936.
59. Talley, K.; Alexov, E. On the pH-optimum of activity and stability of proteins, *Proteins*, **2010**, *78* (12), 2699–2706.

Part IV
Artificial Intelligence in Biophysics and Chemistry

https://doi.org/10.1142/9789819808199_0010

Chapter 10

MLBKFD: Probabilistic Model Methods to Infer Pseudo Trajectories from Single-cell Data*

Changfeng Han[†], Wenjie Cao[‡], Cheng Li[†], Yanbing Guo[§], Yuebin Wang[†], Ya-Zhou Shi[†] and Ben-Gong Zhang[†,¶]

[†]*School of Mathematical & Physical Sciences, Wuhan Textile University, Wuhan 430200, P. R. China*
[‡]*School of Mathematics, Sun Yat-sen University, Guangzhou 510275, P. R. China*
[§]*School of Computer Science and Artificial Intelligence, Wuhan Textile University, Wuhan 430200, P. R. China*
[¶]*bgzhang@wtu.edu.cn*

Cell trajectory inference is very important to understand the details of tissue cell development, state differentiation and gene dynamic regulation. However, due to the high noise and heterogeneity of the single-cell data, it is challenging to infer cell trajectory in complex biological processes. Here, we proposed a new trajectory inference method, called Metric Learning Bhattacharyya Kernel Feature Decomposition (MLBKFD). In MLBKFD, a statistical model was used to infer cell trajectory by calculating the Markov transition matrix and Bhattacharyya kernel matrix between cells. Before that, to expedite the matrix calculation in the statistical model, a deep feedforward neural network was used to perform dimensionality reduction on single-cell data. The MLBKFD was evaluated on four typical datasets as well as seven recent human fetal lung datasets. Comparisons with the two outstanding methods (i.e., DTFLOW and MARGARET) demonstrate that the MLBKFD is capable of accurately inferring cell development and differentiation trajectories from single-cell data with

*To cite this article, please refer to its earlier version published in *J. Comput. Biophys. Chem.* **2024**, *23* (6), 723–739, DOI: 10.1142/S2737416524400015
[¶]Corresponding author.

different sizes and sources. Notably, MLBKFD exhibits nearly twice the speed of DTFLOW while maintaining high precision, particularly when dealing with large datasets. MLBKFD provides accurate and efficient trajectory inference, empowering researchers to gain deeper insights into the complex dynamics of cell development and differentiation.

Keywords: Trajectory inference; single-cell data; probabilistic model; neural network.

1. Introduction

Cell trajectory inference is a powerful tool for analyzing the complexity and variability of cell dynamics in single-cell transcriptomics data.[1–4] It plays a crucial role in improving the performance of gene regulatory network reconstruction and calculating gene expression changes along the trajectory of the system.[5–7] Accurate trajectory inference is essential for identifying cell lineages, analyzing intratumoral heterogeneity, understanding cancer progression, and recognizing gene regulation rules in constructing pseudo-time cell development processes.[8–13]

With the accumulation of single-cell data, several trajectory inference methods have been proposed to understand the developmental processes and cell fate decisions. These methods can be broadly categorized into: transcriptional dynamics-based method (e.g., RNA velocity), Tree/Graph-based methods, methods quantifying cell-fate-bias-based method and transition probabilistic-based model. Each category of methods has its own strengths and limitations.[14–19] For example, methods (e.g., VeTra,[18] CellRank,[20] and UniTVelo[21]) based on RNA velocity, which is the relative ratio between pre-mRNA and mature mRNA reads, can provide valuable information about the direction and speed of movement of individual cells, allowing for the visualization of dynamic transformation of cells without any prior knowledge.[20–22] These methods have proven to be particularly useful in understanding cellular dynamics during development and disease progression, however, estimating RNA velocity vectors are still challenging (due to the limitation single-cell data with pre-/mature mRNA) and computationally intensive.[23]

In contrast, methods based on the minimum spanning tree (MST) or shortest path such as[24–26] TSCAN,[27] GraphDDP,[28] DensityPath[29] and SoptSC,[30] can infer trajectory with faster speed and have shown excellent performances on large datasets.[8,29,31–33] For instance, Monocle2 uses reverse graph embedding to describe multiple fate decisions in a completely unsupervised manner, and PAGA constructs a KNN graph and applies the Louvain algorithm to divide the graph into multiple resolutions.[34] The PAGA algorithm converges faster, and can be extended to the directed graph by using RNA velocity to orient the edges in the PAGA graph. However, it is important to note that these methods often make strong assumptions about the topology of the trajectory and may be susceptible to noise in the dataset.[35] Recently, deep learning technology has shown advantages in single-cell data with large sizes and noise.[36–38] For example, MARGARET introduces a feedforward neural network and can scale to large scRNA-seq datasets making it suitable for analyzing atlas-level scRNA-seq datasets, which shows excellent results in identifying disconnected and multi-branch complex trajectories.[35] Single-cell Deep learning model for ATAC-Seq and RNA-Seq Trajectory integration (scDART) is a scalable deep learning framework that embeds data modalities into a shared low-dimensional latent space that preserves cell trajectory structures in the original datasets.[39] These methods have poor interpretability and are less helpful to understand the differentiation mechanism.[39–44]

Nevertheless, the methods based on the transition probability are interpretable and tractable.[45–54] For example, diffusion pseudo-time (DPT) algorithm defines the diffusion pseudo-time distance between two cells using the accumulated Markov transition matrix and determines the ordering of cells based on the distances between a root cell and all other cells.[55–59] The DTFLOW algorithm uses Bhattacharyya kernel feature decomposition to visualize the underlying dataset structure. DPT is a special case of the random walk with restart (RWR) algorithm (i.e., $p = 1$) in DTFLOW. Both DTFLOW and DPT focus on the need to calculate the transfer matrix between cells.[17] However, these methods require higher computational costs to obtain better results, so dimensionality reduction should be operated.

How to infer cell trajectories quickly and accurately, especially for large datasets, is still a problem to be solved.[19]

In this work, we developed a probabilistic-based model, Metric Learning Bhattacharyya Kernel Feature Decomposition (MLBKFD), to fast infer the cell pseudo-trajectories from single-cell data. Motivated by MARGARET, we constructed a deep feedforward neural network (DFNN) to extract the feature information from scRNA-seq data to reduce the dimensionality. We then applied the interpretable statistical method DTFLOW on this low-dimensional latent space to model the cell trajectories. In comparison to DTFLOW, MLBKFD is more efficient when dealing with large datasets without requiring prior stage information of the cells. Furthermore, the trajectory inference process using MLBKFD is simpler and offers better interpretability compared to MARGARET. The efficiency and accuracy of MLBKFD have been validated through testing on four different types of classical single-cell datasets and seven recent human fetal lung single-cell datasets with different sizes.[17,35,60]

2. Materials and methods

MLBKFD mainly consists of two steps: dimension reduction and trajectory modeling, see Fig. 1. First, the scRNA-seq data are input into DFNN for feature extraction, resulting in a low-dimensional latent space. Second, a probabilistic model is used for stochastic trajectory modeling.

2.1. *Dimensionality reduction to obtain the cell-state latent space*

In order to reduce the dimension of the scRNA-seq data, a DFNN (f) is used to map the relatively high-dimensional initial cell-state space to the low-dimensional latent space, i.e., f: $R^D \rightarrow R^d$ $(D \geq d)$ see Fig. 1(b). As shown in Fig. A.1 in Appendix A, the preprocessed scRNA-seq data with initial clustering label was first fed into a neural network module, which includes two fully connected hidden layers with 128 and 64 nodes, to get a low-dimensional cell-state latent space.[35,61] The neural network was trained through a loss function

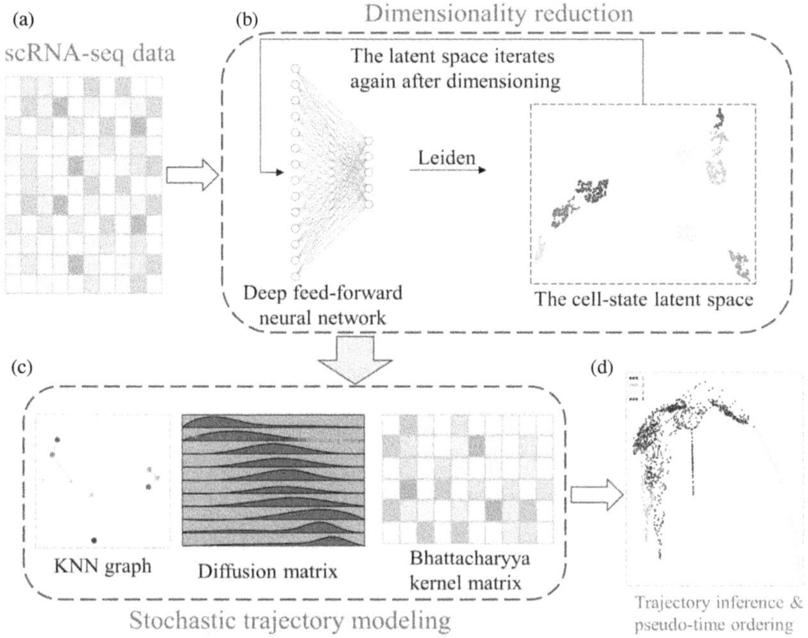

Fig. 1. (Color online) Framework of MLBKFD. (a) The scRNA-seq data are pre-processed into a gene expression matrix. (b) The dimensionality reduction process uses the DFNN iterative feature extraction to reduce the cell-state latent space dimension as much as possible. (c) Stochastic trajectory modeling mainly includes constructing a KNN graph, calculating diffusion matrix and Bhattacharyya kernel matrix. (d) Inferring the trajectory and pseudo-time ordering of datasets through kernel decomposition calculations.

calculated based on the principle that the distance between cells of the same type should be minimized, while the distance between cells of different types should be maximized; see more details in Appendix A. Then, all cells in the latent space (i.e., the output of the trained network) were further clustered using Leiden, which has been widely validated by recent works.[62] Finally, the new clustering labels generated from the latent space were sent back to the neural network module again for iteration until the number of extracted feature dimensions is no longer decreasing[35]; see more details in Appendix A. This iterative process not only effectively maps the data to a low-dimensional space but also achieves good cell clustering simultaneously.

The ordered pair (a_i, b_i) denotes the latent space and cluster labels of cell $i \in N$, where N is the number of cells. After the iterative learning, the low-dimensional representation of the final cell i is f_i^*, which is obtained by $f^*(a_i)$, where f^* indicates that the DFNN is trained to the best condition on the scRNA-seq dataset. At the same time, each cell cluster labels b_i^* in the optimized scRNA-seq dataset is also obtained. The cell-state latent space $\{f_i^*\}_{i=1}^N$ and the cluster labels $\{b_i^*\}_{i=1}^N$ are used in subsequent steps.

2.2. *Trajectory inference using statistical models*

Based on the cell data in low-dimensional latent space, a Markov affinity matrix is constructed. First, for each cell, the k (e.g., 10) most similar cells can be found using K-Nearest Neighbor (KNN) based on pairwise cell–cell Euclidean distances.[17,63] Then, the transition probability between any two adjacent cells is defined by the Gaussian kernel $K(f_i^*, f_j^*) = \exp\left(-\frac{\|f_i^*, f_j^*\|}{2\sigma_i\sigma_j}\right)$, where f_i^* is the low-dimensional representation of the ith cell and σ_i is the local kernel widths of the ith cell (i.e., the distance between ith cell and its kth neighbor cell). Then, the kernel matrix K was further symmetrized through $K = \max(K, K^T)$, where K^T denotes the transpose of K. Subsequently, the Markov transition probability between cells i and j can be defined by

$$M_{ij} = \frac{\tilde{K}(f_i^*, f_j^*)}{\sum_j \tilde{K}(f_i^*, f_j^*)}, \tag{1}$$

where $\tilde{K}(f_i^*, f_j^*) = \frac{K(f_i^*, f_j^*)}{\sqrt{\sum_j K(f_i^*, f_j^*) \sum_i K(f_i^*, f_j^*)}}$ is the normalized $K(f_i^*, f_j^*)$, which is similar to normalization of an adjacency matrix.[17]

Furthermore, considering the stochastic nature of cell differentiation, the transition probability matrix (S) could be modified as

$$S = (1 - \mathcal{P})(I - \mathcal{P}M)^{-1}, \tag{2}$$

where $\mathcal{P}$ is a random probability of the current cell moving to its neighbors, I is the identity matrix. s_i represents the ith rows of matrix

S, and $\sum_{k=1}^{N} \mathcal{S}_{ik} = 1$, where $\mathcal{S}_{ik}$ is the kth element of $\mathcal{S}_i$. Based on Refs. 17 and 57, $\mathcal{P}$ was set as 0.9 in this work.[17,56]

Based on the modified transition probability matrix S, a Bhattacharyya distance matrix between cells can be calculated by $D = -\log G$, where $G = \sqrt{\mathcal{S}}\sqrt{\mathcal{S}}^T$ is the Bhattacharya kernel matrix. That is, $G_{ij} = \langle\sqrt{\mathcal{S}_i}, \sqrt{\mathcal{S}_j}\rangle$, $\forall\, i, j \in [1, 2, \ldots, N]$ represents the inner product of $\sqrt{\mathcal{S}_i}$ and $\sqrt{\mathcal{S}_j}$. Then, the distance matrix D can be decomposed by singular value decomposition (SVD) method, i.e., $D = V\Sigma V^T$, where V is a unitary matrix satisfying $V^T V = I$ and Σ is a diagonal matrix whose elements are singular values of the matrix D. Since the largest d positive singular value could represent the main characteristics of the matrix D, the low-dimensional (d) embedding of $Y(Y_d)$ can be defined as

$$Y_d = V_d \sum_{d}^{\frac{1}{2}}. \tag{3}$$

Finally, the distance between two cells i and j in d-dimensional space was defined by $D_d(i, j) = \sqrt{-2\log G_{ij}}$, which satisfies the triangular inequality of inner product space. If the root cells r are selected, the distance between the ith cell and root cell is denoted as $D_d(r, i)$, and the vector $T_r = D_d(r, *)$, composed of distances from all cells to the root cell, can effectively represent the order of appearance for each cell. Then, the pseudo-time T (a value between 0 and 1) of each cell could be defined as the min–max normalized T_r to

$$T = \frac{T_r - \min\{T_r\}}{\max\{T_r\} - \min\{T_r\}}. \tag{4}$$

2.3. *Evaluation metrics*

To evaluate the accuracy of the inferred trajectory, the Spearman's rank correlation coefficient (ρ) is used in this work, and it is defined as[46]

$$\rho = \frac{\sum_{i=1}^{N}(\text{true}_i - \overline{\text{true}})(\text{rank}_i - \overline{\text{rank}})}{\sqrt{\sum_{i=1}^{N}(\text{true}_i - \overline{\text{true}})^2}\sqrt{\sum_{i=1}^{N}(\text{rank}_i - \overline{\text{rank}})^2}}, \tag{5}$$

where rank(i) denotes the rank of cell i in the pseudo-time ordering, and true(i) denotes the rank of cell i in the real order from experiments. $\rho \in [-1, 1]$, and the closer the ρ value is to 1, the more consistent the pseudo-time ordering is with the real order.

Although dimensionality reduction can preserve the main features of the data and significantly reduce its complexity, in general, dimensionality reduction tends to result in some loss of information from the original data. We further validate the effect and robustness of this dimensionality reduction method. The adjusted rand index (ARI), which is a statistical measure used to evaluate the similarity between two clusterings, takes into account both the similarity of the data points within the same cluster and the dissimilarity between different clusters. On the other hand, the normalized mutual information (NMI), which is an information theoretical measure used to assess the similarity between two clustering results, measures the amount of shared information between the two clustering, normalized by their individual entropies. The details of the calculation for the ARI and NMI can be found in Appendix A. The ARI provides a value between −1 and 1, while the NMI value ranges from 0 to 1, and both values closer to 1 indicate better clustering results, while 0 indicates random clustering.[64,65]

2.4. *Datasets and data processing*

As shown in Table 1, there are four classical datasets which were widely used by other methods[35,59,66,67] and the seven new human fetal lung datasets with different sizes[60] are used to validate the MLBKFD. In four classical datasets, the Fibroblast and ME datasets are labeled by developmental stage,[17,59] while the Planaria parenchyme and human4 datasets do not have real labels, but they provide the detailed cell type information as well as the prior knowledge of transformation between these cell types instead of the labels of developmental stage.[35,66] The human fetal lung datasets with very high resolution were released in 2022.[60,68] The size distribution of these datasets is wide, e.g., the smallest one includes 649 cells and 16122 genes, and size of the largest one is 9096×22746 (Table 1). Furthermore, the

Table 1. 11 Datasets used in this work.

Dataset[a]	Tissue	Organism	Cells	Genes	Stages	Types	Source
Fibroblast	fibroblasts	mouse	405	16379	5	—	https://www.ncbi.nlm.nih.gov/geo/query/acc.cgi?acc=GSE67310
ME	blastocyst	mouse	438	48	7	—	https://github.com/gcyuan/SCUBA/tree/master/sample_data/guo2010
Planaria parenchyme	—	planarian	1986	4210	—	5	https://zenodo.org/record/5850115
human4	brain	human	1303	16381	—	14	https://www.ncbi.nlm.nih.gov/geo/query/acc.cgi?acc=GSM2230760
PNS	lung	Homo sapiens	649	16122	8	5	https://fetal-lung.cellgeni.sanger.ac.uk/scRNA.html
Cilium	lung	Homo sapiens	1641	19881	5	1	
B cells	lung	Homo sapiens	4138	17560	6	9	
T, NK and ILC	lung	Homo sapiens	4778	17851	7	14	
Endothelium	lung	Homo sapiens	5467	20746	9	9	
Myeloid	lung	Homo sapiens	6947	20014	9	19	
Epithelium	lung	Homo sapiens	9096	22746	8	11	

Note: [a] "—" represents a null value.

datasets contain the detailed information of both cell development stage and cell type.[69,70] In order to facilitate training of the neural network in the subsequent steps, for each dataset, we first performed principal component analysis (PCA), and then, we clustered the cells using K-means method to obtain initial class labels.

3. Results

3.1. *Workflow of MLBKFD*

First, we took PNS dataset as an example to show the workflow of MLBKFD for trajectory inference; see Materials and methods for details. First, based on the pre-processed gene expression matrix, the cell-state latent space and the clustering labels for all cells are obtained through the iterative DFNN. Figure 2(a) shows the clustering results visualized by UMAP, and the dataset is divided into five large clusters, two of which are further subdivided into eight clusters. The PNS dataset is composed of five different cell types with eight different time stages.

Second, based on the data in latent space and clustering label for each cell, the Bhattacharyya distance matrix between cells can be calculated, followed by a matrix decomposition, and the cell development trajectory with the highest probability can be found according to the distances between cells. By specifying the root cells, the pseudo time of each cell on the trajectory is solved and normalized, see Fig. 2. It should be noted that our method currently cannot identify root cells. Therefore, in this study, we designate the central cell of the first-stage cells as the root cell for each dataset. The eight real time stages of the PNS dataset are displayed on the obtained trajectories for easy comparison, as shown in Fig. 2(c).

We also used the DTFLOW method to infer the trajectory for the PNS dataset, taking the real stage label as input; see Fig. 2(d). As shown in Figs. 2(c) and 2(d), the stage information points from MLBKFD are more clear. For example, the 11th week stage points inferred by DFLOW are difficult to distinguish from other stage points, while those from MLBKFD form a separate branch. It can be pointed out that MLBKFD does not require the real stage labels,

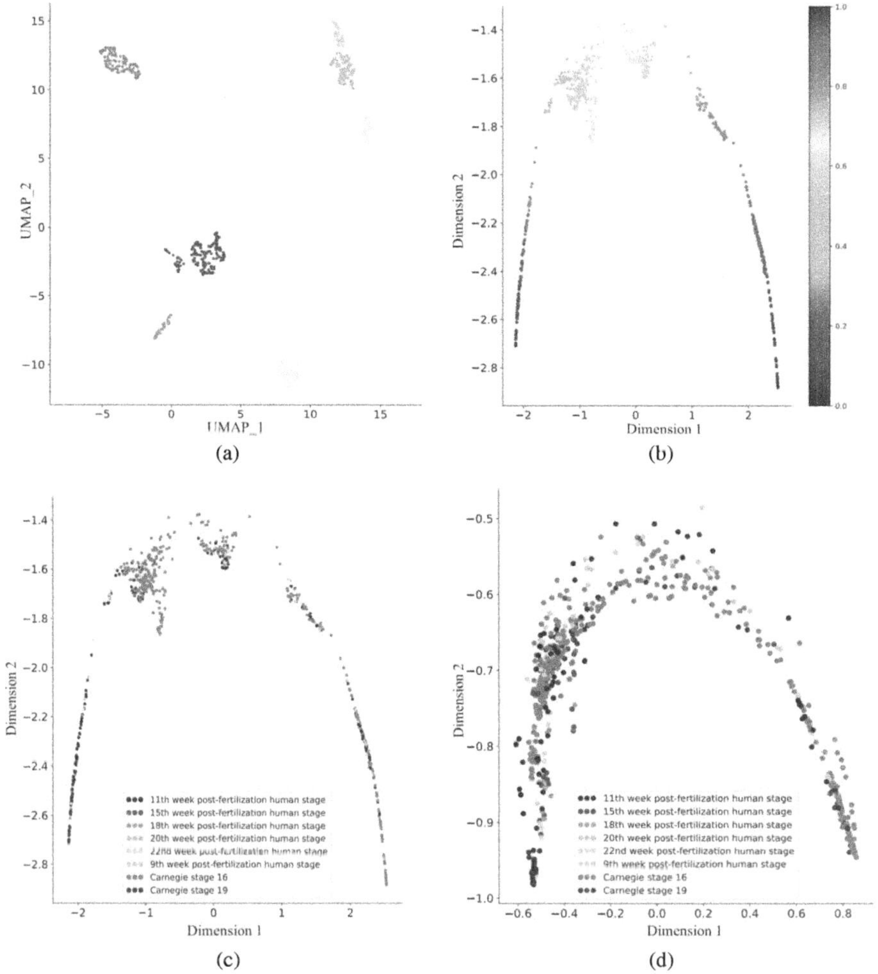

Fig. 2. (Color online) Cell trajectory inference for the PNS dataset. (a) UMAP visualization map of cell state in low-latent space. (b) The pseudo-time corresponds to the differentiation process inferred by MLBKFD. (c) The cell differentiation trajectory inferred by MLBKFD, and different colors represent the real time stage. (d) The cell differentiation trajectory is inferred by DTFLOW.

unlike DTFLOW. In addition, we also performed another inference using MLBKFD with real label as input, and the corresponding results are shown in Fig. S4. For convenience, we used the MLBKFD0 representing our method independent of the real stage label.

In addition, we further calculated the Spearman correlation coefficient between inferred and experimental pseudo-times. For the PNS dataset, the coefficients are 0.31 and 0.36 for MLBKFD0 and MLBKFD, respectively, which are comparable to those of other models (e.g., 0.34 for DTFLOW, and 0.39 for MARGARET), suggesting that it is effective for MLBKFD in pseudo-time inference.

3.2. *Verifying the information loss in the cell-state latent space*

To further verify the extent of information loss between the cell-state latent space and the original data space, we used UMAP to visualize the original data space and cell-state latent space, as shown in Figs. 3(a) and 3(b), respectively. It is clear that the edge between the clusters is much clearer and the compactness within the cluster is better from visualization viewpoint. This means that the information in the original data space contains a lot of noise, and the latent space of cell state only contains the most critical feature information of each cell.

Additionally, to further validate the robustness of this dimensionality reduction method, we also conducted clustering experiments using the K-means, Leiden, and Louvain algorithms on five widely-used clustering datasets with ground-truth labels (Table S1).[62,71–74] We performed the clustering analysis in both the original data space and the latent space obtained through dimensionality reduction. As shown in Fig. 3(c), the mean value of ARI (or NMI) of clustering results (0.20/0.48) on the five datasets in cell-state latent space is only slightly smaller than that (0.28/0.59) in original data space, due to the loss of a small amount of data information in the latent space. However, reducing the data dimensionality to a latent space can effectively improve the speed of subsequent trajectory inference (see Sec. 3.3). Moreover, since dimensionality reduction can effectively preserve the main features of the data, it can enhance the visualization of clustering. The above comparison results suggest that our method can compress scRNA-seq data with minimal loss, allowing for accelerated cell trajectory inference by subsequent statistical models.

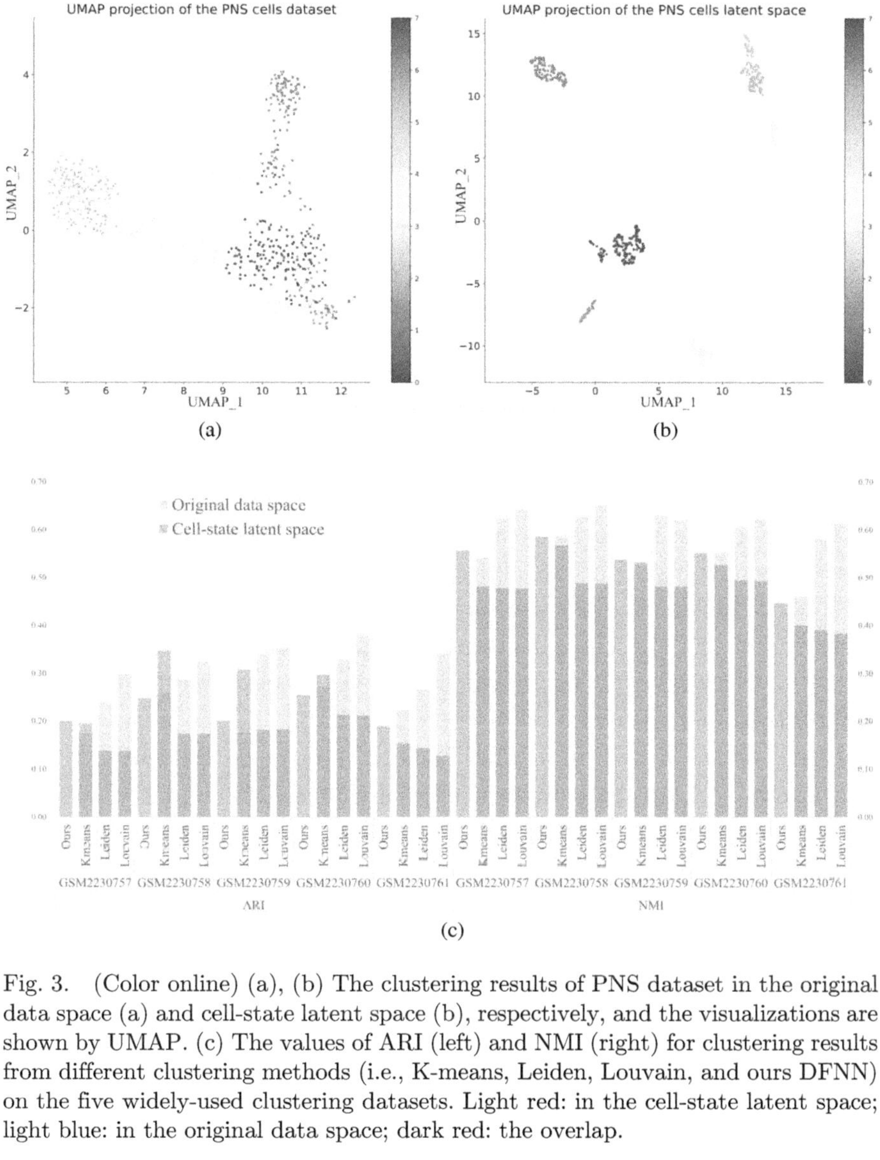

Fig. 3. (Color online) (a), (b) The clustering results of PNS dataset in the original data space (a) and cell-state latent space (b), respectively, and the visualizations are shown by UMAP. (c) The values of ARI (left) and NMI (right) for clustering results from different clustering methods (i.e., K-means, Leiden, Louvain, and ours DFNN) on the five widely-used clustering datasets. Light red: in the cell-state latent space; light blue: in the original data space; dark red: the overlap.

3.3. *Effectiveness and robustness of MLBKFD*

Furthermore, to validate the effectiveness of MLBKFD, we conducted extensive testing on 11 different datasets, and as shown in Table 1, these datasets with different sizes encompass cells of various types and features, covering a wide range of cellular states and variations.[35,59,60,66,67] To provide a visual comparison, we also employed a well-established method DTFLOW for trajectory inference on these datasets; see Figs. 4 and 5. The detailed results of cell trajectory inferred by MLBKFD and DTFLOW for all datasets can also be found in Figs. S1–S10. As shown in Figs. 2 and 4, on small datasets (e.g., less than 10000 genes or 2000 cells), the cell differentiation trajectories inferred by MLBKFD without using the true cell-type labels are consistent with those inferred by DTFLOW using true labels,[17] and both methods can clearly restore the development process of cells. In the ME dataset, MLBKFD restored the cell development process, which is consistent with the DTFLOW results, as shown in Figs. 4(b) and 4(e). In the Cilium dataset, MLBKFD still clearly identified 20th/22th week stage, while the whole development process began to be partially affected by data noise and intermediate state cells; see Fig. 4(c). As shown in Fig. 4(f), the result inferred by DTFLOW becomes unsatisfactory. However, as the number of cells or genes increases, the computational complexity of DTFLOW significantly increases, resulting in a decrease in the clarity of inferred cell differentiation trajectories. On the other hand, as the data are first mapped to a low-dimensional latent space in MLBKFD, the complexity of trajectory inference hardly increases, and the inferred differentiation trajectories can still maintain relative clarity; see Fig. 5 for results on larger datasets. Nevertheless, for complex datasets, there will be some deviations between the trajectory inferred by MLBKFD and the experimental trajectories. For example, in the Myeloid dataset, the number of cells at the 18th/20th week stage is much more than that at the other development stages. The possible reason is that there is a very small amount of the "Unknown" stage, which affects the inference of the entire trajectory. Subsequent considerations combine more multi-omics data to infer this complex trajectory. Further, on these datasets, we do not use the real stage label of the experiment (i.e., MLBKFD0) to simulate the working effect on the dataset without real stage label. The results of MLBKFD,

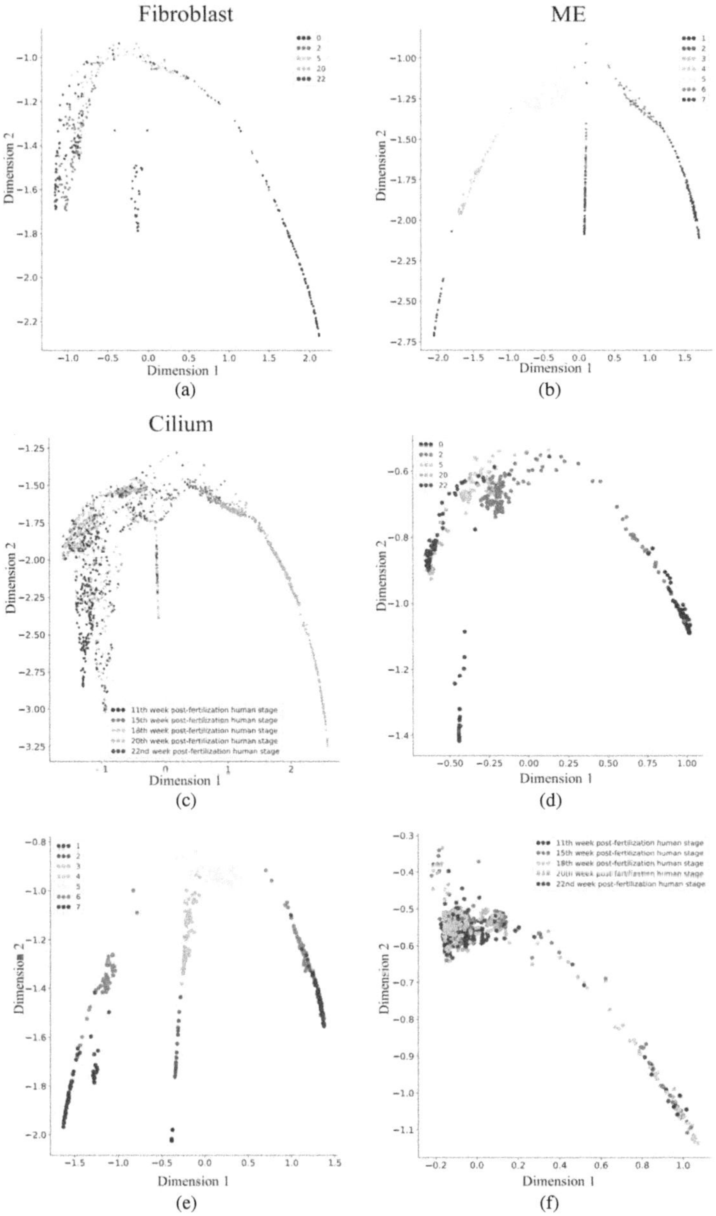

Fig. 4. (Color online) MLBKFD is used to visualize the trajectories of three small datasets. (a)–(c) The results of trajectory inference are visualized by MLBKFD, and the color label in the graph is the real developmental stage. The same (d)–(f) are the results of DTFLOW method.

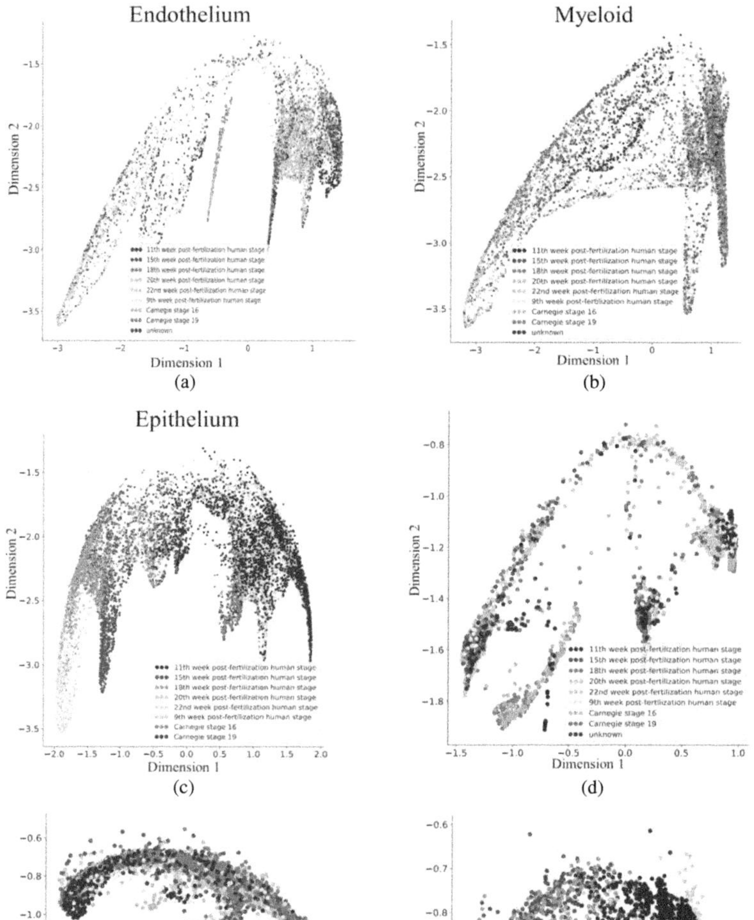

Fig. 5. (Color online) MLBKFD runs on three datasets with more than 20000 genes and more than 5000 cells. (a)–(c) The results of trajectory inference are visualized by MLBKFD, and the color label in the graph is the real developmental stage. The same (d)–(f) are the results of DTFLOW method.

DTFLOW, etc. using real labels are not significantly different from MLBKFD0 on most datasets (see Figs. S6–S10 for details).

Furthermore, we compared the time taken by MLBKFD and DTFLOW for trajectory inference (Fig. 6(a)). As shown in Fig. 6(a), MLBKFD exhibits significantly lower time consumption compared to DTFLOW, especially for larger datasets. For example, on the Epithelium dataset with 9096 cells and 22746 genes, the MLBKFD took $\sim$65 s, which is nearly twice as fast as the DTFLOW method ($\sim$130 s). This is mainly due to the fact that mapping the data to a lower-dimensional latent space significantly reduces the complexity of the data, e.g., for the Epithelium dataset, its dimensionality can be reduced from 22746 to 10, thereby reducing the computation time for cell transition matrix in trajectory inference processes. On large datasets, reducing the computation time for trajectory inference is crucial for downstream cellular processes analysis, which could be significant for advancing the field of cell biology and medicine.

In order to test the prediction accuracy of our method, we further calculated the pseudo-time according to the inferred differentiation trajectory, and calculated the correlation coefficient between it and the time stage given by the experiment (see the section of Materials and Methods). Because the two datasets of Planaria parenchyme and human4 have no experimental stage labels, only nine datasets are used. Any cell was selected from the earliest label in the experimental stage as the root cell of MLBKFD for pseudo-time calculation. For small datasets (i.e., Fibroblast and ME datasets), MLBKFD performed well as other methods, as shown in Fig. 6(b). For example, on the ME dataset, the correlation coefficient obtained by our method is 0.932, which is slightly better than those from the other methods, e.g., 0.926 of DTFLOW, 0.763 of MARGARET, 0.80209 of Monocle. Notably, for the larger Epithelium dataset (containing 9096 cells and 22746 genes), MLBKFD still gives very good results (correlation coefficient is 0.749). However, for other lung datasets, both MLBKFD and DTFLOW did not predict well (the correlation coefficient was less than 0.5). In particular, for B cell and Myeloid datasets, the correlation coefficient was less than 0.3. The reason may be that the number of cells in Epithelium dataset is large and the distribution of

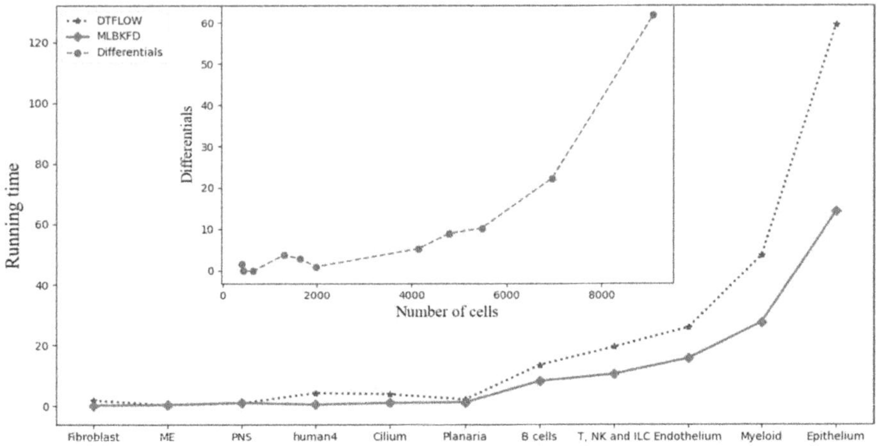

(a)

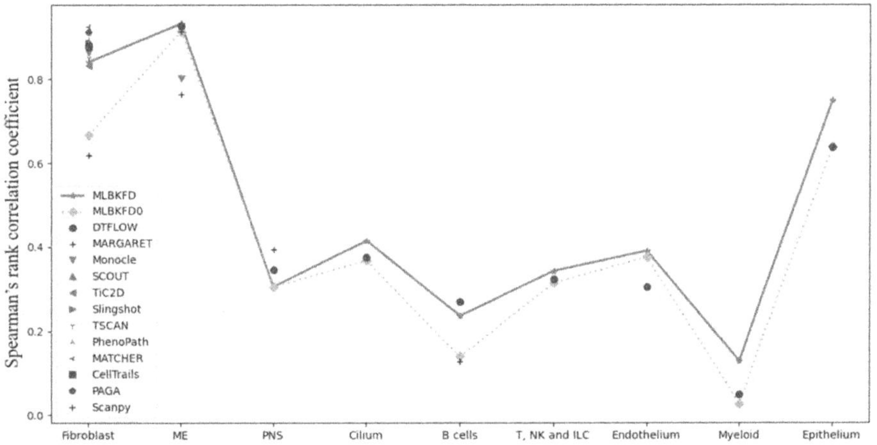

(b)

Fig. 6. (Color online) (a) The running time comparison of DTFLOW and MLBKFD on all datasets. The datasets are arranged from small to large according to the number of cells. (b) The Spearman correlation between the pseudo-time and the true label of various trajectory inference methods on nine datasets with developmental stage labels. The pink line is MLBKFD using developmental stage information, the golden dotted line is MLBKFD0 without a developmental stage, and other methods are displayed as points.

each stage is relatively uniform, while the cells in other datasets are mostly concentrated in some stages, resulting in too few cells in other stages, which has a greater impact on the inference results (see Fig. S11).

Furthermore, we find that the MLBKFD0 method without the label of the experimental stage is not much different from the MLBKFD (or DTFLOW) results in most cases, indicating that our method can be independent of the label of the experiment. For example, Fibroblast contains only 405 cells in five stages except for a few special datasets. The unsatisfactory performance of MLBKFD0 is related to the loss of information after dimensionality reduction of 16379 genes in this dataset. Although our results are not all superior to the existing methods, MLBKFD is based on solving the cell transition probability after learning each data, and has stronger adaptability to the datasets. And PAGA, MATCHER, MARGARET have advantages over pseudo-time correlation coefficients, but MLBKFD presents trajectories in a new way. These results consistently show that MLBKFD is an accurate, fast and robust method for trajectory inference method.

4. Conclusions and discussion

Cell differentiation, a complex biological process involving cell fate transitions, presents challenges in accurately inferring and visualizing developmental trajectories. Existing trajectory clustering and visualization methods often fail to provide comprehensive explanations of cell development inference. To address the limitation, this study introduces a novel framework called MLBKFD, which enhances trajectory inference based on a probabilistic model. MLBKFD leverages a DFNN to learn discriminative features from single-cell data to construct a low-dimensional latent space representing the cell states, and the effectiveness and robustness of the latent space have been validated using four widely-used clustering datasets. Based on the cell data in the latent space and the predicted cell clustering labels, the cell developmental trajectory can be further inferred using a statistical model by calculating the transfer matrix between cells. The tests on 11 datasets of varying sizes and origins demonstrate that

MLBKFD can effectively infer the differentiation trajectories of cells, e.g., it can clearly show the branches of cell differentiation. Although MLBKFD does not show significant superiority over existing methods in terms of pseudo-time inference, it significantly reduces the time required for inference compared to DTFLOW, due to the dimensionality reduction step of DFNN used in MLBKFD. In conclusion, MLBKFD is an efficient, fast, and robust method for inferring and visualizing cell differentiation trajectories from scRNA-seq data. It provides a powerful tool for cellular biology and medical research, aiding in a better understanding and exploration of the process of cell differentiation.

Future work may include processing complex data with many cell types and high noise, and considering combining more multi-omics data (genome, transcriptome and proteome) for mutual verification to finally determine the cell differentiation trajectory. Although calculating the transition probability of cells helps to understand the process of cell differentiation, the transition probability is only affected by some genes in each cell. It is necessary to study the effect of the regulation process of these genes on the probability of intercellular transfer. Starting from the construction of the trajectory, we consider combining the gene regulatory network, that is, constructing a series of gene regulatory networks along the pseudo-time trajectory to form a dynamic gene regulatory network. These will play an important role in the study of genes in cell metastasis analysis and cell fate decisions.

Funding information

This work was supported by the National Natural Science Foundation of China (NSFC) [grant numbers 11971367,12371500].

Conflict of interest

The authors declare that they have no known competing financial interests or personal relationships that could have appeared to influence the work reported in this paper.

Appendix A. Dimensionality reduction and clustering metrics calculation

A.1. *Dimensionality reduction*

DFNN (f) is used to map the relatively high-dimensional initial cell-state space to the low-dimensional latent space, i.e., f: $R^D \to R^d$ $(D \geq d)$, as follows:

Step 1: As shown in Fig. A.1, the preprocessed scRNA-seq data (with the dimension of $N \times 300$) with initial clustering label (i.e., Labels) were first fed into a neural network to reduce the dimension (e.g., to $N \times 10$). The neural network was composed of two fully connected hidden layers of sizes of 128 and 64, followed by an ReLU activation layer. The triple marginal loss function[61] $\mathcal{L}(f_i, f_j, f_k)$ of the network was calculated by

$$\mathcal{L}(f_i, f_j, f_k) = \max\{\|f_i - f_j\|_2 - \|f_i - f_k\|_2 + \text{margin}, 0\}, \quad \text{(A.1)}$$

where f_i denotes the representation of cell i in low-dimensional latent space, i.e.,

$$f_x = f(a_x) \quad x \in \{i, j, k\}. \quad \text{(A.2)}$$

Here, a_i is the initial representation of the ith cell with the label of b_i. The other two cells j and k are randomly selected from cells with the same and different labels as the given ith cell, respectively, i.e., $b_i = b_j \wedge b_i \neq b_k$. For each dataset, five epochs were trained.

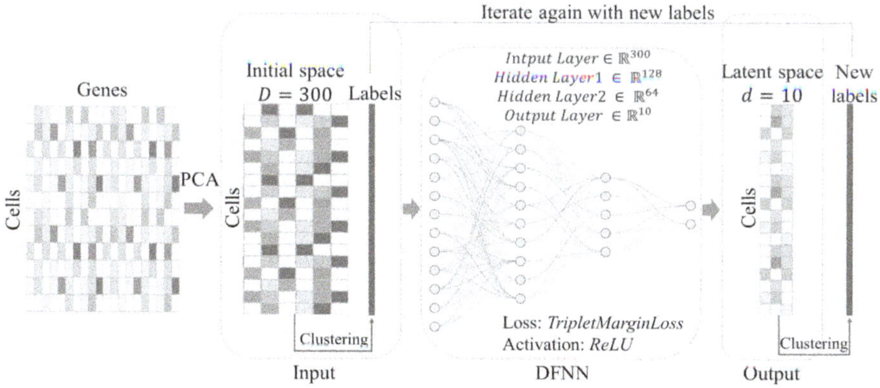

Fig. A.1. (Color online) Details of dimensionality reduction in this work.

Step 2: After the trained network, the cell data in latent space were generated, and then, the cells were clustered again to generate a fine clustering distribution, which was used as a new clustering label of cells.

Step 3: Furthermore, the cell data in the initial space together with the new clustering labels were fed into the neural network again to update the result of dimensionality reduction, and steps 1 and 2 were repeated until the extracted feature dimension was no longer reduced.

Step 4: After training, the cell-state latent space of the final cell i is f_i^*, which is obtained by $f^*(a_i)$, where f^* indicates that the DFNN is trained to the best condition on the scRNA-seq dataset. At the same time, each cell cluster labels in the optimized scRNA-seq dataset is also obtained. The latent space and the cluster labels are used in subsequent steps.

A.2. *Clustering metrics*

To assess the accuracy of clustering, the Adjusted Rand Index (ARI) and Normalized Mutual Information (NMI) metrics are commonly employed. Here, we utilized these metrics to help evaluate the degree of information loss in the cell-state latent space. The ARI can be calculated by[64,65]:

$$\mathrm{ARI} = \frac{\sum_{ij}\binom{n_{ij}}{2} - \left[\sum_i \binom{l_i}{2}\sum_j \binom{m_j}{2}\right]\Big/\binom{n}{2}}{\frac{1}{2}\left[\sum_i \binom{l_i}{2} + \sum_j \binom{m_j}{2}\right] - \left[\sum_i \binom{l_i}{2} + \sum_j \binom{m_j}{2}\right]\Big/\binom{n}{2}}, \tag{A.3}$$

where l_i, m_j and n_{ij} are values from the contingency table which is used to estimate the overlap between two clusterings. The NMI metric between two clusterings C_1 and C_2 can be formulated as follows:

$$\mathrm{NMI} = \frac{\mathbb{I}(C_1, C_2)}{\sqrt{\mathbb{H}(C_1)\mathbb{H}(C_2)}}, \tag{A.4}$$

Table A.1. Details of five clustering datasets in this work.

Dataset	Cells	Genes	Clusters	Source
Baron-human1	1937	16381	14	GSM2230757
Baron-human2	1724	16381	14	GSM2230758
Baron-human3	3605	16381	14	GSM2230759
Baron-human4	1303	16381	14	GSM2230760
Baron-mouse1	822	14878	13	GSM2230761

where $\mathbb{I}$ is the mutual information between the two clusterings C_1 and C_2 and $\mathbb{H}$ is the Shannon-Entropy of the clustering.

This experiment uses five different single-cell data sets with "gold standard" (deemed as true clusters) cluster labels assigned to each single cell.[66] Baron human1, Baron human2, Baron human3 and Baron human4 are four human pancreatic islet datasets, and Baron mouse1 are mouse pancreatic islets. Cells are barcoded by using the in-drop platform and 10X Genomics platform, respectively. Table A.1. lists the basic information of the five benchmark datasets.

Supplementary information

The Supplementary Information are available at: https://www.worldscientific.com/doi/suppl/10.1142/S2737416524400015.

Orcid

Changfeng Han https://orcid.org/0009-0002-5266-8101
Wenjie Cao https://orcid.org/0009-0002-7112-1013
Cheng Li https://orcid.org/0009-0008-3882-7069
Yanbing Guo https://orcid.org/0009-0001-8147-1853
Yuebin Wang https://orcid.org/0009-0004-3756-7390
Ya-Zhou Shi https://orcid.org/0000-0003-2579-3134
Ben-Gong Zhang https://orcid.org/0000-0002-2766-6502

References

1. Campbell, N. V.; Mickael, C.; Kumar, S.; Zhang, H.; Campbell, I. L.; Gillen, A. E.; Trentin, C. O.; Diener, K.; Gao, B.; Kheyfets, V. O.; Gu, S.; Kumar, R.; Phang, T.; Brown, R. D.; Graham, B. B.; Stenmark, K. R. Single-Cell RNA Sequencing and Binary Hierarchical Clustering Define Lung Interstitial Macrophage Heterogeneity in Response to Hypoxia. *Am. J. Physiol. Lung Cell Mol. Physiol.* **2022**, *323* (1), L58–L68, https://doi.org/10.1152/ajplung.00104.2022.
2. Poirion, O. B.; Zhu, X.; Ching, T.; Garmire, L. Single-Cell Transcriptomics Bioinformatics and Computational Challenges. *Front. Genet.* **2016**, *7*.
3. Stegle, O.; Teichmann, S. A.; Marioni, J. C. Computational and Analytical Challenges in Single-Cell Transcriptomics. *Nat. Rev. Genet.* **2015**, *16* (3), 133–145, https://doi.org/10.1038/nrg3833.
4. Yuan, G.-C.; Cai, L.; Elowitz, M.; Enver, T.; Fan, G.; Guo, G.; Irizarry, R.; Kharchenko, P.; Kim, J.; Orkin, S.; Quackenbush, J.; Saadatpour, A.; Schroeder, T.; Shivdasani, R.; Tirosh, I. Challenges and Emerging Directions in Single-Cell Analysis. *Genome Biol.* **2017**, *18* (1), 84, https://doi.org/10.1186/s13059-017-1218-y.
5. Kim, J.; T Jakobsen, S.; Natarajan, K. N.; Won, K.-J. TENET: Gene Network Reconstruction Using Transfer Entropy Reveals Key Regulatory Factors from Single Cell Transcriptomic Data. *Nucleic Acids Res* **2021**, *49* (1), e1, https://doi.org/10.1093/nar/gkaa1014.
6. Matsumoto, H.; Kiryu, H.; Furusawa, C.; Ko, M. S. H.; Ko, S. B. H.; Gouda, N.; Hayashi, T.; Nikaido, I. SCODE: An Efficient Regulatory Network Inference Algorithm from Single-Cell RNA-Seq during Differentiation. *Bioinformatics* **2017**, *33* (15), 2314–2321, https://doi.org/10.1093/bioinformatics/btx194.
7. Qiu, X.; Rahimzamani, A.; Wang, L.; Ren, B.; Mao, Q.; Durham, T.; McFaline-Figueroa, J. L.; Saunders, L.; Trapnell, C.; Kannan, S. Inferring Causal Gene Regulatory Networks from Coupled Single-Cell Expression Dynamics Using Scribe. *Cell Syst.* **2020**, *10* (3), 265–274.e11, https://doi.org/10.1016/j.cels.2020.02.003.
8. Bendall, S. C.; Davis, K. L.; Amir, E. D.; Tadmor, M. D.; Simonds, E. F.; Chen, T. J.; Shenfeld, D. K.; Nolan, G. P.; Pe'er, D. Single-Cell Trajectory Detection Uncovers Progression and Regulatory Coordination in Human B Cell Development. *Cell* **2014**, *157* (3), 714–725, https://doi.org/10.1016/j.cell.2014.04.005.
9. De Micheli, A. J.; Laurilliard, E. J.; Heinke, C. L.; Ravichandran, H.; Fraczek, P.; Soueid-Baumgarten, S.; De Vlaminck, I.; Elemento, O.; Cosgrove, B. D. Single-Cell Analysis of the Muscle Stem Cell Hierarchy Identifies Heterotypic Communication Signals Involved in Skeletal Muscle Regeneration. *Cell Rep.* **2020**, *30* (10), 3583–3595.e5, https://doi.org/10.1016/j.celrep.2020.02.067.
10. Fan, J.; Slowikowski, K.; Zhang, F. Single-Cell Transcriptomics in Cancer: Computational Challenges and Opportunities. *Exp. Mol. Med.* **2020**, *52* (9), 1452–1465, https://doi.org/10.1038/s12276-020-0422-0.

11. Gulati, G. S.; Sikandar, S. S.; Wesche, D. J.; Manjunath, A.; Bharadwaj, A.; Berger, M. J.; Ilagan, F.; Kuo, A. H.; Hsieh, R. W.; Cai, S.; Zabala, M.; Scheeren, F. A.; Lobo, N. A.; Qian, D.; Yu, F. B.; Dirbas, F. M.; Clarke, M. F.; Newman, A. M. Single-Cell Transcriptional Diversity Is a Hallmark of Developmental Potential. *Science* **2020**, *367* (6476), 405–411, https://doi.org/10.1126/science.aax0249.
12. Tran, T. N.; Bader, G. D. Tempora: Cell Trajectory Inference Using Time-Series Single-Cell RNA Sequencing Data. *PLOS Comput. Biol.* **2020**, *16* (9), e1008205, https://doi.org/10.1371/journal.pcbi.1008205.
13. Weinreb, C.; Rodriguez-Fraticelli, A.; Camargo, F. D.; Klein, A. M. Lineage Tracing on Transcriptional Landscapes Links State to Fate during Differentiation. *Science* **2020**, *367* (6479), eaaw3381, https://doi.org/10.1126/science.aaw3381.
14. Lan, T.; Hutvagner, G.; Zhang, X.; Liu, T.; Wong, L.; Li, J. Density-Based Detection of Cell Transition States to Construct Disparate and Bifurcating Trajectories. *Nucl. Acids Res.* **2022**, gkac785, https://doi.org/10.1093/nar/gkac785.
15. Saelens, W.; Cannoodt, R.; Todorov, H.; Saeys, Y. A Comparison of Single-Cell Trajectory Inference Methods. *Nat. Biotechnol.* **2019**, *37* (5), 547–554. https://doi.org/10.1038/s41587-019-0071-9.
16. Van den Berge, K.; Roux de Bézieux, H.; Street, K.; Saelens, W.; Cannoodt, R.; Saeys, Y.; Dudoit, S.; Clement, L. Trajectory-Based Differential Expression Analysis for Single-Cell Sequencing Data. *Nat. Commun.* **2020**, *11* (1), 1201. https://doi.org/10.1038/s41467-020-14766-3.
17. Wei, J.; Zhou, T.; Zhang, X.; Tian, T. DTFLOW: Inference and Visualization of Single-Cell Pseudotime Trajectory Using Diffusion Propagation. *Genom. Proteom. Bioinform.* **2021**, *19* (2), 306–318, https://doi.org/10.1016/j.gpb.2020.08.003.
18. Weng, G.; Kim, J.; Won, K. J. VeTra: A Tool for Trajectory Inference Based on RNA Velocity. *Bioinformatics* **2021**, btab364, https://doi.org/10.1093/bioinformatics/btab364.
19. Sagar; Grün, D. Deciphering Cell Fate Decision by Integrated Single-Cell Sequencing Analysis. *Annu. Rev. Biomed. Data Sci.* **2020**, *3*, 1–22. https://doi.org/10.1146/annurev-biodatasci-111419-091750.
20. Lange, M.; Bergen, V.; Klein, M.; Setty, M.; Reuter, B.; Bakhti, M.; Lickert, H.; Ansari, M.; Schniering, J.; Schiller, H. B.; Pe'er, D.; Theis, F. J. CellRank for Directed Single-Cell Fate Mapping. *Nat. Methods* **2022**, *19* (2), 159–170, https://doi.org/10.1038/s41592-021-01346-6.
21. Gao, M.; Qiao, C.; Huang, Y. UniTVelo: Temporally Unified RNA Velocity Reinforces Single-Cell Trajectory Inference. *Nat. Commun.* **2022**, *13* (1), 6586, https://doi.org/10.1038/s41467-022-34188-7.
22. Liu, R.; Pisco, A. O.; Braun, E.; Linnarsson, S.; Zou, J. Dynamical Systems Model of RNA Velocity Improves Inference of Single-Cell Trajectory, Pseudo-Time and Gene Regulation. *J. Molecul. Biol.* **2022**, 167606, https://doi.org/10.1016/j.jmb.2022.167606.

23. Cannoodt, R.; Saelens, W.; Deconinck, L.; Saeys, Y. Spearheading Future Omics Analyses Using Dyngen, a Multi-Modal Simulator of Single Cells. *Nat. Commun.* **2021**, *12* (1), 3942, https://doi.org/10.1038/s41467-021-24152-2.
24. Setty, M.; Tadmor, M. D.; Reich-Zeliger, S.; Angel, O.; Salame, T. M.; Kathail, P.; Choi, K.; Bendall, S.; Friedman, N.; Pe'er, D. Wishbone Identifies Bifurcating Developmental Trajectories from Single-Cell Data. *Nat. Biotechnol.* **2016**, *34* (6), 637–645, https://doi.org/10.1038/nbt.3569.
25. Welch, J. D.; Hartemink, A. J.; Prins, J. F. SLICER: Inferring Branched, Nonlinear Cellular Trajectories from Single Cell RNA-Seq Data. *Genome Biol.* **2016**, *17*. https://doi.org/10.1186/s13059-016-0975-3.
26. Gan, Y.; Li, N.; Guo, C.; Zou, G.; Guan, J.; Zhou, S. TiC2D: Trajectory Inference from Single-Cell RNA-Seq Data Using Consensus Clustering. *IEEE/ACM Trans. Comput. Biol. Bioinform.* **2022**, *19* (4), 2512–2522, https://doi.org/10.1109/TCBB.2021.3061720.
27. Ji, Z.; Ji, H. TSCAN: Pseudo-Time Reconstruction and Evaluation in Single-Cell RNA-Seq Analysis. *Nucl. Acids Res.* **2016**, *44* (13), e117, https://doi.org/10.1093/nar/gkw430.
28. Costa, F.; Grün, D.; Backofen, R. GraphDDP: A Graph-Embedding Approach to Detect Differentiation Pathways in Single-Cell-Data Using Prior Class Knowledge. *Nat. Commun.* **2018**, *9* (1), 3685, https://doi.org/10.1038/s41467-018-05988-7.
29. Chen, Z.; An, S.; Bai, X.; Gong, F.; Ma, L.; Wan, L. DensityPath: An Algorithm to Visualize and Reconstruct Cell State-Transition Path on Density Landscape for Single-Cell RNA Sequencing Data. *Bioinformatics* **2019**, *35* (15), 2593–2601, https://doi.org/10.1093/bioinformatics/bty1009.
30. Wang, S.; Karikomi, M.; MacLean, A. L.; Nie, Q. Cell Lineage and Communication Network Inference via Optimization for Single-Cell Transcriptomics. *Nucl. Acids Res.* **2019**, *47* (11), e66, https://doi.org/10.1093/nar/gkz204.
31. Anchang, B.; Mendez-Giraldez, R.; Xu, X.; Archer, T. K.; Chen, Q.; Hu, G.; Plevritis, S. K.; Motsinger-Reif, A. A.; Li, J.-L. Visualization, Benchmarking and Characterization of Nested Single-Cell Heterogeneity as Dynamic Forest Mixtures. *Brief Bioinform.* **2022**, *23* (2), bbac017. https://doi.org/10.1093/bib/bbac017.
32. Shin, J.; Berg, D. A.; Zhu, Y.; Shin, J. Y.; Song, J.; Bonaguidi, M. A.; Enikolopov, G.; Nauen, D. W.; Christian, K. M.; Ming, G.; Song, H. Single-Cell RNA-Seq with Waterfall Reveals Molecular Cascades Underlying Adult Neurogenesis. *Cell Stem Cell* **2015**, *17* (3), 360–372, https://doi.org/10.1016/j.stem.2015.07.013.
33. Wei, J.; Zhou, T.; Zhang, X.; Tian, T. SCOUT: A New Algorithm for the Inference of Pseudo-Time Trajectory Using Single-Cell Data. *Comput. Biol. Chem.* **2019**, *80*, 111–120, https://doi.org/10.1016/j.compbiolchem.2019.03.013.
34. Wolf, F. A.; Hamey, F. K.; Plass, M.; Solana, J.; Dahlin, J. S.; Göttgens, B.; Rajewsky, N.; Simon, L.; Theis, F. J. PAGA: Graph Abstraction Reconciles Clustering with Trajectory Inference through a Topology Preserving Map of

Single Cells. *Genome Biol.* **2019**, *20* (1), 59, https://doi.org/10.1186/s13059-019-1663-x.
35. Pandey, K.; Zafar, H. Inference of Cell State Transitions and Cell Fate Plasticity from Single-Cell with MARGARET. *Nucl. Acids Res.* **2022**, gkac412. https://doi.org/10.1093/nar/gkac412.
36. Arisdakessian, C.; Poirion, O.; Yunits, B.; Zhu, X.; Garmire, L. X. DeepImpute: An Accurate, Fast, and Scalable Deep Neural Network Method to Impute Single-Cell RNA-Seq Data. *Genome Biol.* **2019**, *20* (1), 211, https://doi.org/10.1186/s13059-019-1837-6.
37. Eraslan, G.; Simon, L. M.; Mircea, M.; Mueller, N. S.; Theis, F. J. Single-Cell RNA-Seq Denoising Using a Deep Count Autoencoder. *Nat. Commun.* **2019**, *10* (1), 390, https://doi.org/10.1038/s41467-018-07931-2.
38. Xu, C.; Cai, L.; Gao, J. An Efficient ScRNA-Seq Dropout Imputation Method Using Graph Attention Network. *BMC Bioinform.* **2021**, *22* (1), 582, https://doi.org/10.1186/s12859-021-04493-x.
39. Zhang, Z.; Yang, C.; Zhang, X. ScDART: Integrating Unmatched ScRNA-Seq and ScATAC-Seq Data and Learning Cross-Modality Relationship Simultaneously. *Genome Biol.* **2022**, *23* (1), 139, https://doi.org/10.1186/s13059-022-02706-x.
40. Chen, Y.; Zhang, Y.; Li, J. Y. H.; Ouyang, Z. LISA2: Learning Complex Single-Cell Trajectory and Expression Trends. *Front. Genet.* **2021**, *12*, 681206, https://doi.org/10.3389/fgene.2021.681206.
41. Chen, Y.; Zhang, Y.; Ouyang, Z. LISA: Accurate Reconstruction of Cell Trajectory and Pseudo-Time for Massive Single Cell RNA-Seq Data. *Pac. Symp. Biocomput* 2019, 24, 338–349.
42. Jin, K.; Schnell, D.; Li, G.; Salomonis, N.; Prasath, V. B. S.; Szczesniak, R.; Aronow, B. J. CellDrift: Inferring Perturbation Responses in Temporally Sampled Single-Cell Data. *Brief Bioinform.* **2022**, *23* (5), bbac324. https://doi.org/10.1093/bib/bbac324.
43. Song, D.; Li, J. J. PseudotimeDE: Inference of Differential Gene Expression along Cell Pseudotime with Well-Calibrated p-Values from Single-Cell RNA Sequencing Data. *Genome Biol.* **2021**, *22* (1), 124, https://doi.org/10.1186/s13059-021-02341-y.
44. Song, Q.; Wang, J.; Bar-Joseph, Z. ScSTEM: Clustering Pseudotime Ordered Single-Cell Data. *Genome Biol.* **2022**, *23* (1), 150, https://doi.org/10.1186/s13059-022-02716-9.
45. Cannoodt, R.; Saelens, W.; Sichien, D.; Tavernier, S.; Janssens, S.; Guilliams, M.; Lambrecht, B.; Preter, K. D.; Saeys, Y. SCORPIUS Improves Trajectory Inference and Identifies Novel Modules in Dendritic Cell Developmen. Biorxiv. 2016: 079509. https://doi.org/10.1101/079509.
46. Chen, H.; Albergante, L.; Hsu, J. Y.; Lareau, C. A.; Lo Bosco, G.; Guan, J.; Zhou, S.; Gorban, A. N.; Bauer, D. E.; Aryee, M. J.; Langenau, D. M.; Zinovyev, A.; Buenrostro, J. D.; Yuan, G.-C.; Pinello, L. Single-Cell Trajectories Reconstruction, Exploration and Mapping of Omics Data with STREAM. *Nat. Commun.* **2019**, *10* (1), 1903, https://doi.org/10.1038/s41467-019-09670-4.

47. Fischer, D. S.; Fiedler, A. K.; Kernfeld, E. M.; Genga, R. M. J.; Bastidas-Ponce, A.; Bakhti, M.; Lickert, H.; Hasenauer, J.; Maehr, R.; Theis, F. J. Inferring Population Dynamics from Single-Cell RNA-Sequencing Time Series Data. *Nat. Biotechnol.* **2019**, *37* (4), 461–468, https://doi.org/10.1038/s41587-019-0088-0.
48. Lönnberg, T.; Svensson, V.; James, K. R.; Fernandez-Ruiz, D.; Sebina, I.; Montandon, R.; Soon, M. S. F.; Fogg, L. G.; Nair, A. S.; Liligeto, U. N.; Stubbington, M. J. T.; Ly, L.-H.; Bagger, F. O.; Zwiessele, M.; Lawrence, N. D.; Souza-Fonseca-Guimaraes, F.; Bunn, P. T.; Engwerda, C. R.; Heath, W. R.; Billker, O.; Stegle, O.; Haque, A.; Teichmann, S. A. Single-Cell RNA-Seq and Computational Analysis Using Temporal Mixture Modeling Resolves TH1/TFH Fate Bifurcation in Malaria. *Sci. Immunol.* **2017**, *2* (9), eaal2192, https://doi.org/10.1126/sciimmunol.aal2192.
49. Matsumoto, H.; Kiryu, H. SCOUP: A Probabilistic Model Based on the Ornstein–Uhlenbeck Process to Analyze Single-Cell Expression Data during Differentiation. *BMC Bioinform.* **2016**, *17*, https://doi.org/10.1186/s12859-016-1109-3.
50. Reid, J. E.; Wernisch, L. Pseudotime Estimation: Deconfounding Single Cell Time Series. *Bioinformatics* **2016**, *32* (19), 2973–2980, https://doi.org/10.1093/bioinformatics/btw372.
51. Setty, M.; Kiseliovas, V.; Levine, J.; Gayoso, A.; Mazutis, L.; Pe'er, D. Characterization of Cell Fate Probabilities in Single-Cell Data with Palantir. *Nat. Biotechnol.* **2019**, *37* (4), 451–460, https://doi.org/10.1038/s41587-019-0068-4.
52. Street, K.; Risso, D.; Fletcher, R. B.; Das, D.; Ngai, J.; Yosef, N.; Purdom, E.; Dudoit, S. Slingshot: Cell Lineage and Pseudotime Inference for Single-Cell Transcriptomics. *BMC Genom.* **2018**, *19* (1), 477, https://doi.org/10.1186/s12864-018-4772-0.
53. Wang, D.; Gu, J. VASC: Dimension Reduction and Visualization of Single-Cell RNA-Seq Data by Deep Variational Autoencoder. *Genom. Proteom. Bioinform.* **2018**, *16* (5), 320–331, https://doi.org/10.1016/j.gpb.2018.08.003.
54. Weinreb, C.; Wolock, S.; Tusi, B. K.; Socolovsky, M.; Klein, A. M. Fundamental Limits on Dynamic Inference from Single-Cell Snapshots. *Proc. Natl. Acad. Sci.* **2018**, *115* (10), E2467–E2476, https://doi.org/10.1073/pnas.1714723115.
55. Angerer, P.; Haghverdi, L.; Büttner, M.; Theis, F. J.; Marr, C.; Buettner, F. Destiny: Diffusion Maps for Large-Scale Single-Cell Data in R. *Bioinformatics* **2016**, *32* (8), 1241–1243, https://doi.org/10.1093/bioinformatics/btv715.
56. Haghverdi, L.; Büttner, M.; Wolf, F. A.; Buettner, F.; Theis, F. J. Diffusion Pseudotime Robustly Reconstructs Lineage Branching. *Nat. Methods* **2016**, *13* (10), 845–848, https://doi.org/10.1038/nmeth.3971.
57. Haghverdi, L.; Buettner, F.; Theis, F. J. Diffusion Maps for High-Dimensional Single-Cell Analysis of Differentiation Data. *Bioinformatics* **2015**, *31* (18), 2989–2998, https://doi.org/10.1093/bioinformatics/btv325.
58. van Dijk, D.; Sharma, R.; Nainys, J.; Yim, K.; Kathail, P.; Carr, A. J.; Burdziak, C.; Moon, K. R.; Chaffer, C. L.; Pattabiraman, D.; Bierie, B.; Mazutis, L.; Wolf, G.; Krishnaswamy, S.; Pe'er, D. Recovering Gene Interactions from Single-Cell

Data Using Data Diffusion. *Cell* **2018**, *174* (3), 716–729.e27, https://doi.org/10.1016/j.cell.2018.05.061.

59. Gan, Y.; Guo, C.; Guo, W.; Xu, G.; Zou, G. Entropy-Based Inference of Transition States and Cellular Trajectory for Single-Cell Transcriptomics. *Brief. Bioinform.* **2022**, *23* (4), bbac225, https://doi.org/10.1093/bib/bbac225.
60. He, P.; Lim, K.; Sun, D.; Pett, J. P.; Jeng, Q.; Polanski, K.; Dong, Z.; Bolt, L.; Richardson, L.; Mamanova, L.; Dabrowska, M.; Wilbrey-Clark, A.; Madissoon, E.; Tuong, Z. K.; Dann, E.; Suo, C.; Goh, I.; Yoshida, M.; Nikolić, M. Z.; Janes, S. M.; He, X.; Barker, R. A.; Teichmann, S. A.; Marioni, J. C.; Meyer, K. B.; Rawlins, E. L. A Human Fetal Lung Cell Atlas Uncovers Proximal-Distal Gradients of Differentiation and Key Regulators of Epithelial Fates. *Cell* **2022**, *185* (25), 4841–4860.e25, https://doi.org/10.1016/j.cell.2022.11.005.
61. Balntas, V.; Riba, E.; Ponsa, D.; Mikolajczyk, K. Learning Local Feature Descriptors with Triplets and Shallow Convolutional Neural Networks. In *British Machine Vision Conference*, Vol. 1, 2016, p. 3.
62. Traag, V. A.; Waltman, L.; van Eck, N. J. From Louvain to Leiden: Guaranteeing Well-Connected Communities. *Sci. Rep.* **2019**, *9* (1), 5233, https://doi.org/10.1038/s41598-019-41695-z.
63. Rohrdanz, M. A.; Zheng, W.; Maggioni, M.; Clementi, C. Determination of Reaction Coordinates via Locally Scaled Diffusion Map. *J. Chem. Phys.* **2011**, *134* (12), 124116, https://doi.org/10.1063/1.3569857.
64. Santos, J. M.; Embrechts, M. On the Use of the Adjusted Rand Index as a Metric for Evaluating Supervised Classification. In *Artificial Neural Networks — ICANN 2009*, Eds., Alippi, C., Polycarpou, M., Panayiotou, C., Ellinas, G., Lecture Notes in Computer Science; Springer: Berlin, Heidelberg, 2009, pp 175–184. https://doi.org/10.1007/978-3-642-04277-5_18.
65. Modat, M.; Vercauteren, T.; Ridgway, G. R.; Hawkes, D. J.; Fox, N. C.; Ourselin, S. Diffeomorphic Demons Using Normalized Mutual Information, Evaluation on Multimodal Brain MR Images. In *Medical Imaging 2010: Image Processing*; SPIE, Vol. 7623, 2010, pp 800–807, https://doi.org/10.1117/12.843962.
66. Cao, W.; Shi, Y.-Z.; Qiu, H.; Zhang, B. SS-RNN: A Strengthened Skip Algorithm for Data Classification Based on Recurrent Neural Networks. *Front. Genet.* **2021**, *12*.
67. Guo, G.; Huss, M.; Tong, G. Q.; Wang, C.; Li Sun, L.; Clarke, N. D.; Robson, P. Resolution of Cell Fate Decisions Revealed by Single-Cell Gene Expression Analysis from Zygote to Blastocyst. *Develop. Cell* **2010**, *18* (4), 675–685, https://doi.org/10.1016/j.devcel.2010.02.012.
68. Cao, S.; Feng, H.; Yi, H.; Pan, M.; Lin, L.; Zhang, Y. S.; Feng, Z.; Liang, W.; Cai, B.; Li, Q.; Xiong, Z.; Shen, Q.; Ke, M.; Zhao, X.; Chen, H.; He, Q.; Min, M.; Cai, Q.; Liu, H.; Wang, J.; Pei, D.; Chen, J.; Ma, Y. Single-Cell RNA Sequencing Reveals the Developmental Program Underlying Proximal-Distal Patterning of the Human Lung at the Embryonic Stage. *Cell Res.* **2023**, *33* (6), 421–433, https://doi.org/10.1038/s41422-023-00802-6.
69. Lim, K.; Donovan, A. P. A.; Tang, W.; Sun, D.; He, P.; Pett, J. P.; Teichmann, S. A.; Marioni, J. C.; Meyer, K. B.; Brand, A. H.; Rawlins, E. L. Organoid

Modeling of Human Fetal Lung Alveolar Development Reveals Mechanisms of Cell Fate Patterning and Neonatal Respiratory Disease. *Cell Stem Cell* **2023**, *30* (1), 20–39.e9, https://doi.org/10.1016/j.stem.2022.11.013.

70. Li, J.; Tang, N. Empowering Human Lung Development. *Cell Stem Cell* **2023**, *30* (1), 5–6, https://doi.org/10.1016/j.stem.2022.12.012.
71. Blondel, V. D.; Guillaume, J.-L.; Lambiotte, R.; Lefebvre, E. Fast Unfolding of Communities in Large Networks. *J. Stat. Mech.* **2008**, *2008* (10), P10008, https://doi.org/10.1088/1742-5468/2008/10/P10008.
72. Fortunato, S.; Barthélemy, M. Resolution Limit in Community Detection. *Proc. Natl. Acad. Sci.* **2007**, *104* (1), 36–41, https://doi.org/10.1073/pnas.0605965104.
73. Ikotun, A. M.; Ezugwu, A. E.; Abualigah, L.; Abuhaija, B.; Heming, J. K-Means Clustering Algorithms: A Comprehensive Review, Variants Analysis, and Advances in the Era of Big Data. *Inform. Sci.* **2023**, *622*, 178–210, https://doi.org/10.1016/j.ins.2022.11.139.
74. MacQueen, J. Some Methods for Classification and Analysis of Multivariate Observations. *Proceedings of the fifth Berkeley symposium on mathematical statistics and probability.* 1967, 1(14), 281–297.

https://doi.org/10.1142/9789819808199_0011

Chapter 11

Inferring Transcriptional Bursting Kinetics Using Gene Expression Model with Memory and Crosstalk from scRNA-seq Data*

Mengyuan Wang [†], Wenjie Cao [‡], Yanbing Guo [§], Guilin Wang [¶], Jian Jiang [¶], Huahai Qiu [¶,||,††] and Ben-Gong Zhang [¶,**,††]

[†]*School of Mathematical & Physical Sciences, Wuhan Textile University, Wuhan 430200, P. R. China*
[‡]*School of Mathematics, Sun Yat-sen University, Guangzhou 510275, P. R. China*
[§]*School of Computer Science and Artificial Intelligence, Wuhan Textile University, Wuhan 430200, P. R. China*
[¶]*School of Mathematical & Physical Sciences, Wuhan Textile University, Wuhan 430200, P. R. China*
[||]*qiuhuahai2006@163.com*
[**]*bgzhang@wtu.edu.cn*

Genetically identical cell populations growing in the same environment show a large degree of heterogeneity in gene expression profiles, resulting in significant phenotypic consequences. One of the important sources of this heterogeneity is transcriptional bursting. In recent years, single cell transcriptome sequencing (scRNA-seq) data have been widely used to infer the kinetics of transcriptional bursting. Moreover, the increasing

*To cite this article, please refer to its earlier version published in the *J. Comput. Biophys. Chem.* **2024**, *23* (6), 765–779, DOI: 10.1142/S2737416524400040
[††]Corresponding author.

evidence showed that genes are jointly regulated by several competitive pathways when activated by external signals, and there is molecular memory in the process of gene state switching. Therefore, a gene expression model considering both gene activation pathway and molecular memory can better reflect the nature of transcriptional bursting. In this paper, we used an approximate Bayesian computation algorithm called ABC-PRC combined with a gene expression model that considered crosstalk and memory, to infer the kinetics of transcriptional bursting. We analyze the internal relationship between transcriptional bursting and gene expression using scRNA-seq data obtained from mouse embryonic stem cells and mouse embryonic fibroblasts. Both model analysis and data simulations demonstrate that the bimodal coefficient increases with the increase of burst size, and the noise intensity decreases with the increase of burst frequency. Additionally, we show that some previously proposed models can be simplified as special cases of the gene expression model and inference algorithm proposed here under certain circumstances.

Keywords: Transcriptional bursting; single-cell data; gene expression model; parameter inference.

1. Introduction

In most biological systems, isogenic organisms and cells growing in a common environment exhibit significant phenotypic differences, including the expression of individual genes.[1,2] This difference stems from gene expression variability, which produces phenotypic diversity to help organisms adapt to an unpredictable environment.[3,4] Transcriptional bursting is not only a fundamental mode of gene expression process but also a pivotal source of gene expression variability. Revealing the features of transcriptional bursting to understand the origins of phenotypic variation is a basic goal of biology.[5] In previous studies, scientists usually quantify mRNA production on time series by real-time quantitative PCR to analyze the dynamics of transcriptional bursting, or directly observe the number and distribution of specific mRNA molecules in cells by fluorescence *in situ* hybridization (FISH), so as to study the dynamic changes in the process of transcription. Mathematicians use mathematical models to simulate and understand the dynamic behavior of transcriptional bursting, and explore parameters to understand the dynamic characteristics of transcriptional bursts.[6–9] Many theoretical models have been proposed to explain the phenomenon of transcriptional bursting. For example, according to the classical telegraph model (CTM), the gene switches between ON states and OFF states and transcribes only in ON states. The switches between ON states and OFF states are Markovian (memoryless), meaning that the waiting time between states is distributed exponentially.[10] The stochastic telegraph model is widely used because of its simplicity. In recent years, more and more experimental data have shown that gene transcription is a complex multi-step biochemical process that can produce memory between different reaction events, this means that the traditional memoryless two-state model is not enough to characterize complex transcriptional mechanisms.[11–17] Therefore, the researchers proposed the polymorphism model to represent a more real transcription process. This polymorphism not only provides a convenient way to maintain Markov properties, but also has the opportunity to reveal the details of the underlying biological mechanism behind promoter switching.[18–20] More specifically, the expressions of most genes are regulated in a complex way, involving several repressors, transcription factors and mediators, as well as chromatin remodeling or superhelix changes. These processes can be modeled by non-exponential waiting time distributions.[21] Researchers have developed a generalized telegraph model (GTM) based on the CTM, which uses the non-exponential waiting time distribution between state transitions to simulate molecular memory.[22]

With the development of single-cell transcriptome sequencing technology (scRNA-seq), scientists began to infer the dynamic transcriptional bursting process from single-cell transcriptome static snapshots data. In the beginning, the simple CTM was used to first infer genome-wide transcriptional bursting kinetics from scRNA-seq data, and then reveal the characteristics and regulatory mechanisms of burst kinetics.[23–27] The characteristics of transcriptional bursting are described by burst size (BS), burst frequency (BF), average ON states dwell time and average OFF states dwell time. The regulatory mechanisms include static promoter structure, dynamic E-P communication (enhancer-promoter communication), histone modification and other molecular factors.[23,28–30] In recent years, Luo *et al.* have developed a statistical inference framework to infer the kinetics of transcriptional bursting by combining scRNA-seq data with non-Markov GTM, and analyze how molecular memory affects the inference of BS and frequency.[22] However, there is a potential assumption in the above gene expression model that gene is activated through a single pathway. But experiments have shown that many inducible genes for stress response, stem cell renewal, development and immunity, when activated by a signal, are often co-regulated by several competitive pathways, and there are multi-step intermediate reactions in each pathway. That is, there is molecular memory. Therefore, the gene expression model which integrates the two crosstalk signal pathways and molecular memory into the process of gene state switches can more truly reflect the process of transcriptional bursting.[31,32] In 2021, Zhang *et al.* introduced a modified version of the random telegraph model in Ref. 33, which takes into account both gene activation pathways and molecular memory.[33] They solved the model analytically and derived the arbitrarily high-order steady-state binomial moments for mRNAs. However, they did not consider applying this model to the inference of transcriptional bursting kinetics.

For further study, we present an inference procedure to obtain the characteristics of transcriptional bursting kinetics from scRNA-seq data by combining a gene expression model that simultaneously considers both crosstalk and molecular memory, as shown in Fig. 1. We use a parameter inference algorithm called approximate Bayesian computation-partial rejection control (ABC-PRC). Then we apply the algorithm to the scRNA-seq data produced by mouse embryonic fibroblasts (MEF) and mouse embryonic stem cells (ESC) to analyze the effect of transcriptional bursting on gene expression. We find that both model analysis and data simulations show that the bimodal coefficient increases with the increase of BS, and the noise intensity decreases with the increase of BF. Additionally, some previously proposed models can be simplified as special cases of the gene

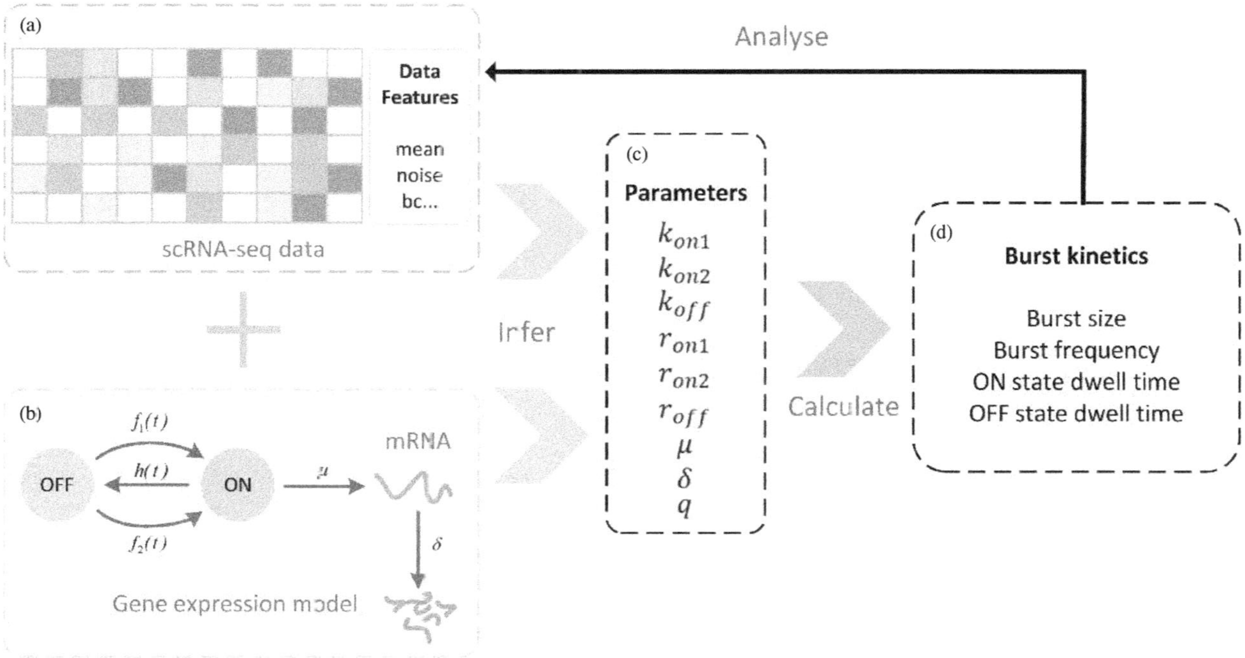

Fig. 1. (Color online) A schematic diagram illustrating the inference of transcriptional bursting kinetics from scRNA-seq data. (a) scRNA-seq data, it contains the distribution characteristics of gene expression level, such as mean, noise, bimodal coefficient, etc. (b) The gene expression model considers memory and crosstalk simultaneously and the gene is activated by one of the activation pathways with a certain probability and transcribed in active states. The molecular memory is simulated using non-exponential waiting time distribution $f_1(t)$, $f_1(t)$ and $h(t)$. (c) The model parameters, these parameters are obtained through the inference algorithm. (d) The kinetics characteristics of transcriptional bursting, including BS, BF, average ON states dwell time and average OFF states dwell time.

expression model and inference algorithm proposed here under certain circumstances.

2. Material and methods

2.1. *Model description*

Here, we employ a gene expression model that incorporates both molecular memory and crosstalk. The schematic diagram of the model is shown in Fig. 1(b). The model assumes that the gene switches randomly between active and inactive states. The gene is activated by the first pathway selected with a constant probability q, the second pathway is selected with probability $1-q$. The gene is inactivated by a single pathway. The waiting time of genes in the three pathways is represented by random variables τ_{off1}, τ_{off2}, τ_{on}, that follow the distribution function $f_1(t)$, $f_2(t)$, $h(t)$, respectively. Because the promoter states transition involves a series of intermediate reaction steps, molecular memory is created in between state switches. Here, to simulate the presence of molecular memory between gene state switches, we use the gamma waiting time distribution instead of the exponential waiting time distribution used in the CTM.[34]

The activation pathway:

$$f_1(t) = \frac{r_{\text{on1}}^{\text{kon1}} t^{k_{\text{on1}}-1} e^{-r_{\text{on1}} t}}{\Gamma(k_{\text{on1}})}, \tag{1}$$

$$f_2(t) = \frac{r_{\text{on2}}^{\text{kon2}} t^{k_{\text{on2}}-1} e^{-r_{\text{on2}} t}}{\Gamma(k_{\text{on2}})}. \tag{2}$$

The inactivation pathway:

$$h(t) = \frac{r_{\text{off}}^{\text{koff}} t^{k_{\text{off}}-1} e^{-r_{\text{off}} t}}{\Gamma(k_{\text{off}})}, \tag{3}$$

where $r_S (S \in \{\text{on1}, \text{on2}, \text{off}\})$ is the rate parameter of gene switching from OFF state to ON state (from ON state to OFF state), and $k_S (S \in \{\text{on1}, \text{on2}, \text{off}\})$ is the number of intermediate reaction steps of gene switching from OFF state to ON state (from ON state to OFF state), also known as memory index. In addition, the parameter μ is used to represent the synthesis rate of mRNA and δ is the degradation rate of mRNA. These constant rates mean that the model assumes that the synthesis and degradation of mRNA is Markov.

In this model, it is known that when the value of q is 1, the model reduces to a GTM without crosstalk. And when k_S is also equal to 1, the model simplified to a CTM without crosstalk and memory. In this manuscript, our focus is on the case where q and k_S not equal 1.

The kinetic characteristics of transcriptional bursting are commonly described by BS, BF, average ON state dwell time and average OFF state dwell time, as shown in Fig. 2. The average cycle time is defined as the sum of the average OFF state dwell time and the average ON state dwell time. BS represents the average number of

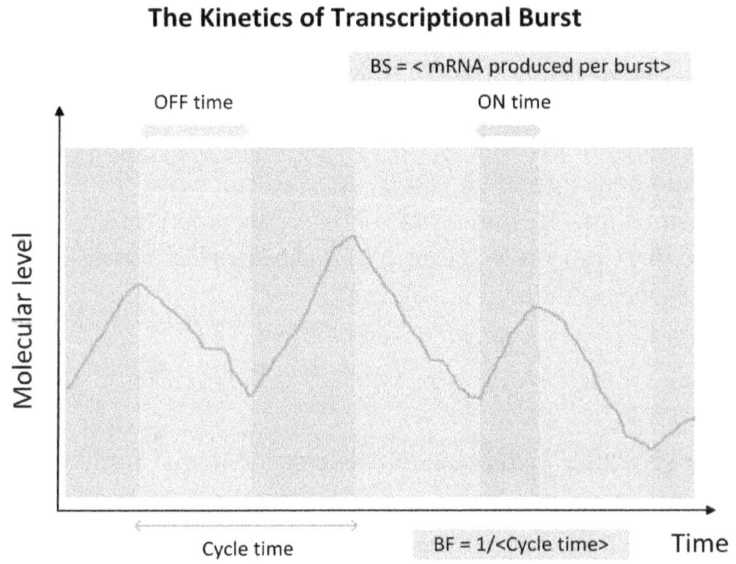

Fig. 2. (Color online) A schematic diagram of transcriptional burst kinetics. The blue curve illustrates the fluctuation of mRNA molecular level over time, while the blue shadow represents the waiting time of the gene in the OFF state and the pink shadow represents the waiting time of the gene in the ON state. The cycle time is defined as the sum of ON state time and OFF state time within a burst. The BS is defined as the average number of mRNA molecules produced per burst, and the BF indicates the number of bursts per unit time.

mRNA molecules produced per burst, and the BF represents the average number of bursts per unit time, that is, the reciprocal of the average cycle time. Make the average dwell time of the OFF state as $\langle\tau_{\text{off}}\rangle = q\langle\tau_{\text{off1}}\rangle + (1-q)\langle\tau_{\text{off2}}\rangle$, the formulas for BS and frequency are as follows:

$$\text{BS} = \mu\langle\tau_{\text{on}}\rangle, \tag{4}$$

$$\text{BF} = \frac{1}{\langle\tau_{\text{off}}\rangle + \langle\tau_{\text{on}}\rangle}, \tag{5}$$

where the symbol $\langle\rangle$ indicates expectation value. And based on the given distribution $f_1(t)$, $f_2(t)$, $h(t)$, the·expectation of τ_{off1}, τ_{off2}, τ_{on} is $\frac{k_{\text{on1}}}{r_{\text{on1}}}$, $\frac{k_{\text{on2}}}{r_{\text{on2}}}$, $\frac{k_{\text{on}}}{r_{\text{off}}}$, respectively.

Theoretically, the master equation can capture the random information of the gene expression biochemical system. But in general, it is difficult to obtain the analytical solution, and the numerical simulation is very time-consuming. Therefore, it is a meaningful topic to develop the theoretical analysis method and numerical solution method of the master equation. Among them, the binomial moment approach can be used to calculate the steady-state distribution of mRNA and its statistical characteristics.[35]

For the distribution function $P(n)$, the K-order binomial moment (denoted by b_k) is defined as follows:

$$b_k = \sum_{n \geq k} \binom{n}{k} P(n), \quad k = 0, 1, 2, \ldots, \tag{6}$$

where the symbol $\binom{n}{k}$ represents the combinatorial number.

Binomial moments have good properties, such as: binomial moments can be reconstructed by the following formula:

$$P(n) = \sum_{k \geq n} \binom{k}{n} b_k, \quad k = 0, 1, 2, \ldots, \tag{7}$$

at the same time, when the order of the binomial moment tends to infinite, the binomial moment converges to 0. Moreover, the binomial moments can be expressed as a linear combination of the raw moments or central moments. The characteristic statistics concerned in this paper can be given by binomial moments.

Motivated by Ref. 33, we obtain the first three binomial moments of the model:

$$b_1 = \frac{\mu\langle\tau_{\text{on}}\rangle}{\langle\tau_{\text{on}}\rangle + \langle\tau_{\text{off}}\rangle}, \tag{8}$$

$$b_2 = \frac{\langle\tau_{\text{on}}\rangle}{\langle\tau_{\text{on}}\rangle + \langle\tau_{\text{off}}\rangle}\frac{\mu^2}{2} - \frac{\nu_1\tilde{H}(1)}{\langle\tau_{\text{on}}\rangle + \langle\tau_{\text{off}}\rangle}\frac{\mu^2}{2}, \tag{9}$$

$$b_3 = \frac{c\mu^3}{6}\langle\tau_{\text{on}}\rangle - \frac{c\nu_1\mu^3}{6}\tilde{H}(1) - \frac{c\nu_2\mu^3}{6}\tilde{H}(2) + \frac{c\nu_1\nu_2\mu^3}{6}[\tilde{h}(1) - \tilde{h}(2)], \tag{10}$$

where $c = \frac{1}{\langle\tau_{\text{on}}\rangle + \langle\tau_{\text{off}}\rangle}$, $\nu_k = \frac{1-\tilde{f}(k)}{1-\tilde{f}(k)\tilde{h}(k)}$, $\tilde{f}(k) = q\tilde{f}_1(k) + (1-q)\tilde{f}_2(k)$, $k = 1, 2, \ldots$, $H(x) = \int_x^\infty h(t)dt$, is the survival function of the ON state. The wavy line represents the Laplace transform of the function.[33]

By using the first third-order binomial moments, we can easily obtain the mean, variance, noise intensity (coefficient of variation), Fano factor and other features of the mRNA distribution.

In fact, the gene expression model used in this paper can be simplified to the CTM and the GTM when we set the memory index k_{on} and k_{off} of the inactivation path to 1 and fix the probability q of choosing the first activation path to 1. Here we set k_{off} and q equal to 1. This simplification means that our model is reduced to a two-state model without crosstalk, where molecular memory only exists in the activation pathway. Then we analyze how the presence of molecular memory affects gene expression. We simulate the effect of molecular memory on gene expression using the stochastic simulation algorithm, and the results are shown in Fig. 3. We observe that, except for the mean of gene expression level, all other indicators (noise strength, Fano factor, skewness, kurtosis, bimodal coefficient) increase with the increase of the memory index which is the same with the results of Luo *et al.* And the results also imply that the summary statistics are sensitive to the memory index.[22]

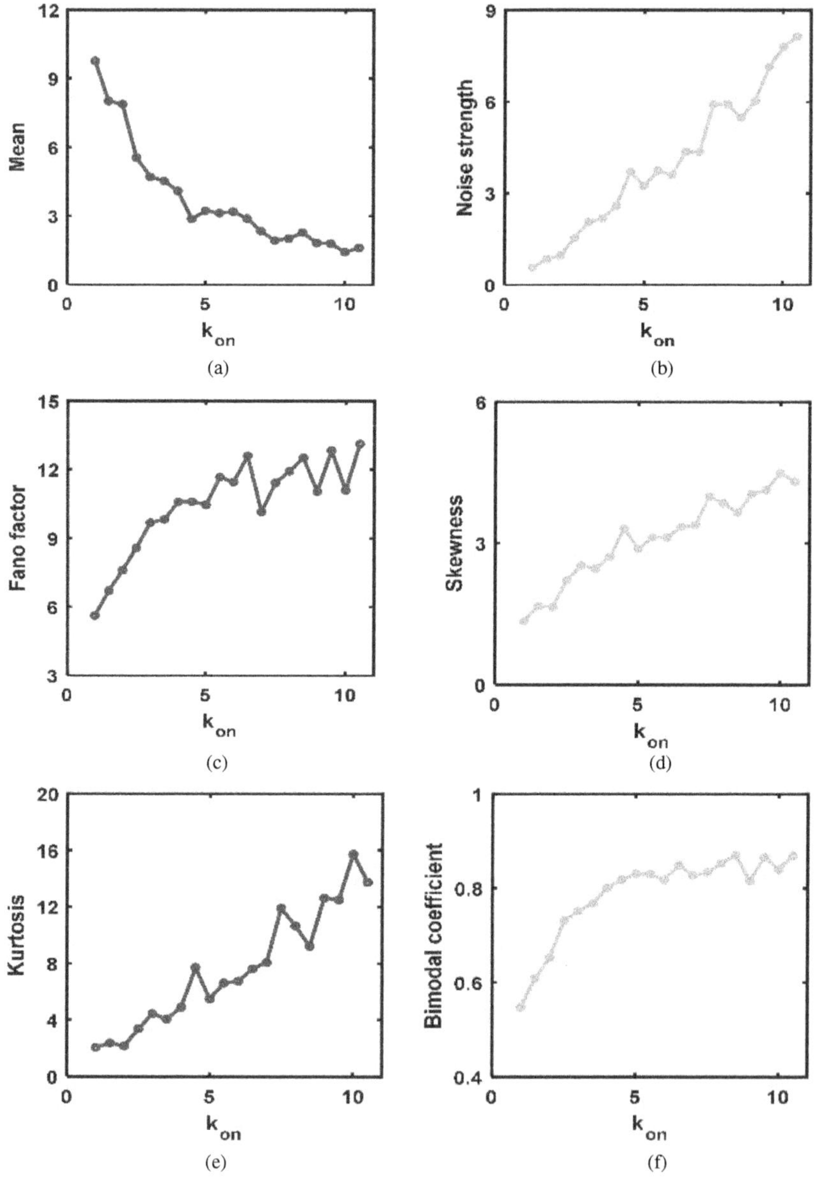

Fig. 3. (Color online) The effect of the memory index of the activated pathway on the mean (a), noise strength (b), Fano factor (c), skewness (d), kurtosis (e) and bimodal coefficient (e) of gene expression. The points represent the results of stochastic simulation algorithm in fixed parameters $k_{\text{off}} = 1$, $r_{\text{on1}} = 0.5$, $r_{\text{off}} = 0.5$, $\mu = 20$, $\delta = 1$, $q = 1$ and varying increment of k_{on}.

2.2. Parameter inference algorithm

Our goal is to infer the posterior distribution of the parameters in the model from scRNA-seq data. When the likelihood function is difficult to deal with in analysis or calculation, approximate Bayesian computation (ABC) offers a method to estimate the posterior distribution. ABC has gained popularity across various fields such as epidemiology, evolutionary biology and molecular genetics.[36–44] The objective of ABC is to find *a posteriori* distribution $f(\theta \mid x_0) \propto f(x_0 \mid \theta)\pi(\theta)$, that is, the probability distribution of the parameters of the observation data x_0, given the likelihood function $f(x_0 \mid \theta)$ and the prior distribution $\pi(\theta)$ on the parameter space Θ. In order to avoid directly evaluating likelihood and perform the above tasks, the ABC algorithm typically follows the procedure outlined below:[45]

For the candidate parameter vector θ^* sampled from a certain density, the simulation data x^* is generated from the likelihood function $f(x|\theta^*)$ with θ^* as the condition. If the simulation data x^* and the observation data x_0 are sufficiently close, the parameter vector θ^* will be accepted. However, due to the large size and high dimensionality of general data, directly calculating their proximity is challenging. Therefore, it is necessary to employ summary statistics $S()$ to project the simulation data and observation data into a lower-dimensional space, and then use a distance metric ρ to measure their proximity. If the distance between the two summary statistics is within a given tolerance threshold ϵ, they are considered sufficiently close. According to this way, the ABC algorithm samples from the joint distribution $f(\theta, x \mid \rho(S(x^*), S(x_0)) \leq \epsilon)$, and we are interested in the marginal distribution $f(\theta|\rho(S(x^*), S(x_0)) \leq \epsilon)$. If the summary statistic is close enough and ϵ is very small, then $f(\theta \mid \rho(S(x^*), S(x_0)) \leq \epsilon)$ is a reasonable approximation of the posterior distribution $f(\theta \mid x_0)$.

As is widely known, the efficiency of most of the ABC algorithms for sampling from the posterior distribution is typically low. Such as the methods based on rejection sampling or Markov chain Monte Carlo (MCMC) simulation.[22,45,46] To address the issue of low sampling efficiency, we use as ABC sampling algorithm ABC-PRC. The algorithm combines partials rejection control with sequential

Monte Carlo to effectively overcome the problem of low efficiency. The specific algorithm flow is as follows:

(1) Initialize a sequence of tolerance threshold $\epsilon_1, \ldots, \epsilon_t$, and given the prior distribution of samples $\pi(\theta)$. Set the population indicator $t = 1$.
(2) Set particle indicator $i = 1$.
   (2.1) If $t = 1$, sample $\theta^{**} \sim \pi(\theta)$.
   If $t > 1$, sample θ^* from the previous population with weights and perturb the particle to $\theta^{**} \sim K_t \ (\theta \mid \theta^*)$ according to a Markov transition kernel K_t. Generate a data set $x^{**} \sim f\ (x \mid \theta^{**})$. If $\rho(S(x^{**}), S(x_0)) \geq \epsilon_t$ then go to 2.1.
   (2.2) Set $\theta_t^{(i)} = \theta^{**}$ and

$$W_t^{(i)} = \begin{cases} 1, & t = 1, \\ \dfrac{\pi(\theta_t^{(i)})}{\sum_{j=1}^{N} W_{t-1}(\theta_{t-1}^{(j)})K_t(\theta_t^{(i)} \mid \theta_{t-1}^{(j)})}, & t > 1, \end{cases}$$

   If $i < N$, increment $i = i + 1$ and go to 2.1.
   (1) If $t < T$, increment $t = t + 1$ and go to 2.

Next, we will explain the details of the above algorithm. The first step is the selection of the tolerance threshold series. To improve computational efficiency, we use a larger threshold to quickly accept a series of coarse samples in the initial iteration. For the tth iteration, we select the median of the discrepancies of the results produced by the $(t-1)$th iteration as the tolerance threshold of this iteration.

For the selection of prior distribution in the initial sampling, according to the model description, we need to infer the parameter vector $\theta = [k_{\text{on1}}, k_{\text{on2}}, k_{\text{off}}, r_{\text{on1}}, r_{\text{on2}}, r_{\text{off}}, \mu, \delta, q]$. As in previous studies, we set that the degradation rate δ of mRNA is 1 to normalize the other kinetic parameters so that they are independent of time.[22,23] For other parameters, we set k_S follow the uniform distribution $U[0,5]$ and r_S follow the logarithmic uniform distribution from interval $[-1,1]$ based on 10. In addition, we set the prior distribution of mRNA synthesis rate μ to $U[0,50]$ and set the prior distribution of probability q of the first path to $U[0,1]$.

In the procedure of the above inference algorithm, we need to perturb the samples taken from the previous population slightly. Therefore, we set the Markov transition kernel to a lognormal distribution in this process. Additionally, we define the distance metric ρ used to measure the proximity between the observed data and the model simulation data in the following form:

$$\rho(S(x^{**}), S(x_0)) = \sqrt{\sum_{i=1}^{n} (\log S_{**}^{i} - \log S_0^{i})^2}, \tag{11}$$

where $S_{**} = S(x^{**})$, $S_0 = S(x_0)$, i represents the component index of the summary statistics vector.

In order to compare the proximity between data, it is necessary to project high-dimensional data into low-dimensional summary statistics. By utilizing the first three binomial moments b_n, $(n = 1, 2, 3)$ of mRNA abundance in the gene expression model, as calculated in Ref. 33 and the relationship between binomial moments and several common moment statistics, we obtain the following[33]:

$$\text{Mean} = b_1, \tag{12}$$

$$\text{Noise} = \frac{2b_2 + b_1 - b_1^2}{b_1^2}, \tag{13}$$

$$\text{Fanofactor} = \frac{2b_2 + b_1 - b_1^2}{b_1}, \tag{14}$$

$$\text{Skewness} = \frac{6b_3 + 6b_2 + b_1 - 3b_1(2b_2 + b_1) + 2b_1^3}{(2b_2 + b_1 - b_1^2)^{\frac{3}{2}}}. \tag{15}$$

We utilize the aforementioned moment statistics, which describe the characteristics of the data, as summary statistics to measure the proximity between the data.

Finally, to generate simulation data based on the gene expression model with given parameters, we employ a stochastic simulation algorithm based on the Gillespie algorithm. The algorithm follows these steps:

Input:

(1) The number of mRNA molecules $N(t) = n_0 \geq 0$.
(2) The gene state $U(t) \in \{0, 1\} = \{OFF, ON\}$.
(3) The OFF state dwell time distribution $f_1(t)$ of the first activation pathway, The OFF state dwell time distribution $f_1(t)$ of the second activation pathway and the ON state dwell time distribution $h(t)$.
(4) The mRNA synthesis rate μ, the degradation rate δ and probability q of selecting the first activation pathway.

Sampling:

(1) Initialize $t = t_0 = 0$, $N(t) = n_0 = 0$, $U(t) = 0$. For $t = 0$ to T do repeat:

(1.1) If $U(t) = 0$,

(a) Sample one random value r from the uniform distribution U(0,1);
(b) If $r < q$, sample the OFF state dwell time τ_{off} from distribution $f_1(t)$, otherwise sample the OFF state dwell time τ_{off} from distribution $f_2(t)$;
(c) Set $\tau_{\text{off}}^{\text{end}} = t + \tau_{\text{off}}$;
(d) While $t < \tau_{\text{off}}^{\text{end}}$ do repeat:

(d1) Set the reaction propensity function of mRNA degradation $a_2 = \delta N(t)$;
(d2) Sample one random value r_1 from the uniform distribution U(0,1);
(d3) Set $\tau = -\ln r_1 / a_2$, and $t = t + \tau$;
(d4) If $t < \tau_{\text{off}}^{\text{end}}$, set $N(t) = N(t) - 1$;

(e) Set $t = \tau_{\text{off}}^{\text{end}}$, $U(t) = 1$

(1.2) If $U(t) = 1$

(a) Sample the ON state dwell time τ_{on} from distribution $h(t)$;
(b) Set $\tau_{\text{on}}^{\text{end}} = t + \tau_{\text{on}}$;

(c) While $t < \tau_{\text{on}}^{\text{end}}$ do repeat:

(c1) Set the reaction propensity function of mRNA synthesis $a_1 = \mu$, and the reaction propensity function of mRNA degradation $a_2 = \delta N(t)$;

(c2) Sample two random value r_1, r_2 from the uniform distribution $U(0,1)$;

(c3) Set $\tau = -\ln r_1/a_1 + a_2$, and $t = t + \tau$;

(c4) Choose $p \in \{1, 2\}$ to satisfy $\sum_{i=1}^{p-1} a_i \leq r_2(a_1 + a_2) \leq \sum_{i=1}^{p} a_i$;

(c5) If $t < \tau_{\text{on}}^{\text{end}}$, if $\mathcal{P} = 1$, set $N(t) = N(t) + 1$; If $\mathcal{P} = 2$, set $N(t) = N(t) - 1$;

(d) Set $t < \tau_{\text{on}}^{\text{end}}$, $U(t) = 0$

Output:

Time series $(t, U(t), N(t))$.

3. Results

3.1. *Validation on the synthetic datasets*

To demonstrate the reliability of the above process, we first give a set of parameters, and use the stochastic simulation algorithm mentioned above to generate synthetic mRNA distribution data based on this parameter. Subsequently, we utilize the above inference process to obtain the transcriptional bursting kinetic parameters from the synthetic data. In essence, we consider the pre-assigned as ground truth and compare them with the inferred parameters to illustrate the accuracy of our inference process.

Figure 4(a) displays a comparison between the characteristic line chart (green) of the synthetic data and the characteristic line chart (orange) generated using the inferred parameters when the parameter $\theta = [3, 6, 5, 0.5, 0.7, 0.5, 20, 1, 0.4]$ is provided. It can be observed that the data inferred by the inference algorithm closely aligns with the data under the real labels. The Euclidean distance between the features of these two datasets is 0.5880. This shows that the data characteristics of the two sets of data are very similar. Figure 4(b) shows the distribution of 1000 parameter samples obtained through

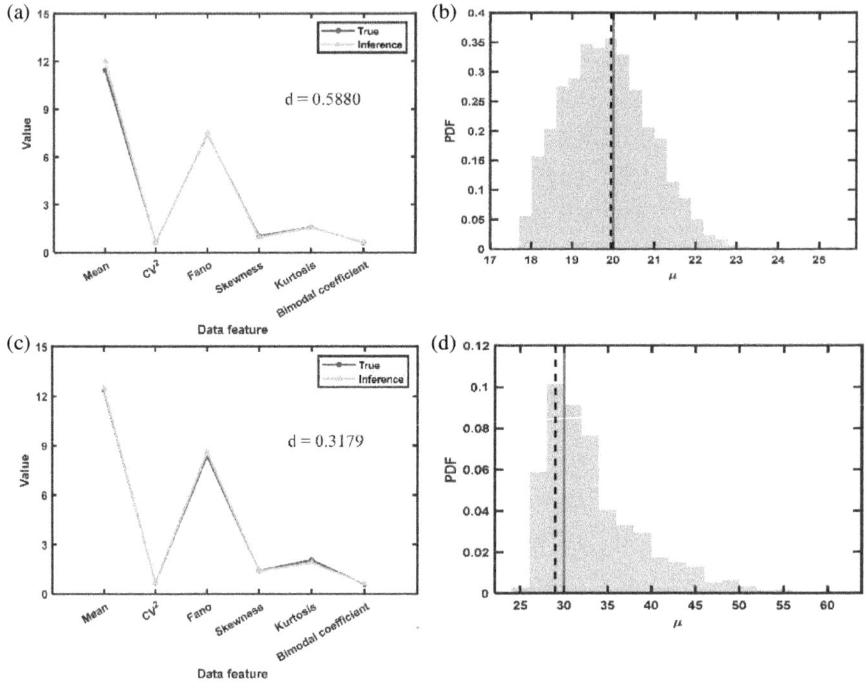

Fig. 4. (Color online) The effectiveness of the inference algorithm is verified on the simulated datasets. (a) The green line represents the line chart of mean, noise intensity, Fano factor, skewness, kurtosis and bimodal coefficient of the data generated by the simulation algorithm under a given parameter $\theta = [3, 6, 5, 0.5, 0.7, 0.5, 20, 1, 0.4]$. The orange line represents the same characteristics of the data generated by the parameters of bursting kinetics inferred from the simulated data. (b) The blue histogram represents the μ distribution of value obtained from the 1000 parameter samples inferred by our inference algorithm. The black solid line represents the mode of the inferred $1000\,\mu$ values, and the red solid line is the given real parameter. d is the Euclidean distance. (c), (d) The inference results are presented under another set of given parameters $\theta = [1, 3, 5, 0.8, 0.7, 3, 30, 1, 0.6]$. The color scheme and letters have the same meaning as the panels (a) and (b).

the inference process, with a focus on the mode of the inferred 1000 mRNA synthesis rate in comparison to real label. In Figs. 4(c) and 4(d), we present another inference conducted with a given set of parameters $\theta = [1, 3, 5, 0.8, 0.7, 3, 30, 1, 0.6]$. It is evident that our inference algorithm accurately estimates the actual parameter values. At the same time, it is also proved that the summary statistics selected above can fully reflect the data characteristics.

3.2. *Validation on the real datasets*

In order to further validate the effectiveness of the algorithm, we apply the parameter inference program to real datasets obtained from MEF cells and mouse ESC. The datasets were obtained by Larsson *et al.* in 2018 from single-cell transcriptome sequencing technology and used them to infer transcriptional burst kinetics at the allelic scale.[25] To enhance the algorithm's performance, we delete genes expressed in less than 40 cells and cells with less than 2000 genes. Additionally, genes with an average expression level of less than 2 are excluded for quality control purposes. Finally, we employ these two datasets to infer the kinetics of transcriptional bursting.

To evaluate the accuracy of our inference algorithm, we randomly select the genes, 1700123O20Rik, Acsl5, Adipor2 and Mrpl15 from the MEF dataset and compare the actual mRNA distribution of these genes with the mRNA distribution generated using the parameters obtained from our inference algorithm. The specific parameter values are shown in Table 1, and more parameter inference results are shown in the github link: https://github.com/kylinwwwm/tranburstQM. The results, shown in Fig. 5, demonstrate that the mRNA distribution data generated by the inferred parameters closely align with the real gene expression distribution. This indicates that our algorithm effectively infers the gene bursting kinetics and accurately reconstructs the real gene expression at the single cell level.

The above experiments demonstrate the accurate inference of parameters in the gene expression model with memory and crosstalk using our algorithm. Furthermore, the transcriptional bursting kinetics

Table 1. The transcriptional bursting parameters of four randomly selected genes were inferred. The degradation rate of mRNAs is 1.

Gene	k_{on1}	k_{on2}	r_{on1}	r_{on2}	k_{off}	r_{off}	μ	q
1700123O20Rik	2.73	1.93	1.10	0.63	5.61	1.37	5.97	0.45
Acsl5	3.51	3.50	0.57	0.70	1.47	0.21	4.33	0.58
Adipor2	4.52	4.05	0.89	0.45	2.13	0.45	4.79	0.76
Mrpl15	2.14	6.88	2.52	1.59	3.52	0.88	7.33	0.23

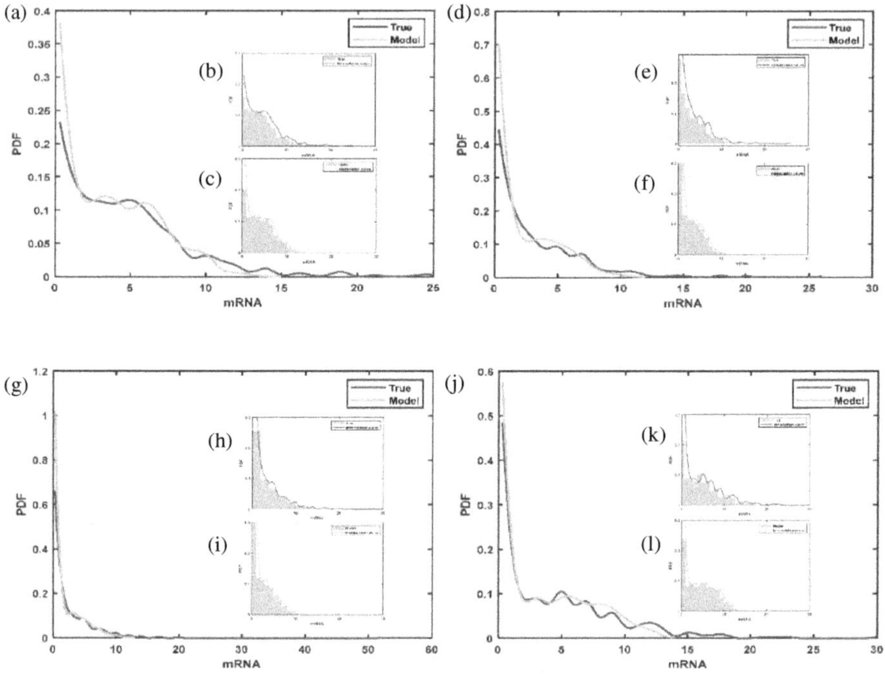

Fig. 5. (Color online) The effectiveness of the inference algorithm is verified on real datasets. (a) The real distribution curve of gene 1700123O20Rik (green) is compared with the distribution curve generated by the inferred parameters (orange). (b) The blue histogram represents the real mRNA distribution of the gene Acsl5, while the green line shows the smooth distribution curve obtained by an interpolation algorithm. (c) The blue histogram represents the data distribution of the model generated by the gene Acsl5 using the inferred bursting parameters. The orange line is the smooth distribution curve obtained using an interpolation algorithm. (d), (l) The real distribution of gene Acsl5, Adipor2 and Mrpl15 is compared with the distribution generated by the inferred parameters, respectively. The color scheme is consistent with panels (a) and (c).

of the gene can be analyzed. Additionally, the inference process also encompasses the case of CTM and GTM.

3.3. *Analytic results*

Next, we will analyze the results of transcriptional bursting kinetics inferred from the above two datasets. We observe that in the scRNA-seq data produced by MEF, genes with higher expression levels tend to have higher BS or frequency, and their average ON state dwell time is also relatively high. This observation aligns with our intuition. Interestingly, we notice a negative correlation between gene BS and BF, the Spearman correlation coefficient is −0.6035, while the average ON state dwell time is positively correlated with the average OFF state dwell time. Their Spearman correlation coefficient is 0.8465. This indicates that gene expression in organisms is finely regulated by balancing BS, BF, ON state dwell time and OFF state dwell time. A similar pattern is observed scRNA-seq data produced by mouse ESC, as depicted in Fig. 6. At the same time, we use the maximum likelihood algorithm to infer the transcriptional burst kinetics of CTM on the same datasets. If we assume that the rate of gene switching from ON state to OFF state in CTM is k_{OFF}, the rate of switching from OFF state to ON state is k_{ON}, and the transcription rate of gene in ON state is s. Then, according to previous work, it is easy to obtain the BS and frequency of CTM, which is $\frac{s}{k_{\mathrm{OFF}}}$, $\frac{1}{\frac{1}{k_{\mathrm{OFF}}}+\frac{1}{k_{\mathrm{ON}}}}$ respectively.[23,25] On this basis, we infer the kinetic characteristics of transcriptional burst based on CTM from the same datasets, as shown in Figs. 6(a) and 6(c). The color spots in the figure represent the inferred results of our model, and the gray spots represent the inferred results of CTM. The pictures show that the CTM-based BS and frequency scatter mostly coincide with our model-based BS and frequency scatter but tend to move to the lower right. This shows that compared with our model, the results inferred by CTM overestimate the BF and underestimate the BS.

Next, we examine the inferred gene BF from scRNA-seq data and divide it into different intervals based on its magnitude. We observe that as the BF increases, the gene expression noise decreases. This suggests that BF plays a role in reducing gene expression noise to some extent, as shown in Fig. 7(a). In the gene expression data of mouse ESC, the bimodal coefficient tends to increase with increasing BS. Bimodality coefficient can be used to describe the bimodal distribution, which is usually an important feature in a dynamical system. The values of the bimodality coefficient range from 0 to 1, and values greater than 5/9 may indicate a bimodal and multimodal

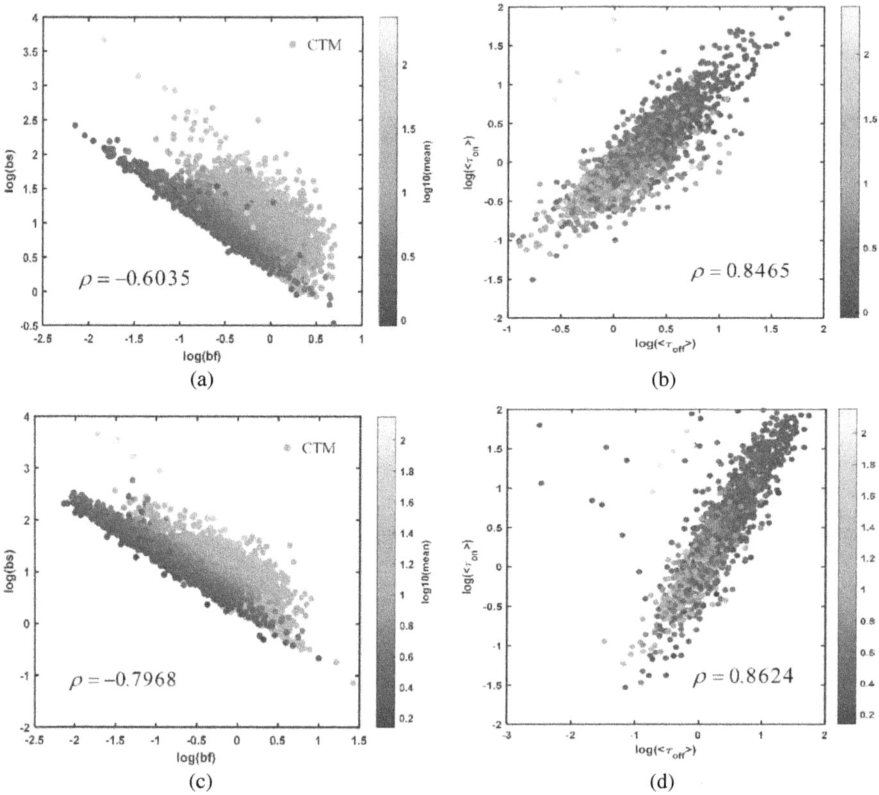

Fig. 6. (Color online) The inferred kinetic characteristics of transcriptional bursting from scRNA-seq data. (a) A scatter plot illustrating the relationship between BS and BF of the genome-wide, inferred from the scRNA-seq data produced by MEF. The color spots represent the inferred results of our model, the gray spots represent the inferred results of CTM. The color represents the average level of gene expression. (b) A scatter plot illustrating the relationship between average ON state dwell time and average OFF state dwell time of the genome-wide, inferred from the scRNA-seq data produced by MEF. The color represents average gene expression level. ρ is the Spearman correlation coefficient. (c), (d) Inferred results from the scRNA-seq data produced by mouse ESC. The color scheme and the Greek alphabet have the same meaning as the panels (a) and (b).

distribution. The coefficient of bimodality is given by the formula $\frac{\text{skewness}^2+1}{\text{kurtosis}}$.[47] The result indicates that higher BS can increase the probability of observing bimodal gene expression distribution to some extent, as shown in Fig. 7(b).

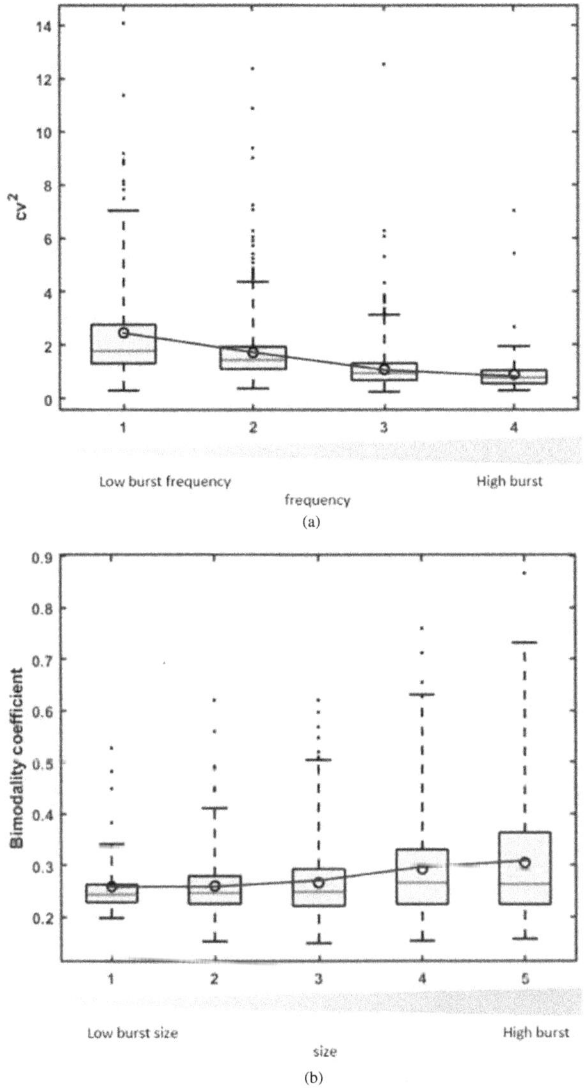

Fig. 7. (Color online) The relationship between the kinetic characteristics of transcriptional bursts and gene expression. (a) Boxplots showing the gene BF and gene expression noise in MEF. The BF is divided into several intervals from small to large, with the gene expression noise plotted on the y-axis. The circles represent the average value of each data group. (b) Boxplots showing the BS and bimodal coefficient in mouse ESC. The BS is divided into several intervals from small to large, with the bimodal coefficient of gene expression plotted on the y-axis. The circles represent the average value of each data group.

Then, in order to investigate the effect of molecular memory on the distribution of gene expression, we randomly select the genes with inferred memory index of activation pathway approximately 1, 6 and 10, and observe the trend of gene expression distribution with the increase of memory index (This is because we infer that the memory index range of genes is about 1 to 10). We observed that, consistent with the theoretical analysis in Ref. 33, with the increase of molecular memory, indicating an increase in the number of intermediate reaction steps from inactive to active gene states, the bimodality of mRNA distribution showed a trend of disappearance.[33] As shown in Fig. 8.

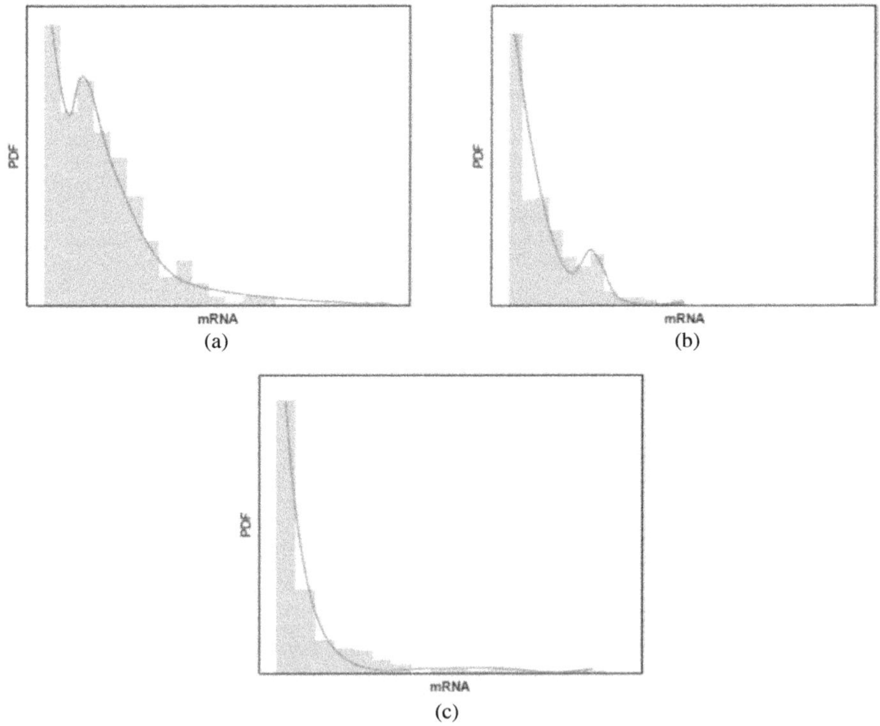

Fig. 8. (Color online) Influence of the molecular memory of the activation pathway on the mRNA distribution. Panels (a), (b) and (c) represent the mRNA distribution of genes with inferred memory index of activation pathway is about 1, 6 and 10, respectively.

4. Conclusions and discussion

The transcription of genetic information from DNA to RNA is one of the most important cellular activities. Gene transcription exhibits two seemingly contradictory characteristics: precision and randomness.[48] On one hand, the timing and rate of gene transcription are tightly controlled and precisely executed. On the other hand, even with constant regulatory signals, the process of gene expression is dynamic and stochastic. Increasing evidence indicates that transcription, whether in prokaryotic genes or higher eukaryotic genes, occurs in bursts rather than following a smooth Poisson process.[13,49–54] The advancement of scRNA-seq technology has provided researchers with great opportunities to study the kinetic mechanisms of genome-wide transcriptional bursting. In previous studies, researchers have often combined gene expression models with scRNA-seq data to infer the kinetics of transcriptional bursting. Strating from simple Markov telegraph models to the polymorphic models, and then to the non-Markov telegraph model, researchers have been gradually exploring the fundamental mechanism of the transcription process step by step.

In this paper, we utilize a gene expression model with memory and crosstalk to infer the transcriptional bursting kinetics from scRNA-seq data. We validate the effectiveness of our algorithm using synthetic datasets and real datasets respectively. We obtain genome-wide transcriptional bursting kinetics and analyze the effects of BS and frequency on gene expression through the set of inference processes. Both model analysis and data simulations demonstrate that the bimodal coefficient increases with increasing BS, while the noise intensity decreases with increasing BF. At the same time, we observe that an increase in molecular memory, which corresponds to an increase in the number of intermediate reaction steps in the gene switch from inactive to active states, leads to a decrease in the bimodality of mRNA distribution. Furthermore, we point out that some previously proposed models can be regarded as special cases of the gene expression model and inference algorithm proposed in this study under certain circumstances.

However, the model analyzed in this paper has some limitations, that is, it is assumed that there are only two activation pathways in the process of gene expression. We can extend the model to more activation pathways in the same way, but there are difficulties in calculating binomial moments. In the following work, we will further improve the method to generalize the gene expression model to the case of multiple activation pathways. Meanwhile, the model implicitly assumes that the transcription rate is constant. In reality, gene expression is influenced by various regulatory factors. And even when regulated by a constant signal, the transcription rate can still exhibit dynamics. Therefore, considering the fluctuating transcription rate can provide a better reflection of biological reality.

Furthermore, with the advancements in sequencing technology, the availability of large-scale single-cell multi-group data offers a tremendous opportunity to study the complex gene expression and regulation processes in organisms. However, due to the complexity of such data, future work may include processing complex data with multiple cell types and high noise, and considering combining more multi-omics data (genome, transcriptome and proteome) for mutual validation. Further, we will soon try to incorporate more biological factors into the model to more fully describe the cell differentiation process. It is necessary to develop more sophisticated mathematical models and statistical inference algorithms to effectively analyze them. For example, interactions between cells and the impact of environmental factors on cell fate can be considered. This will help us gain a deeper understanding of the regulatory mechanisms of cell fate. This presents a challenge in uncovering dynamic molecular mechanisms from static data. In the near future, we aim to incorporate these factors into the gene expression model and data analysis to reveal more insights into the transcriptional dynamics hidden within phenotype.

Statement of usage of artificial intelligence

This study did not use artificial intelligence.

Data availability

The scRNA-seq data used in this manuscript analysis study are available in the EBI European Nucleotide Archive (ENA) under the accession numbers E-MTAB-6362, E-MTAB-6385 and E-MTAB-7098. The code for transcriptional kinetic inference and analysis in this study is provided through Github (https://github.com/kylinwwwm/tranburstQM).

Author contributions

Writing—original draft, M.W. and W.C.; writing—review & editing, M.W., H.-h.Q. and B.-g.Z.; visualization, Y.G., G.W.; supervision, J.J., B.-g.Z.; project administration, B.-g.Z.; funding acquisition, H.-h.Q. and B.-g.Z. All authors have read and agreed to the published version of the manuscript.

Conflict of interest

The authors declare that they have no known competing financial interests or personal relationships that could have appeared to influence the work reported in this paper.

Funding information

This work was supported by the National Natural Science Foundation of China (NSFC) [Grant numbers. 11971367, 12371500 and 12271416].

Orcid

Mengyuan Wang https://orcid.org/0009-0000-1227-607X
Wenjie Cao https://orcid.org/0009-0002-7112-1013
Yanbing Guo https://orcid.org/0009-0001-8147-1853
Guilin Wang https://orcid.org/0009-0000-4223-6837
Jian Jiang https://orcid.org/0000-0002-5994-292x
Huahai Qiu https://orcid.org/0000-0002-5695-7851
Ben-Gong Zhang https://orcid.org/0000-0002-2766-6502

References

1. Burga, A.; Lehner, B. Predicting Phenotypic Variation from Genotypes, Phenotypes and a Combination of the Two. *Curr. Opin. Biotechnol.* **2013**, *24* (4), 803–809, https://doi.org/10.1016/j.copbio.2013.03.004.
2. Raj, A.; van Oudenaarden, A. Nature, Nurture or Chance: Stochastic Gene Expression and Its Consequences. *Cell* **2008**, *135* (2), 216–226, https://doi.org/10.1016/j.cell.2008.09.050.
3. Casanueva, M. O.; Burga, A.; Lehner, B. Fitness Trade-Offs and Environmentally Induced Mutation Buffering in Isogenic C. Elegans. *Science* **2012**, *335* (6064), 82–85, https://doi.org/10.1126/science.1213491.
4. Kussell, E.; Leibler, S. Phenotypic Diversity, Population Growth, and Information in Fluctuating Environments. *Science (New York, N.Y.)* **2005**, *309* (5743), https://doi.org/10.1126/science.1114383.
5. Faure, A. J.; Schmiedel, J. M.; Lehner, B. Systematic Analysis of the Determinants of Gene Expression Noise in Embryonic Stem Cells. *Cell Syst.* **2017**, *5* (5), 471–484.e4, https://doi.org/10.1016/j.cels.2017.10.003.
6. Heid, C. A.; Stevens, J.; Livak, K. J.; Williams, P. M. Real Time Quantitative PCR. *Genom. Res.* **1996**, *6* (10), 986–994.
7. Ståhlberg, A.; Kubista, M. The Workflow of Single-Cell Expression Profiling Using Quantitative Real-Time PCR. *Exp. Rev. Mol. Diag.* **2014**, *14* (3), 323–331, https://doi.org/10.1586/14737159.2014.901154.
8. Levsky, J. M.; Singer, R. H. Fluorescence *in situ* Hybridization: Past, Present and Future. *J. Cell Sci.* **2003**, *116*, 2833–2838.
9. Chong, S.; Chen, C.; Ge, H.; Xie, X. S. Mechanism of Transcriptional Bursting in Bacteria. *Cell* **2014**, *158* (2), 314–326.
10. Ko, M. S. Induction Mechanism of a Single Gene Molecule: Stochastic or Deterministic? *Bioessays* **1992**, *14* (5), 341–346, https://doi.org/10.1002/bies.950140510.
11. Zhang, J.; Zhou, T. Markovian Approaches to Modeling Intracellular Reaction Processes with Molecular Memory. *Proc. Natl. Acad. Sci. USA* **2019**, *116* (47), 23542–23550, https://doi.org/10.1073/pnas.1913926116.
12. Harper, C. V.; Finkenstädt, B.; Woodcock, D. J.; Friedrichsen, S.; Semprini, S.; Ashall, L.; Spiller, D. G.; Mullins, J. J.; Rand, D. A.; Davis, J. R. E.; White, M. R. H. Dynamic Analysis of Stochastic Transcription Cycles. *PLoS Biol.* **2011**, *9* (4), e1000607, https://doi.org/10.1371/journal.pbio.1000607.
13. Suter, D. M.; Molina, N.; Gatfield, D.; Schneider, K.; Schibler, U.; Naef, F. Mammalian Genes are Transcribed with Widely Different Bursting Kinetics. *Science* **2011**, *332* (6028), 472–474, https://doi.org/10.1126/science.1198817.
14. Nishinari, K.; Okada, Y.; Schadschneider, A.; Chowdhury, D. Intracellular Transport of Single-Headed Molecular Motors KIF1A. *Phys. Rev. Lett.* **2005**, *95* (11), 118101, https://doi.org/10.1103/PhysRevLett.95.118101.
15. Stumpf, P. S.; Smith, R. C. G.; Lenz, M.; Schuppert, A.; Müller, F.-J.; Babtie, A.; Chan, T. E.; Stumpf, M. P. H.; Please, C. P.; Howison, S. D.; Arai, F.; MacArthur, B. D. Stem Cell Differentiation as a Non-Markov Stochastic

Process. *Cell Syst.* **2017**, *5* (3), 268–282.e7, https://doi.org/10.1016/j.cels.2017.08.009.
16. Guérin, T.; Bénichou, O.; Voituriez, R. Non-Markovian Polymer Reaction Kinetics. *Nat. Chem.* **2012**, *4* (7), 568–573.
17. Barabási, A.-L. The Origin of Bursts and Heavy Tails in Human Dynamics. *Nature* **2005**, *435* (7039), 207–211, https://doi.org/10.1038/nature03459.
18. Schwabe, A.; Rybakova, K. N.; Bruggeman, F. J. Transcription Stochasticity of Complex Gene Regulation Models. *Biophys. J.* **2012**, *103* (6), 1152–1161.
19. Zhang, J.; Zhou, T. Promoter-Mediated Transcriptional Dynamics. *Biophys. J.* **2014**, *106* (2), 479–488.
20. Jia, C.; Li, Y. Analytical Time-Dependent Distributions for Gene Expression Models with Complex Promoter Switching Mechanisms. *SIAM J. Appl. Math.* **2023**, *83* (4), 1572–1602, https://doi.org/10.1137/22M147219X.
21. Zhang, J.; Zhou, T. Computation of Stationary Distributions in Stochastic Models of Cellular Processes with Molecular Memory. *bioRxiv* **2019**, 521575, https://doi.org/10.1101/521575.
22. Luo, S.; Zhang, Z.; Wang, Z.; Yang, X.; Chen, X.; Zhou, T.; Zhang, J. Inferring Transcriptional Bursting Kinetics from Single-Cell Snapshot Data Using a Generalized Telegraph Model. *R. Soc. Open Sci.* **2023**, *10* (4), 221057, https://doi.org/10.1098/rsos.221057.
23. Kim, J. K.; Marioni, J. C. Inferring the Kinetics of Stochastic Gene Expression from Single-Cell RNA-Sequencing Data. *Genom. Biol.* **2013**, *14* (1), R7, https://doi.org/10.1186/gb-2013-14-1-r7.
24. Jiang, Y.; Zhang, N. R.; Li, M. SCALE: Modeling Allele-Specific Gene Expression by Single-Cell RNA Sequencing. *Genom. Biol.* **2017**, *18* (1), 74, https://doi.org/10.1186/s13059-017-1200-8.
25. Larsson, A. J. M.; Johnsson, P.; Hagemann-Jensen, M.; Hartmanis, L.; Faridani, O. R.; Reinius, B.; Segerstolpe, Å.; Rivera, C. M.; Ren, B.; Sandberg, R. Genomic Encoding of Transcriptional Burst Kinetics. *Nature* **2019**, *565* (7738), 251–254, https://doi.org/10.1038/s41586-018-0836-1.
26. Luo, S.; Wang, Z.; Zhang, Z.; Zhou, T.; Zhang, J. Genome-Wide Inference Reveals that Feedback Regulations Constrain Promoter-Dependent Transcriptional Burst Kinetics. *Nucl. Acids Res.* **2023**, *51* (1), 68–83, https://doi.org/10.1093/nar/gkac1204.
27. Han, C.; Cao, W.; Li, C.; Guo, Y.; Wang, Y.; Shi, Y.-Z.; Zhang, B.-G. MLBKFD: Probabilistic Model Methods to Infer Pseudo Trajectories from Single-Cell Data. *J. Comput. Biophys. Chem.* **2023**, 1–17, https://doi.org/10.1142/S2737416524400015.
28. Ozbudak, E. M.; Thattai, M.; Kurtser, I.; Grossman, A. D.; van Oudenaarden, A. Regulation of Noise in the Expression of a Single Gene. *Nat. Genet.* **2002**, *31* (1), 69–73, https://doi.org/10.1038/ng869.
29. Sanchez, A.; Golding, I. Genetic Determinants and Cellular Constraints in Noisy Gene Expression. *Science* **2013**, *342* (6163), 1188–1193, https://doi.org/10.1126/science.1242975.
30. das Neves, R. P.; Jones, N. S.; Andreu, L.; Gupta, R.; Enver, T.; Iborra, F. J. Connecting Variability in Global Transcription Rate to Mitochondrial

Variability. *PLoS Biol.* **2010**, *8* (12), e1000560, https://doi.org/10.1371/journal.pbio.1000560.
31. Yu, J.; Sun, Q.; Tang, M. The Nonlinear Dynamics and Fluctuations of mRNA Levels in Cross-Talking Pathway Activated Transcription. *J. Theor. Biol.* **2014**, *363*, 223–234, https://doi.org/10.1016/j.jtbi.2014.08.024.
32. Jiao, F.; Sun, Q.; Lin, G.; Yu, J. Distribution Profiles in Gene Transcription Activated by the Cross-Talking Pathway. *DCDS-B* **2019**, *24* (6), 2799–2810, https://doi.org/10.3934/dcdsb.2018275.
33. Zhang, Z.; Deng, Q.; Wang, Z.; Chen, Y.; Zhou, T. Exact Results for Queuing Models of Stochastic Transcription with Memory and Crosstalk. *Phys. Rev. E* **2021**, *103* (6–1), 062414, https://doi.org/10.1103/PhysRevE.103.062414.
34. Pedraza, J. M.; Paulsson, J. Effects of Molecular Memory and Bursting on Fluctuations in Gene Expression. *Science* **2008**, *319* (5861), 339–343, https://doi.org/10.1126/science.1144331.
35. Zhang, J.; Nie, Q.; Zhou, T. A Moment-Convergence Method for Stochastic Analysis of Biochemical Reaction Networks. *J. Chem. Phys.* **2016**, *144* (19), 194109.
36. Marjoram, P.; Molitor, J.; Plagnol, V.; Tavaré, S. Markov Chain Monte Carlo without Likelihoods. *Proc. Natl. Acad. Sci. USA* **2003**, *100* (26), 15324–15328, https://doi.org/10.1073/pnas.0306899100.
37. Bortot, P.; Coles, S. G.; Sisson, S. A. Inference for Stereological Extremes. *J. Amer. Stat. Associat.* **2007**, *102* (477), 84–92, https://doi.org/10.1198/016214506000000988.
38. Beaumont, M. A.; Zhang, W.; Balding, D. J. Approximate Bayesian Computation in Population Genetics. *Genetics* **2002**, *162* (4), 2025–2035, https://doi.org/10.1093/genetics/162.4.2025.
39. Pritchard, J. K.; Seielstad, M. T.; Perez-Lezaun, A.; Feldman, M. W. Population Growth of Human Y Chromosomes: A Study of Y Chromosome Microsatellites. *Mol. Biol. Evol.* **1999**, *16* (12), 1791–1798, https://doi.org/10.1093/oxfordjournals.molbev.a026091.
40. Marjoram, P.; Tavaré, S. Modern Computational Approaches for Analysing Molecular Genetic Variation Data. *Nat. Rev. Genet.* **2006**, *7* (10), 759–770, https://doi.org/10.1038/nrg1961.
41. Tanaka, M. M.; Francis, A. R.; Luciani, F.; Sisson, S. A. Using Approximate Bayesian Computation to Estimate Tuberculosis Transmission Parameters from Genotype Data. *Genetics* **2006**, *173* (3), 1511–1520, https://doi.org/10.1534/genetics.106.055574.
42. Leman, S. C.; Chen, Y.; Stajich, J. E.; Noor, M. A. F.; Uyenoyama, M. K. Likelihoods from Summary Statistics: Recent Divergence between Species. *Genetics* **2005**, *171* (3), 1419–1436, https://doi.org/10.1534/genetics.104.040402.
43. Thornton, K.; Andolfatto, P. Approximate Bayesian Inference Reveals Evidence for a Recent, Severe Bottleneck in a Netherlands Population of Drosophila Melanogaster. *Genetics* **2006**, *172* (3), 1607–1619, https://doi.org/10.1534/genetics.105.048223.

44. Butler, A.; Glasbey, C. A Latent Gaussian Model for Compositional Data with Zeros. *J. R. Stat. Soc. Ser. C (Appl. Stat.)* **2008**, *57* (5), 505–520.
45. Sisson, S. A.; Fan, Y.; Tanaka, M. M. Sequential Monte Carlo without Likelihoods. *Proc. Natl. Acad. Sci. USA* **2007**, *104* (6), 1760–1765, https://doi.org/10.1073/pnas.0607208104.
46. Tavaré, S.; Balding, D. J.; Griffiths, R. C.; Donnelly, P. Inferring Coalescence Times from DNA Sequence Data. *Genetics* **1997**, *145* (2), 505–518, https://doi.org/10.1093/genetics/145.2.505.
47. Ellison, A. M. Effect of Seed Dimorphism on the Density–Dependent Dynamics of Experimental Populations of Atriplex Triangularis (Chenopodiaceae). *Amer. J. Botany* **1987**, *74* (8), 1280–1288, https://doi.org/10.1002/j.1537-2197.1987.tb08741.x.
48. Wang, Y.; Ni, T.; Wang, W.; Liu, F. Gene Transcription in Bursting: A Unified Mode for Realizing Accuracy and Stochasticity. *Biol. Rev. Camb. Philos. Soc.* **2019**, *94* (1), 248–258, https://doi.org/10.1111/brv.12452.
49. Rodriguez, J.; Larson, D. R. Transcription in Living Cells: Molecular Mechanisms of Bursting. *Ann. Rev. Biochem.* **2020**, *89*, 189–212, https://doi.org/10.1146/annurev-biochem-011520-105250.
50. Tunnacliffe, E.; Chubb, J. R. What Is a Transcriptional Burst? *Trends Genet.* **2020**, *36* (4), 288–297, https://doi.org/10.1016/j.tig.2020.01.003.
51. Chubb, J. R.; Trcek, T.; Shenoy, S. M.; Singer, R. H. Transcriptional Pulsing of a Developmental Gene. *Curr. Biol.* **2006**, *16* (10), 1018–1025, https://doi.org/10.1016/j.cub.2006.03.092.
52. Golding, I.; Paulsson, J.; Zawilski, S. M.; Cox, E. C. Real-Time Kinetics of Gene Activity in Individual Bacteria. *Cell* **2005**, *123* (6), 1025–1036, https://doi.org/10.1016/j.cell.2005.09.031.
53. Raj, A.; Peskin, C. S.; Tranchina, D.; Vargas, D. Y.; Tyagi, S. Stochastic mRNA Synthesis in Mammalian Cells. *PLoS Biol.* **2006**, *4* (10), e309, https://doi.org/10.1371/journal.pbio.0040309.
54. Gorin, G.; Vastola, J. J.; Fang, M.; Pachter, L. Interpretable and Tractable Models of Transcriptional Noise for the Rational Design of Single-Molecule Quantification Experiments. *Nat. Commun.* **2022**, *13* (1), 7620, https://doi.org/10.1038/s41467-022-34857-7.

https://doi.org/10.1142/9789819808199_0012

Chapter 12

Investigating Enzyme Biochemistry by Deep Learning: A Computational Tool for a New Era*

Milad Rayka, Morteza Mirzaei[†,§], Gholamreza Farnoosh and Ali Mohammad Latifi[‡]

Applied Biotechnology Research Center, Baqiyatallah University of Medical Sciences, Tehran 1435916471, Iran
[†]*mraga85@gmail.com*
[‡]*amlatifi290@gmail.com*

Enzymes are protein molecules that play a crucial role in various biological processes in living organisms. They function as catalysts in biological reactions such as digestion, metabolism, DNA replication and other physiological processes. Furthermore, enzymes are widely used in food production, pharmaceuticals and biofuel production. In these industries, they accelerate desired chemical reactions as biocatalysts. Therefore, applying computational methods and data-driven algorithms to predict enzyme properties is essential. Over the past decade, deep learning has made remarkable advancements in science and technology. Deep learning is a subset of machine learning algorithms that rely on artificial neural networks. These algorithms can be employed for supervised, semi-supervised and unsupervised learning. Here, to provide an update on the current literature, we provide an overview of various deep learning algorithms and recent advancements in their application to enzyme science. These applications can generally be categorized into diverse subjects: function prediction, enzyme kinetic parameters prediction, enzyme-substrate identification, condition optimization, thermophilic property prediction, enzyme catalytic site prediction and enzyme design. In conclusion, we discuss

*To cite this article, please refer to its earlier version published in *J. Comput. Biophys. Chem.* **2024**, *23* (6), 781–799, DOI: 10.1142/S2737416524400052
§Corresponding author.

the convergence of enzyme science and deep learning, highlighting the potential opportunities and challenges.

Keywords: Enzyme; deep learning; bioinformatics; artificial intelligence.

1. Introduction

Enzymes are protein biomolecules that function as biological catalysts to lower the activation energy of the transition state of chemical reactions in organisms. Also, enzymes are one of the most significant and extensive groups of proteins since approximately 50% of all proteins exhibit enzymatic activities.[1] The molecules that enzymes act on them are called substrates.[2] The catalytic process happens in the specific amino acid residues of the enzyme, which are called the active site. The active site consists of binding and catalytic sites. In the binding site, the orientation of the substrate is modified, and in the catalytic site, the activation energy is reduced. The specificity and catalytic power of enzymes is their most important feature. This property relates to the precise interaction between the substrate and the enzyme. Factors such as the complementarity of the shape of the substrate and the binding site, the type of electric charge, hydrophobicity and hydrophilicity are effective in this interaction.[3]

Enzymes require particular conditions to perform their role. For example, other environmental molecules can decrease or increase enzyme activity.[4] These molecules are called inhibitors and activators, respectively.[5] Enzyme activity is the number of molecules produced per unit of time. If the enzyme concentration is known, the specific activity can be derived by dividing the enzyme activity by enzyme concentration. Among other factors impacting the activity of enzymes, we can mention the temperature (T) and pH of the environment.[6] The catalytic activity of many enzymes relies on small molecules called cofactors. Metals and coenzymes are two broad categories of cofactors. The first class consists of metal ions such as Zn^{2+} and Ni^{2+} or metal elements such as Mn and Se. The second class is small organic molecules based on derivatives of vitamins.[7] For example, monoamine oxidase, which catalyzes the oxidation reaction of monoamines, has a coenzyme called flavin adenine dinucleotide (FAD), composed of vitamin B2.[2]

According to the mentioned properties, it is evident that enzymes play a critical role in many chemical reactions. This diversity makes enzymes primary tools to solve crucial problems in various industries such as chemical, pharmaceutical, biological and food industries.[8–10] However, the pervasiveness of these applications requires advanced biotechnology techniques, especially in the fields of recombinant genes, protein engineering and directed evolution,[11] which are costly and specialist-demanding.[12] In order to overcome these challenges, a range of computational tools have been introduced and created.[13–15] Data-driven approaches like machine and deep learning (DL) algorithms are among computational methods that have attracted considerable attention recently.[16–18]

Deep learning is a subset of machine learning algorithms that can learn from available data without explicit programming. These algorithms are based on artificial neural networks (ANNs) designed in the multiple layers scheme. These layers extract progressively higher-level features from data. This kind of feature learning, which automatically realizes the representations required for feature detection or classification from data, is called representation learning. DL can be applied to tasks involving continuous quantities or discrete class labels for regression and classification.[19]

In the last decade, we witnessed a proliferation of DL in various subjects.[20] This progress can be attributed to several reasons: (1) The advances in learning algorithms, research,[19,21] and availability of open source packages for DL such as PyTorch[22] and TensorFlow.[23] (2) The availability of enormous amounts of data for training, especially in the technology related subjects like machine vision and natural language processing (NLP).[19,21] (3) The immense increase in computer processing capabilities, like the advent of the graphics processing unit (GPU)[24] for parallel computation and specific design chips for artificial intelligence, such as the tensor processing unit (TPU).[25]

Suitable areas for applying DL are scientific and technological problems that need tremendous computational and financial resources or there are no analytic solutions to explain them. As an example, the prediction of the 3D structure of a protein from its amino acid sequence is one of the challenges in structural bioinformatics.[26] A diverse range of approaches is introduced for solving this problem, from physics-based simulations like molecular dynamics to homology methods, which utilize the information encoded in sequence and structural databases. However, these techniques require significant computational power or do not meet the quality standards.[27,28] This is one of the issues in which DL algorithms can play a decisive role by training on experimental data like the X-ray crystallography structure of the proteins. For instance, in the critical assessment of methods of protein structure prediction or CASP14, AlphaFold2, by applying DL for protein structure prediction, accomplished a median score of 87.0 GDT (global distance test) in the free-modeling category.[29,30] It is worth mentioning that although training these types of algorithms can be computationally expensive, once the learning process is complete, the cost of using them is minimal, making them suitable for use on a standard computer. By this premise, DL is widely applied as a new computational tool for predicting and depicting phenomena in chemistry,[31] physics,[32] biology,[33] medicine,[34] mathematics[35] and other scientific domains.[36–38]

This review paper has attempted to investigate the applications of DL algorithms in the fields of enzyme engineering, design and properties prediction. In Sec. 2, we define DL algorithms, their basic terminologies and the most popular architectures which are usually applied in the enzyme field. Section 3 is devoted to applications of DL algorithms from enzyme function prediction to enzyme design. In Sec. 4, we conclude our overview and state the current problems and challenges in this field.

2. Deep learning overview

2.1. *Feed-forward neural network*

The essence of a DL algorithm can be summarized by the following Tom M. Mitchell's quote: "A computer program is said to learn from experience E with respect to some task T and some performance measure P, if its performance on T, as measured by P, improves with experience E".[31] It conveys that a computer program does not necessarily need an explicit programming scheme to cover every possible situation. We can create an algorithm that learns from experience to perform our desired task. The experiences are called train sets. Each data or input instance is associated with a label. The procedure of converting data instances to numerical arrays, which describes it before usage, is called feature engineering, and the result is a feature vector. The algorithm's performance is tracked during training by using a validation set. After the training stage, the generalization capability of the algorithm is checked by employing an independent test set.[39] For instance, our objective is to construct an algorithm to predict the LogP of molecules (specify the preference of a molecule to dissolve in either water or an organic solvent). We can show the training set mathematically as a set of ordered-pair $\mathbb{X} = \{(\boldsymbol{x}^{(i)}, y^{(i)})\}_{i=1}^{N}$ in which $\boldsymbol{x}^{(i)}$ and $y^{(i)}$ are the feature vector and LogP value of the ith molecule in the $\mathbb{X}$ set with size N. Therefore, a proper label is a numerical LogP value. Molecules can be featurized in numerous ways, like physicochemical descriptors and 3D features.[40]

Artificial neural networks, inspired by biological neural, are employed to estimate or approximate functions that correlate input data to the output labels or targets. One of the most popular architectures of ANN is a feed-forward neural network (FFNN). FFNN contains artificial neurons (units or nodes) organized in layers with varying widths (number of neurons in a layer). The first and last layers are the input and output layers. Other layers are referred to as the hidden layers. Each neuron is connected to all neurons in the successive and previous layers to send and receive signals. Because these neurons are fully connected, their corresponding layers are called dense layers. The strength of these connections, represented as weight parameters, is adjusted during training. These weights are initialized randomly before learning. Each neuron accepts input values from the preceding layer and projects them to the new space by multiplying them by the weight parameters (depicted as solid lines in Fig. 1). First, the projected values are summed, and then a nonlinear activation function is applied to add the nonlinearity characteristic. These nonlinear transformations allow FFNN to extract intricate and

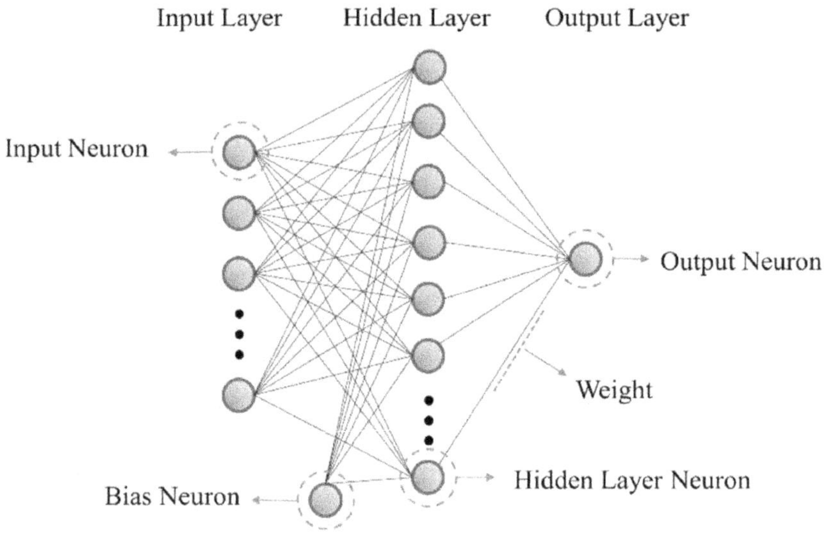

Fig. 1. (Color online) Minimalist picture of a simple artificial neural network.

abstract features. The output of this function is utilized as the input for the next layer until it reaches the last layer. The final output corresponds to the predicted target or label. Figures 1 and 2 depict a picture of an ANN and diagrams of some popular activation functions, respectively.[21,41,42]

Besides weight and bias parameters, which are variable during training, FFNN has hyperparameters that are determined prior to the training. The number of hidden layers, the number of neuron units in each layer and the type of activation function are some examples of hyperparameters. An example of FFNN with a hidden layer can be written mathematically as the following equation:

$$\hat{y} = \sum_{j=1}^{n_2} \phi\left(\sum_{i=1}^{n_1} x^{(i)} \times w_{i,1} + b_1\right) \times w_{j,2}, \quad (1)$$

$x^{(i)}$ is the ith index of the feature vector ($\boldsymbol{x} \in \mathbb{R}^D$) and $\hat{y}$ is the predicted label value. $w_{i,1}$ and $w_{j,2}$ are weight parameters for input and hidden layers with n_1 and n_2 neurons, respectively. b_1 is the bias parameter. ϕ is a typical activation function.[21,41]

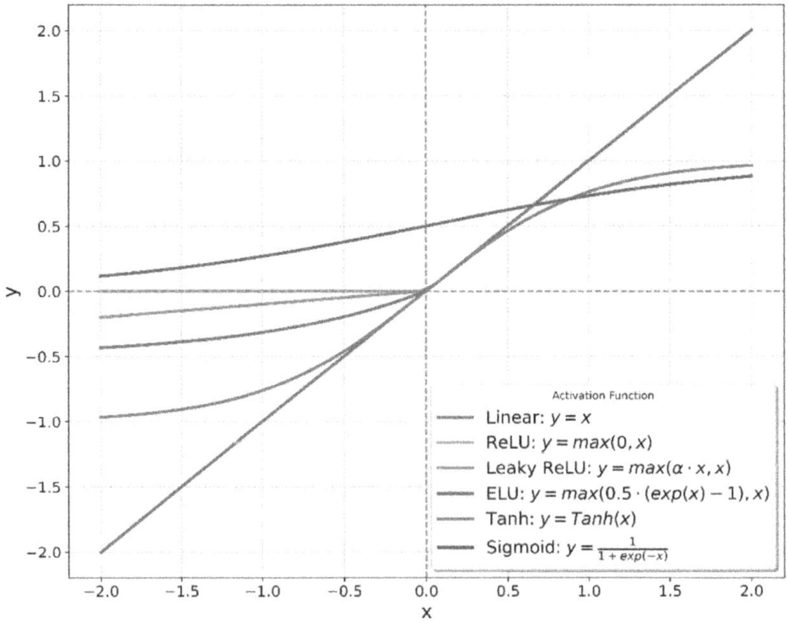

Fig. 2. (Color online) The diagram represents different activation functions alongside their mathematical formulas. Rectified linear unit and exponential linear unit are abbreviated as ReLU and ELU, respectively. Max is a function that accepts two variables and returns the greater one. Tanh is tangent hyperbolic.

The object of the training is to discern the relationship between the training data and labels, which FFNN weight and bias parameters change during that. At first, during forward propagation, each training input data is loaded into the ANN to predict the corresponding label. Then, the predicted value is compared to the ground truth value through a loss function. For example, the following equation shows a mean squared error (MSE) loss function:

$$\mathrm{MSE} = \frac{1}{N}\sum_{i=1}^{N}(y^{(i)} - \hat{y}^{(i)})^2, \tag{2}$$

which $\hat{y}^{(i)}$ is dependent on FFNN weight and bias parameters.[39,43] In the learning procedure, the loss function is minimized using the backpropagation technique and an optimization algorithm like gradient descent. For MSE, we can write a gradient descent step as the following equation:

$$\theta_{i+1} = \theta_i - \eta \times \nabla_\theta \mathrm{MSE}, \quad (3)$$

which θ is the union of weight and bias parameters and η (learning rate) determines the step size of the gradient descent. The above equation displays that FFNN parameters iteratively update until they reach the local minimum. This technique ensures that FFNN acquires suitable weight and bias parameters to predict output with better accuracy. The entire training set may be loaded several times, with every complete pass of data being called an epoch.[39,42]

2.2. *Convolutional neural network*

Another popular neural network, loosely based on the visual cortex, is the convolutional neural network (CNN). CNN, which has translation-equivariant properties, is typically applied to scrutinize visual imagery.[19] It is important to note that CNN input data is a tensor — a suitable mathematical to represent multidimensional arrays — in contrast to the feature vector in FFNN. An image, for instance, is a tensor ($\mathbb{R}^{n \times m \times d}$) with a rank of three, where the first (n) and second dimensions (m) represent the intensity of each pixel and the final dimension (d) shows the color (in the RGB format). CNN consists of several convolutional layers (Fig. 3).

Contrary to FFNN, convolutional layer neurons are only connected to certain pixels in their receptive fields. Inputs are transformed by weight and bias parameters and applying an activation function. A filter is a set of weights and biases. Each layer consists of several filters. For example, one filter detects vertical lines while the other recognizes horizontal lines. The parameters of each filter are shared with other neurons, which reduces memory consumption.

The first layer converts the input data to feature maps. Several convolutional layers apply a similar procedure to extract low-level to high-level features. Like the first layer, the other layer's neurons accept input from only a local region of the previous layer (typically a square area). In order to reduce the number of network computations required, some local pooling layers gradually reduce the input image's

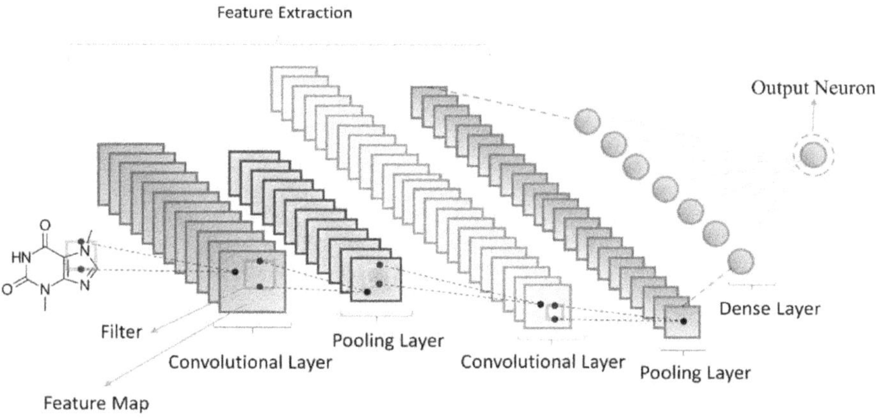

Fig. 3. (Color online) Simple view of a CNN with convolutional and pooling layers for feature extraction. Each color square is a feature map, and each smaller square represents a filter. After global pooling, several dense layers are applied to make a prediction.

spatial width and height. Further, global pooling applies average or max functions to each feature map to convert the 3D tensor ($\mathbb{R}^{\acute{n}\times\acute{m}\times\acute{f}}$) to a feature vector (with dimension $\acute{f}$) suitable for the dense layer. Eventually, a series of dense (fully connected) layers apply nonlinear transformations to make a prediction based on the generated feature vector from the global pooling layer.[21,44,45]

2.3. *Recurrent neural network*

A recurrent neural network (RNN) is another well-known DL algorithm. RNN usually applies to sequential or time series data like natural language translation or fluctuations in heartbeat rate through time (Fig. 4).

It is a reasonable assumption to presume that a data point in a time series is not independent of prior data points. For instance, the meaning of a word relies upon other words in a sentence, its location, and usage context. To learn the contextual information, RNN utilizes a memory notion by taking details from prior inputs to influence the current input and output. Long short-term memory (LSTM) and gated recurrent unit (GRU) are among the attempts to employ the

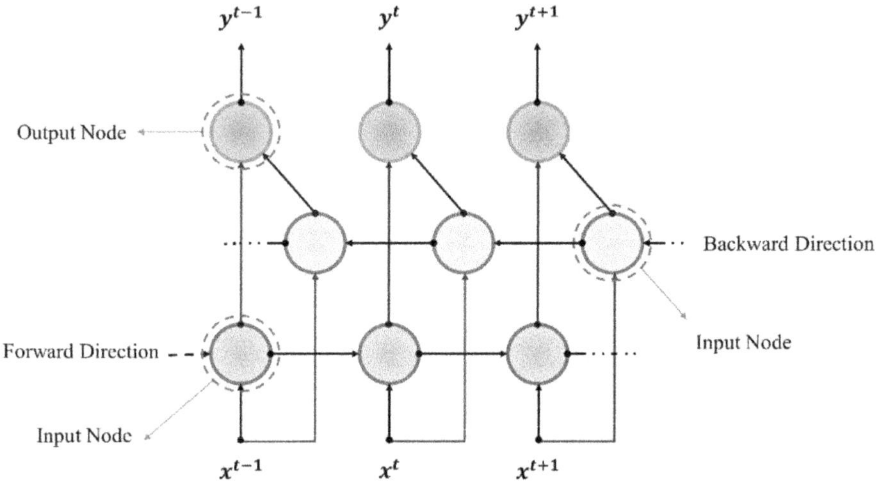

Fig. 4. (Color online) A bidirectional RNN. x and y are input and predicted labels, respectively. t determines the location of x or y in the sequence.

memory concept.[46,47] These units utilize a feedback mechanism through input, output and forget gates to grasp and retain valuable information from preceding inputs. These computations are carried out in the forward direction from the beginning to the end of a sequence, which ignores the influence of subsequent inputs on preceding ones. The bidirectional layer is introduced to circumvent this obstacle by capturing past and future contexts.[21,48]

2.4. *Transformer*

The aim of the transformer, as an encoder-decoder model, is similar to RNN. Both want to capture contextual information between constituents of sequential data (Fig. 5). Instead of memory-based units like LSTM, the transformer applies self-attention to detect subtle relationships between sequence elements. Each sequence element is provided iteratively for RNN, which makes RNN computationally demanding and unable to maintain long dependencies between each segment. To overcome this drawback, at the beginning, a positional encoder is applied to encode the position of each element.

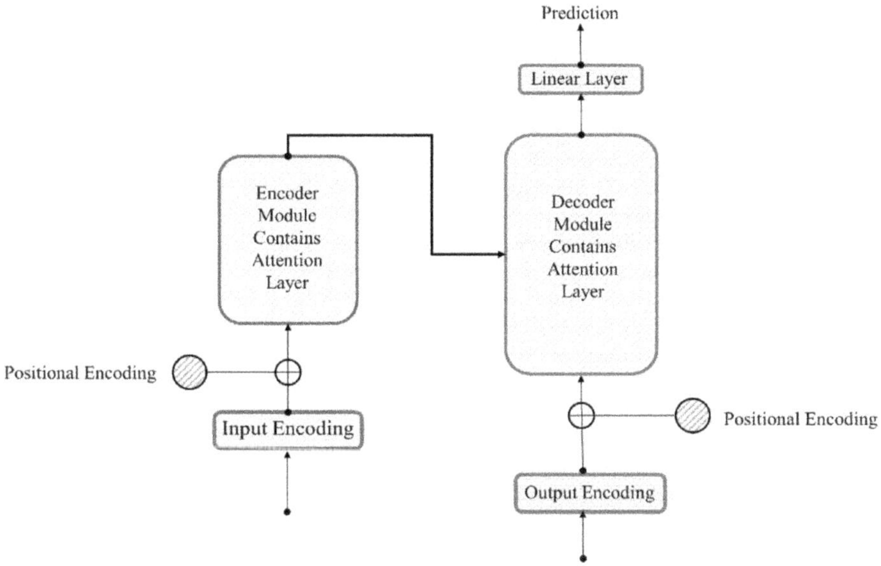

Fig. 5. (Color online) A simple view of a transformer with encoder and decoder modules.

This technique makes it possible for the transformer to process the entire input simultaneously and parallelly. Then, the attention layer is used to capture the latent connections between different segments. As before, after several attention layers, the extracted features are flattened to make them ready for processing by dense layers and final prediction.[49] These capabilities indicated the transformer as a multi-purpose method in the NLP field. Pre-trained models like bidirectional encoder representations from transformers (BERT)[50] and generative pre-trained transformer (GPT)[51] are among the numerous applications of the transformer.[52–54]

2.5. *Graph neural network*

In the past few years, by increasing their capability and expressive power, graph neural networks (GNN) have evolved into a practical tool for learning the representation of graph data structure without any feature engineering (Fig. 6).[55] A natural application of GNN is to apply them to problems that contain molecular graphs like

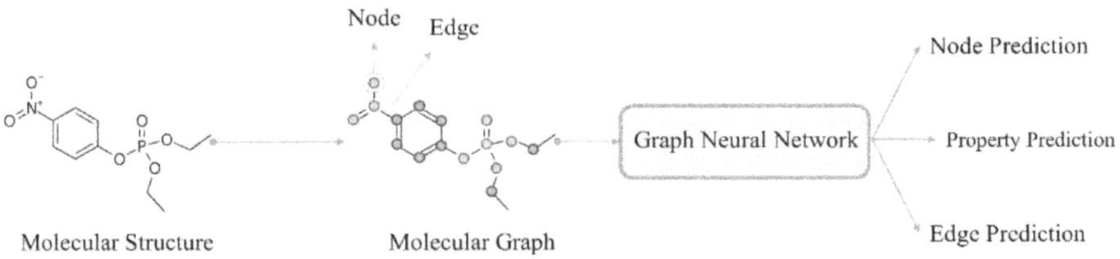

Fig. 6. (Color online) Graph neural networks accept graph inputs and do different tasks like node, edge and property prediction.

protein-ligand binding affinity prediction[56,57] and quantum chemistry.[58–60] A graph (G) typically is a tuple of sets of vertices (V) and edges (E) as $G = (V, E)$. Specific layers in GNN can accept a graph as an input and, after several operations, make an embedding of the whole graph. For example, in the message-passing scheme, these operations consist of three steps. In the first step, for each node of the graph, all neighborhood messages (features of each node) are gathered. Next, messages aggregate using an aggregate function. In the final step, new features (embedding) for each node are generated using an update function. By stacking message-passing GNN layers together, distance nodes can affect each other and incorporate information from across the entire graph. These networks can be used to do different tasks like node, edge and property prediction.

2.6. *Generative models*

In addition to prediction problems, the mentioned DL algorithms can be utilized for generation tasks in which the purpose of training a DL model is to generate new data instances with the same characteristics as the train set. Here, the model infers the distribution of the observed data and tries to emulate that. Two widespread schemes for generation tasks are variational autoencoder (VAE)[61] and generative adversarial network (GAN).[62] VAE is composed of encoder and decoder parts. At first, the encoder maps input from the features space to the latent space. This new space consists of parameters of distribution, which the decoder maps its value to features space and generates a data instance. Like VAE, GAN is composed of two components, which are called generator and discriminator. In practice, the objectives of the generator and discriminator are to produce new data instances and to distinguish them from reference data, which the generator tries to approximate its output distribution to the reference distribution of the observed data. By integrating the GAN scheme and aforementioned neural networks, several variants of GAN are proposed.[63,64] For example, in transformer GAN, both parts are based on transformer architecture instead of the original implementation of CNN.[65] Self-attention GAN enforces the self-attention mechanism for the generator and discriminator,[66] while GraphGAN accepts graph representation as the input.[67]

3. Applications of deep learning in enzymes

We have categorized DL's applications in enzyme science into several classes (Fig. 7). The first class deals with predicting enzyme function, which is typically described by an EC number. In the second class, DL predicts enzyme kinetic parameters. The third and fourth classes focus on condition optimization, such as temperature (T) and pH, as well as predicting the thermophilic properties of enzymes. The fifth and sixth classes involve predicting catalytic residues in an enzyme and its enzymatic reaction. We conclude this section by discussing the design of new enzymes using DL algorithms.

Up to now, various traditional machine learning approaches, such as nearest neighbor methods (kNN) and support vector machines (SVM), have been used for enzyme engineering, as documented in Refs. 16,18 and 68–74. In this study, we focus solely on research that applies DL algorithms. It's worth mentioning that the application of

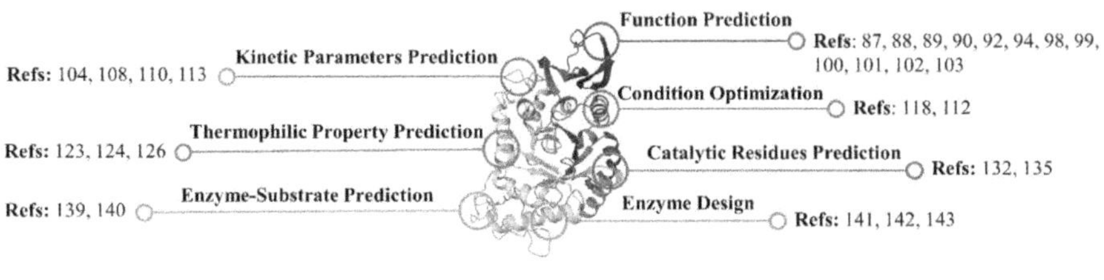

Fig. 7. (Color online) Application of DL in enzyme science. Refs indicate references to the reviewed paper. The displayed enzyme is an amylase with 1AMY PDB ID, which is generated by the RCSB PDB website (www.rcsb.org).

DL algorithms in enzyme science is a vast discipline, and we omitted the topics that are related to drug discovery, such as enzyme inhibition and allosteric site prediction.

3.1. *Enzyme function prediction*

Identifying an enzyme's function is critical to understanding its catalytic behavior,[75] proposing new metabolic pathways[76] and annotating a vast amount of genomic sequence data.[77] Additionally, some enzymes have multi-functional capability, which comprises a significant portion of all enzymes.

Enzymes are classified based on their selective catalytic reactions. To assign each enzyme a unique identifier, the International Union of Biochemistry and Molecular Biology (IUBMB) has proposed the enzyme commission (EC) number.[78] This number consists of four digits like $x.y.z.k$. The first digit specifies the general mechanism of the enzymes and divides them into seven categories: (1) Oxidoreductases, (2) Transferases, (3) Hydrolases, (4) Lyases, (5) Isomerases, (6) Ligases, (7) Translocases. The next three digits divide enzymes into distinct subcategories based on different elements involved in the reaction, such as the type of bond or substrate. For instance, the enzyme with EC 3.4.11.4 is called tripeptide aminopeptidase, and this is an enzyme that belongs to the family of hydrolases (EC 3), which hydrolyses the peptide bonds (EC 3.4) at the end of (EC 3.4.11) a tripeptide (EC 3.4.11.4) and releases an amino group.[79]

The conventional method of identifying an EC number is through experimental efforts, which are expensive and time-consuming. Therefore, devising a computational procedure to characterize enzyme capabilities is an alternative solution. In the past few decades, several bioinformatic approaches based on structure[80] and sequence similarity[81] (homology-based) have been introduced. In these methods, researchers use BLAST,[82,83] or similar techniques, to compare enzyme sequences against databases to indicate their functions by finding another similar enzyme whose function is experimentally known. Nonetheless, BLAST-based methods may be erroneous[84] and lack appropriate sensitivity when databases are growing in size and diversification.[75] In their large-scale evaluation, Radivojac *et al.* demonstrated that 40% of the annotated enzymes based on computational tools are incorrectly annotated.[85] Additionally, their efficacy is diminished when it comes to proteins that have not been extensively researched or those with unclear functions or multiple activities, i.e., promiscuous enzymes.[86]

The CNN has gained popularity in predicting EC numbers due to its exceptional ability to identify functional regions and domains in protein sequences. DeepEC[87] is a CNN-based model that predicts EC numbers based on a protein sequence as input, using three independent CNNs for the prediction task. The first CNN, also known as CNN-1, categorizes the input protein sequence as an enzyme or nonenzyme protein. On the other hand, CNN-2 and CNN-3 are dedicated to predicting third and fourth-level EC numbers and can assign a promiscuous enzyme to different classes. It is worth mentioning that these CNNs are trained and validated separately, and each protein sequence is represented by using an on-hot encoding scheme.

mlDEEPre, an enhanced version of DEEPre,[88] is a multi-functional enzyme predictor with two input layers. The first layer accepts one-hot encoded and position-specific-scoring-matrix (PSSM) obtained from evolutionary information of protein sequences, and the second layer accepts functional domain encoding of the protein — a vector composed of 0s and 1s. The first layer processes the input data through CNN, while the second layer extracts information using fully-connected layers. Finally, the outputs are concatenated, and the label is predicted based on this combined vector.[89] The EPP-HMCNF model utilizes information from the BRENDA database and is based on the hierarchical multi-label classification network and FFNN.[90] This data included natural and nonnatural substrates that interacted with enzymes and was applied to predict enzyme function.[91] Huang *et al.* compiled protein data from RCSB PDB and filtered it by CD-HIT[92] to reduce bias and selected only enzyme proteins. The researchers utilized several DL algorithms, including CNN, LSTM, GRU, BiLSTM, and an attention-based bidirectional-LSTM model (ABLE), along with the synthetic minority oversampling technique

(SMOTE) oversampling method to handle the enzyme class imbalance issue by creating new data points for minority classes. In the end, ABLE with SMOTE sampling demonstrated the best performance, achieving the highest recall and F-Score.[93]

Contrastive learning-enabled enzyme annotation (CLEAN)[86] is a newly published sequence-based function annotator that uses a contrastive learning scheme[94] to predict the likelihood of enzyme functions. To achieve this, CLEAN employs the UniProt database[95] and ESM-1b,[96] a pre-trained transformer-based language model to generate protein sequence embedding. During contrastive learning, CLEAN adjusts the embedding space to minimize the Euclidean distance between amino acid sequences with the same EC number and maximize the distance for those with different EC numbers. Similar to CLEAN, DeEPn[97] gathers protein sequence data from UniProt that meets several criteria, such as having a certain length. Each sequence is mapped to a high-dimensional vector, which serves as input to a series of convolutional layers for feature extraction. The final dense layer performs a classification task by assigning one of eight classes (seven EC number classes and one nonenzyme class) to the input sequence. Unlike the mentioned methods, which are solely based on sequential or structural features, Gao *et al.*[98] proposed a fusion model based on different biological information for EC number prediction. To this end, a dataset was compiled from RCSB PDB enzymes that belonged to six EC number classes. Three sets of features were generated to represent each enzyme from different perspectives. The first set consists of a 2D histogram representation of the PSSM derived by PSI-BLAST. PSSM represents the probability of mutating amino acids at each sequence position. The second and third sets, structure-based features, were derived by expressing the 3D structure of the enzyme using the two torsion angles of the polypeptide chain and the distance between pairwise amino acids, respectively. Three parallel CNNs were trained based on the three sets of enzyme representations for EC number prediction. The results showed that combining sequential and structural information led to better predictions than using either type of information alone.

Bio2Rxn used a set of six types of previously published EC number predictor tools, such as DeepEC, to perform a consensus-based EC number classification task.[99] Hierarchical enzyme classification network (HECNet), a hybrid approach, consists of a single network and a Siamese triplet network (STNet). HECNet is designed to handle the abundance of enzyme data at higher EC number levels and the scarcity of data at lower levels. The dataset was gathered from the Swiss-Prot database with annotated EC numbers up to the fourth level. Several categories of features (including the sequence itself, PSSM, disordered regions, secondary structure, solvent accessibility, amino acid composition and functional domains) were used to represent enzymes.[100] In their later publication, they developed mlHECNet, a multi-functional enzyme predictor with a similar architecture to HECNet.[101]

Strodthoff *et al.* introduced a different approach for the enzyme classification task. First, they developed a universal deep sequence model (UDSMProt) using self-supervised learning on unlabeled protein sequences from Swiss-Prot. They applied RNN architecture to help the model to learn the implicit representation of a protein sequence. Second, they fine-tuned the pre-trained model to predict enzyme EC number. The final model achieved better performance compared to the baseline models, which were trained from scratch or using PSSM features.[102]

3.2. *Enzyme kinetic parameters prediction*

Enzyme performance can be assessed using parameters such as k_{cat} and k_M. k_M or Michaelis constant is the required substrate concentration for an enzyme to operate at half of its maximum velocity. A higher k_M value indicates that the enzyme has a lower affinity for the substrate. On the other hand, k_{cat}, or the turnover number, represents the maximum number of substrate conversions per second that a single active site can facilitate at a given enzyme concentration. Understanding the values of k_M and k_{cat} is crucial for comprehending enzyme-substrate interactions and catalytic behavior.[2] However, it is difficult and time-consuming to obtain accurate estimates of these parameters experimentally, resulting in a lack of data for many enzymes.[103] For example, only about 10% of the k_{cat} values for enzyme-catalyzed reactions in *Escherichia coli* organisms are currently known.[104]

Kroll *et al.*[103] developed a general model for predicting k_M values. They collected data points from BRENDA,[90] including the organism name, EC number, UniProt ID, amino acid sequence, substrate information and associated k_M values. Substrate representation was derived from a task-specific GNN. UniRep,[105] a pre-trained LSTM model, produced a numerical representation of enzyme sequence. Two generated vectors were concatenated to create the final enzyme-substrate representation. Gradient boosting trees (GBT)[106] were used for the learning procedure. The results showed that the hybrid model based on the enzyme-substrate representation outperformed the pure enzyme or substrate models.

In a separate study, Kroll *et al.*[107] developed a model called turnover number prediction (TurNuP) for predicting the k_{cat}. They compiled a dataset from BRENDA, Sabio-RK,[108] and UniProt, which included information about enzyme sequences, reactant IDs, reaction equations and k_{cat} values. During the data preprocessing step, they removed nonwild-type enzymes, nonnatural reactions, identical data points, incomplete reactions and enzyme information. They transformed the reaction information into numerical reaction fingerprints and used a fine-tuned version of ESM-1b to generate an embedding of the enzyme sequences. The concatenated representation of two generated vectors was used to train a GBT algorithm. The final model, TurNuP, outperformed models based solely on reaction information or enzyme representation.[107]

The DLKcat[109] model utilizes substrate structures and enzyme sequences as inputs to predict k_{cat} values for various organisms. The DLKcat architecture consisted of a GNN for substrates and a CNN for enzyme amino acid sequences. Similar to Kroll *et al.*, the training, validation and test data were collected from the BRENDA and SABIO-RK databases. Additionally, DLKcat demonstrated the ability to distinguish between preferred and alternative substrates and identify enzyme promiscuity. Moreover, by implementing a neural attention mechanism, DLKcat provided a quantitative assessment of the importance of amino acids for the predicted enzyme activity. The creators of DLKcat have recently published the GotEnzymes database,[110] which encompasses over 25.7 million enzyme-compound pairs across 8,099 organisms, with their respective k_{cat} values predicted by DLKcat. It is worth mentioning that a recently published pre-print on bioRxiv shows that DLKcat does not have generalization ability. Its predictions are only accurate for cases where enzymes have highly similar analogs in the training set.[111]

A pre-trained language model-based k_{cat} predictor, named PreKcat, was developed using DLKcat data.[112] PreKcat utilized a concatenated representation vector of both the protein and substrate. The protein representation was derived from the ProtT5-XLUniRef50 model,[113] while the substrate representation was generated using a pre-trained SMILES transformer.[114] To train PreKcat, an extremely randomized trees (ERT) algorithm[115] was employed, resulting in a 20% improvement over DLKcat. To incorporate environmental factors such as temperature (T) and pH, a two-layer framework called EF-PreKcat was proposed. The base layer of this framework consisted of two individual models: PreKcat and Revised PreKcat. The Revised PreKcat model was trained using a smaller dataset that considered pH or temperature. The input of the revised PreKcat model comprised a concatenated representation vector of the protein and substrate along with the pH or temperature value. The meta-layer of the framework was a linear regression model that incorporated the predicted k_{cat} values from both the PreKcat and Revised PreKcat models. By utilizing information from both datasets, with and without environmental information, the accuracy of the final predicted k_{cat} value was significantly improved.[112]

3.3. *Condition optimization*

Enzymes that remain active at high temperatures (T) or far from the optimal pH are preferable for catalyzing chemical reactions in industrial and biotechnological processes. The traditional approach to finding enzymes with desirable pH and T is based on sequence homology with characterized enzymes.[116] However, this procedure is inadequate due to the possibility of sparse or absent representation of sequences with enzymatic activity, optimal pH and T.

EpHod (enzyme pH optimum prediction with DL) predicts the optimal pH of enzymes based on their protein sequences.[117] For this study, a primary dataset of enzymes with known optimal pH values was extracted from the BRENDA database. Additionally, a secondary dataset of secreted bacterial proteins with recorded environmental pH values for optimal growth was compiled from BacDive[118] and NCBI[119] databases. This secondary dataset was used for pre-training purposes. ESM-1v,[120] a pre-trained language model similar to ESM-1b, generated embedding inputs from the protein sequences. For the training and testing stage, a residual light attention architecture, consisting of CNN and fully connected layers, was employed in combination with transfer learning.

DeepET[121] adopted CNN layers to predict optimal catalytic temperatures (T_{opt}) from protein amino acid sequences. Since T_{opt} data are limited, DeepET employed transfer learning. DeepET was trained on optimal growth temperature (OGT) data, almost 3 million enzyme sequences from microorganisms with known OGTs, then fine-tuned on the T_{opt} dataset, 1,902 enzyme sequences with known T_{opt}. Both datasets were collected from the BRENDA database.

3.4. *Thermophilic property prediction*

The capability of enzymes to preserve their chemical and spatial structures without denaturation at high temperatures is called thermal stability. Enzymes with the mentioned ability are called thermophilic.[72,122] Thermal stability can be attributed to biological aspects of enzymes, such as amino acid composition, salt bridges and hydrogen and disulfide bonds. Thermophilic enzymes have an outstanding advantage in industrial production due to their increased catalytic efficiency at higher temperatures. However, differentiating between thermophilic and mesophilic proteins through experimental techniques is a time-consuming, labor-intensive and costly process.[72,122]

iThermo[123] was developed using a meticulously collected dataset of protein sequences in order to predict thermophilic and nonthermophilic proteins. The authors classified proteins in microorganisms with an OGT above 60°C as thermophilic and those with an OGT below 30°C as mesophilic. The protein sequences were transformed into numerical vectors using the iFeature package.[124] Redundant features were removed, and an MLP was chosen as the learning algorithm. By implementing a fusion strategy to incorporate valuable information from all generated features, iThermo achieved an appropriate performance.

BertThermo[125] is a DL model that uses the iThermo dataset to predict whether proteins are thermophilic or nonthermophilic based on their amino acid sequences. Unlike traditional methods that manually generate features, BertThermo employed deep representation learning to extract relevant features. It utilized a pre-trained BERT model to embed the protein sequence and processed the embeddings using a transformer. Finally, it employed a logistic function to classify the protein as either thermophilic or nonthermophilic. In 5-fold cross-validation and independent testing, BertThermo achieved high accuracy.

DeepTP[122] is a DL model designed to distinguish between thermophilic and mesophilic proteins. The initial dataset used for this model was obtained from the Li *et al.* paper[126] and was further processed to create a balanced dataset. Proteins with OGT above 60°C or below 37°C were classified as thermophilic or mesophilic, respectively. Each input sequence was represented using two embedding techniques and one biological feature generation method. These representations were processed through multiple layers, including CNN, BiLSTM and Self-Attention. The extracted representations were combined by concatenation and employed to make predictions by an MLP. DeepTP outperformed other state-of-the-art models.

3.5. *Enzyme catalytic residues prediction*

Understanding which residues in an enzyme are involved in catalytic activity is crucial for studying enzyme sequence, structure and function. Also, this characterization has crucial applications in drug design and enzyme engineering. However, experimental methods for identifying these catalytic residues, like before, are time-consuming, expensive and labor-intensive. Furthermore, bioinformatics methods that utilize a sequence approach can have high false negative rates for novel proteins.[127–130]

Torng and Altman[128] introduced 3DCNNs (3D CNNs) trained on the PROSITE database for structure-based enzyme site detection. 98.89% and 92.88% of nitric oxide synthase and TRYPSIN-like enzyme sites were detected in the Catalytic Site Atlas by 3DCNNs. While 3DCNNs only use structural information of enzymes, adaptive edge-gated graph attention neural network (AEGAN),[131] a recently published model, incorporates sequential and structural characteristics of proteins alongside manually annotated and verified enzyme catalytic data of UniProt and M-CSA[132] databases to predict enzyme catalytic site.

3.6. *Enzyme-substrate prediction*

Synthesizing organic compounds by using enzymes as biocatalysts is challenging because some compounds may not be converted by enzymes. Therefore, a complete characterization of an enzyme's function and the reactions it catalyzes is essential. Kreutter *et al.*[133] introduced a transformer model as computer-assisted synthetic planning (CASP). At first, they trained their model on a general chemical reaction dataset and then transferred it to a specialized dataset of enzymatic reactions of the Reaxys database. The resulting transformer model predicted whether an organic compound is a substrate for the applied enzyme and its enzymatic transformation outcome.

SEP-DNN, developed by Watanabe *et al.*,[134] is a dense neural network that is trained by collected reaction data from the Kyoto encyclopedia of genes and genomes (KEGG).[135] Each data instance consists of substrate-enzyme-product (SEP) information. ProtVec[136] and SMILESVec[137] are used for the numerical representation of enzymes and compounds, respectively. SEP-DNN achieved a suitable performance when applied to unknown enzymatic reactions that are not included in the training data.

Enzymatic link prediction (ELP)[138] is a unique method for predicting enzymatic transformations. ELP is based on the idea that enzymes and compounds in a chemical interaction are similar to those found in previously cataloged enzymatic reactions. To use ELP, known enzymatic reactions are mapped onto a graph structure, where compounds and reactions are represented as nodes and edges, respectively. This allows the prediction of a new substrate for an enzyme by simply predicting a link between the relevant nodes on the graph.

The enzyme substrate prediction (ESP) model, suggested by Kroll *et al.*,[139] is a general machine-learning model that predicts the enzyme-substrate pairs. Substrate and enzyme embeddings were produced using task-specific GNN and modified ESM-1b, respectively. A gradient-boosted decision tree model was trained on the combined enzyme and substrate representations, and testing it on an independent and diverse test set resulted in a good performance.

3.7. *Enzyme design*

Natural enzymes do not perform well in industrial conditions. Therefore, designing new enzymes with desired catalytic activity, substrate specificity, enantioselectivity, or thermostability is a crucial task.[140–142] Iterative cycles of mutagenesis and screening, also known as directed evolution, are a common protein engineering technique. But, for an enzyme with a length of N, the sequence space is composed of 20^N amino acid sequences, which makes exploring this enormous multidimensional space experimentally impossible and computationally intractable. To speed up the discovery procedure of a new enzyme, a generative modeling approach, like GAN, can be used to learn sequence space and the underlying data distribution to generate new samples.[73,74]

ProteinGAN,[143] a GAN-based generative model, learns the hidden pattern of protein sequences by adopting a temporal convolutional network to scrutinize local and global sequence features and self-attention mechanisms to focus on catalytic residues. ProteinGAN has been trained on filtered and processed data of bacterial malate dehydrogenase (MDH) sequences of the UniProt database. In testing, ProteinGAN generated more diverse sequences compared to training data containing novel mutations and structural domains, and their MDH catalytic activity was verified experimentally. Evolutionary context-integrated neural network or ECNet[144] is a DL model that maps protein sequence to its functional measurement, which is derived from mutational scanning. The specific protein's function often depends on residue interdependencies, which were incorporated into the supervised model using evolutionary information. Additionally, global protein structure was integrated into the model from an unsupervised protein language model. An LSTM model was used to learn the sequence-to-function relation of proteins. During testing, ECNet demonstrated better performance with respect to other traditional machine learning methods and captured interdependencies between mutations within a protein sequence. ECNet was used in a real experiment to engineer TEM-1 β-lactamase and identify variants with enhanced ampicillin resistance. Subsequently, these variants underwent experimental validation, which confirmed the capability of ECNet.

Ramírez-Palacios *et al.*[145] used a different approach to enzyme design. They employed graph convolutional neural networks (GCNNs) to explore the fitness landscape of ω-transaminase from *Vibrio fluvialis* (Vf-TA), which has industrial application in the production of chiral amines. The GCNNs were trained by a set of randomly mutated Vf-TA enzymes, whose binding energy between each variant and a constant substrate was calculated by the Rosetta suite. It is worth mentioning that each enzyme variant was represented as a graph formed by protein residues (nodes) with pairwise connections (edges). GCNNs showed an appropriate accuracy in the highly mutated dataset, which implied that GCNNs learned patterns for combining mutants from a few examples.[145]

4. Challenges and concluding remarks

Here, we reviewed recent advancements in DL applications in the enzyme field. Enzymes are biocatalysts with wide application in various industries. After reviewing DL concepts and algorithms, we investigated their applications in several categories: (1) Enzyme function prediction, (2) Kinetic parameters prediction, (3) Condition Optimization, (4) Thermophilic property prediction, (5) Catalytic residues prediction, (6) Enzyme-substrate prediction, (7) Enzyme design. Reviewed papers are gathered in Table 1.

These studies predominantly employed enzyme amino acid sequences, to derive a suitable representation for utilized algorithms and ignored the 3D structural information of enzymes. This limitation arises from the fact that the majority of enzyme databases only include sequential information, which, typically, is obtained from high-throughput screening experiments.

One additional limitation can be attributed to imbalanced and biased datasets. When enzymes with a specific EC number are abundant, or there is a hidden bias within the dataset, it can result in overfitting and a lack of generalization in the suggested model.[111] Gathering and reporting a negative control dataset, e.g., mutated enzyme sequences that lost their thermophilic property, using contrastive learning and sampling methods, are promising approaches to deal with the severe class imbalance. Also, applying realistic data splitting instead of random splitting prevents algorithms from overfitting.

While considerable developments have occurred in DL applications, the potential for enzyme engineering is still far from being fully explored. For example, predicting the effects of co-factors, co-solvents, or additives on enzyme behavior has not yet been thoroughly investigated. This drawback can be related to the scarcity of this kind of data in the established databases. Also, not all enzyme parameters, such as optimum temperature and pH or kinetic parameters, are reported. The reporting of experiment results in a specific format in scientific papers, which can be easily read by text-mining algorithms, might help solve data scarcity problems.[146]

Another way to deal with insufficient data is leveraging large amounts of unlabeled data to make a pre-trained model, which can be fine-tuned to specific tasks.[147] Also, the multimodal representations of

Table 1. Reviewed papers, their employed DL algorithms, and their utilized data are gathered in this table. GANN is a graph attention neural network. GCNN is a graph convolutional neural network. The rest of the abbreviations and acronyms are presented in the main text. "Various" means that data is compiled from different published papers.

Entry	Name	Application	DL Algorithm	Data	Reference
1	CLEAN	Function prediction	FFNN, Transformer	UniProt	86
2	DeepEC	Function prediction	CNN	Swiss-Prot, TrEMBL	87
3	DEEPre	Function prediction	CNN, LSTM	ENZYME, Swiss-Prot, PDB	88
4	mlDEEPre	Function prediction	CNN	UniProt	89
5	EPP-HMCNF	Function prediction	FFNN	BRENDA	91
6	ABLE	Function prediction	Attention based biLSTM	PDB	93
7	DeEPn	Function prediction	CNN	Swiss-Prot, TrEMBL	97
8	Gao *et al.*	Function prediction	CNN	PDB	98
9	Bio2Rxn	Function prediction	Consensus	Various	99
10	HECNet	Function prediction	CNN, LSTM, FFNN	Swiss-Prot	100
11	mlHECNet	Function prediction	CNN, LSTM, FFNN	Swiss-Prot	101

Table 1. (*Continued*)

Entry	Name	Application	DL Algorithm	Data	Reference
12	UDSMProt	Function prediction	LSTM	Various	102
13	Kroll *et al.*	Kinetic parameters prediction	GBT, LSTM, GNN	BRENDA	103
14	TurNuP	Kinetic parameters prediction	GBT, Transformer	BRENDA, Sabio-RK, UniProt	107
15	DLKcat	Kinetic parameters prediction	CNN, GNN	BRENDA, Sabio-RK	109
16	PreKcat	Kinetic parameters prediction	ERT, Transformer	BRENDA, Sabio-RK	112
17	EpHod	Condition Optimization	FFNN, CNN, Transformer	BRENDA, BacDive, NCBI	117
18	DeepET	Condition Optimization	CNN	BRENDA	121
19	iThermo	Thermophilic property prediction	FFNN	UniProt	123
20	BertThermo	Thermophilic property prediction	Transformer	UniProt	125
21	DeepTP	Thermophilic property prediction	FFNN, BiLSTM, CNN	Li *et al.*[126]	122
22	3DCNNs	Catalytic residues prediction	CNN	PROSITE	128

(*Continued*)

Table 1. (*Continued*)

Entry	Name	Application	DL Algorithm	Data	Reference
23	AEGAN	Catalytic residues prediction	GANN	UniProt, M-CSA	131
24	SEP-DNN	Enzyme-substrate prediction	FFNN	KEGG	134
25	ELP	Enzyme-substrate prediction	GNN	KEGG	138
26	ESP	Enzyme-substrate prediction	GBT, GNN, Transformer	UniProt	139
27	ProteinGAN	Enzyme design	GAN, CNN, Attention	UniProt	143
28	ECNet	Enzyme design	BiLSTM, Transformer	Various	144
29	Ramírez-Palacios *et al.*	Enzyme design	GCNN	—	145

input data and adopting a multi-task learning approach could further help to circumvent the data scarcity problem.[148]

However, in our opinion, the main challenges of this field are the lack of benchmark datasets and the validation of predicted values through experiments. These obstacles hinder the further development and application of DL algorithms. Finally, we hope that our overview paper on the current application of DL algorithms in enzyme science, which is far from an exhaustive and systematic review, will draw the attention of scientists to this field and help them to solve its current challenges and problems.

Supplementary information

Not applicable.

Statement of usage of artificial intelligence

The authors declare that no assistance from artificial intelligence was utilized in the preparation of this paper. The paper presented here is the result of our research without the involvement of any AI tools.

Data availability

Not applicable.

Authors contributions

M.R. wrote the main text. M.M., G.F. and A.M.L. read the paper, rewrote parts of it, and made some corrections.

Conflict of interest

The authors declare no conflict of interests.

Funding information

Funding is not applicable.

Acknowledgements

Not applicable.

Publisher's note

As per author request, corresponding authors updated as Morteza Mirzae and Ali Mohammad Latifi, which were not in the earlier published version.

Orcid

Milad Rayka https://orcid.org/0000-0002-4378-9609
Morteza Mirzaei https://orcid.org/0000-0001-5568-2422
Gholamreza Farnoosh https://orcid.org/0000-0001-8951-892X
Ali Mohammad Latifi https://orcid.org/0000-0002-8952-5174

References

1. Rost, B. Enzyme Function Less Conserved than Anticipated. *J. Mol. Biol.* **2002**, *318* (2), 595–608.
2. Berg, J.; Tymoczko, J.; Stryer, L. *Biochemistry*, 7th edn.; W.H. Freeman and Company, England, 2012.
3. Nelson, L.; Cox, M. *Lehninger Principles of Biochemistry*, 7th edn.; W.H. Freeman, New York, 2017.
4. Mirzaei, M.; Khorshahi, H.; Tebyanian, H.; Moradi, R. S.; Rastegar, M.; Panahi, S.; Sariri, R.; Latifi, A. M. The Effect of Deep Eutectic Solvents as Co-Solvent on Organophosphorus Hydrolase Targeting Engineering Enzyme-Catalyzed. *Biointerface Res. Appl. Chem.* **2020**, *10* (5), 6488–6497.
5. Saboury, A. Enzyme Inhibition and Activation: A General Theory. *J. Iran. Chem. Soc.* **2009**, *6*, 219–229.
6. Grahame, D.; Bryksa, B.; Yada, R. Factors Affecting Enzyme Activity, in *Improving and Tailoring Enzymes for Food Quality and Functionality*, Elsevier, 2015, pp. 11–55.
7. Fischer, J. D.; Holliday, G. L.; Thornton, J. M. The CoFactor Database: Organic Cofactors in Enzyme Catalysis. *Bioinformatics* **2010**, *26* (19), 2496–2497.
8. Vellard, M. The Enzyme as Drug: Application of Enzymes as Pharmaceuticals. *Curr. Opin. Biotechnol.* **2003**, *14* (4), 444–450.
9. Dinmukhamed, T.; Huang, Z.; Liu, Y.; Lv, X.; Li, J.; Du, G.; Liu, L. Current Advances in Design and Engineering Strategies of Industrial Enzymes. *Syst. Microbiol. Biomanuf.* **2021**, *1*, 15–23.

10. Tarafdar, A.; Sirohi, R.; Gaur, V. K.; Kumar, S.; Sharma, P.; Varjani, S.; Pandey, H. O.; Sindhu, R.; Madhavan, A.; Rajasekharan, R. Engineering Interventions in Enzyme Production: Lab to Industrial Scale. *Bioresour. Technol.* **2021**, *326*, 124771.
11. Heckmann, C. M.; Paradisi, F. Looking Back: A Short History of the Discovery of Enzymes and How they Became Powerful Chemical Tools. *ChemCatChem* **2020**, *12* (24), 6082–6102.
12. Kirk, O.; Borchert, T. V.; Fuglsang, C. C. Industrial Enzyme Applications. *Curr. Opin. Biotechnol.* **2002**, *13* (4), 345–351.
13. Ebert, M. C.; Pelletier, J. N. Computational Tools for Enzyme Improvement: Why Everyone Can–and Should–Use Them. *Curr. Opin. Chem. Biol.* **2017**, *37*, 89–96.
14. Damborsky, J.; Brezovsky, J. Computational Tools for Designing and Engineering Enzymes. *Curr. Opin. Chem. Biol.* **2014**, *19*, 8–16.
15. Damborsky, J.; Brezovsky, J. Computational Tools for Designing and Engineering Biocatalysts. *Curr. Opin. Chem. Biol.* **2009**, *13* (1), 26–34.
16. Feehan, R.; Montezano, D.; Slusky, J. S. Machine Learning for Enzyme Engineering, Selection and Design. *Protein Eng. Des. Sel.* **2021**, *34*, gzab019.
17. Siedhoff, N. E.; Schwaneberg, U.; Davari, M. D. Machine Learning–Assisted Enzyme Engineering. *Methods Enzymol.* **2020**, *643*, 281–315.
18. Mazurenko, S.; Prokop, Z.; Damborsky, J. Machine Learning in Enzyme Engineering. *ACS Catal.* **2019**, *10* (2), 1210–1223.
19. LeCun, Y.; Bengio, Y.; Hinton, G. Deep Learning. *Nature* **2015**, *521* (7553), 436–444.
20. Shinde, P. P.; Shah, S.A Review of Machine Learning and Deep Learning Applications. In *2018 Fourth Int. Conf. Computing Communication Control and Automation (ICCUBEA)*, IEEE, 2018, pp. 1–6.
21. Goodfellow, I.; Bengio, Y.; Courville, A. *Deep Learning*, MIT Press, 2016.
22. Paszke, A.; Gross, S.; Chintala, S.; Chanan, G.; Yang, E.; DeVito, Z.; Lin, Z.; Desmaison, A.; Antiga, L.; Lerer, A. Automatic Differentiation in Pytorch. In *31st Conf. Neural Information Processing Systems NeurIPS 2017*, Long Beach, CA, USA, 2017, pp. 1–4.
23. Abadi, M.; Agarwal, A.; Barham, P.; Brevdo, E.; Chen, Z.; Citro, C.; Corrado, G. S.; Davis, A.; Dean, J.; Devin, M. TensorFlow: Large-Scale Machine Learning on Heterogeneous Distributed Systems. arXiv:1603.04467.
24. Pandey, M.; Fernandez, M.; Gentile, F.; Isayev, O.; Tropsha, A.; Stern, A. C.; Cherkasov, A. The Transformational Role of GPU Computing and Deep Learning in Drug Discovery. *Nat. Mach. Intell.* **2022**, *4* (3), 211–221.
25. Wang, Y. E.; Wei, G.-Y.; Brooks, D. M. Benchmarking TPU, GPU, and CPU Platforms for Deep Learning. arXiv:1907.10701.
26. Dill, K. A.; Ozkan, S. B.; Shell, M. S.; Weikl, T. R. The Protein Folding Problem. *Annu. Rev. Biophys.* **2008**, *37*, 289–316.
27. Al-Lazikani, B.; Jung, J.; Xiang, Z.; Honig, B. Protein Structure Prediction. *Curr. Opin. Chem. Biol.* **2001**, *5* (1), 51–56.
28. Kuhlman, B.; Bradley, P. Advances in Protein Structure Prediction and Design. *Nat. Rev. Mol. Cell Biol.* **2019**, *20* (11), 681–697.

29. Kryshtafovych, A.; Schwede, T.; Topf, M.; Fidelis, K.; Moult, J. Critical Assessment of Methods of Protein Structure Prediction (CASP) — Round XIV. *Proteins Struct. Funct. Bioinf.* **2021**, *89* (12), 1607–1617.
30. Jumper, J.; Evans, R.; Pritzel, A.; Green, T.; Figurnov, M.; Ronneberger, O.; Tunyasuvunakool, K.; Bates, R.; Žídek, A.; Potapenko, A. Highly Accurate Protein Structure Prediction with AlphaFold. *Nature* **2021**, *596* (7873), 583–589.
31. Mater, A. C.; Coote, M. L. Deep Learning in Chemistry. *J. Chem. Inf. Model.* **2019**, *59* (6), 2545–2559.
32. Tanaka, A.; Tomiya, A.; Hashimoto, K. *Deep Learning and Physics*, Springer, 2021.
33. Iqbal, M. S.; Ahmad, I.; Khan, T.; Khan, S.; Ahmad, M.; Wang, L. Recent Advances of Deep Learning in Biology, in *Deep Learning for Unmanned Systems*, Springer, 2021, pp. 709–732.
34. Wang, F.; Casalino, L. P.; Khullar, D. Deep Learning in Medicine — Promise, Progress, and Challenges. *JAMA Intern. Med.* **2019**, *179* (3), 293–294.
35. Lu, P.; Qiu, L.; Yu, W.; Welleck, S.; Chang, K.-W. A Survey of Deep Learning for Mathematical Reasoning. arXiv:2212.10535.
36. Zhang, J.; Lei, Y.-K.; Zhang, Z.; Chang, J.; Li, M.; Han, X.; Yang, L.; Yang, Y. I.; Gao, Y. Q. A Perspective on Deep Learning for Molecular Modeling and Simulations. *J. Phys. Chem. A* **2020**, *124* (34), 6745–6763.
37. Su, C.; Xu, Z.; Pathak, J.; Wang, F. Deep Learning in Mental Health Outcome Research: A Scoping Review. *Transl. Psychiatry* **2020**, *10* (1), 116.
38. Chen, L.; Wang, L.; Miao, J.; Gao, H.; Zhang, Y.; Yao, Y.; Bai, M.; Mei, L.; He, J. Review of the Application of Big Data and Artificial Intelligence in Geology. *J. Phys.: Conf. Ser.* **2020**, *1684*, 012007.
39. Géron, A. *Hands-on Machine Learning with Scikit-Learn, Keras, and Tensorflow*, O'Reilly Media, 2022.
40. Todeschini, R.; Consonni, V. *Handbook of Molecular Descriptors*, John Wiley & Sons, 2008.
41. Marsland, S. *Machine Learning: An Algorithmic Perspective*, CRC Press, 2015.
42. Goh, G. B.; Hodas, N. O.; Vishnu, A. Deep Learning for Computational Chemistry. *J. Comput. Chem.* **2017**, *38* (16), 1291–1307.
43. Aggarwal, C. C. *Neural Networks and Deep Learning*, Springer, 2018.
44. O'Shea, K.; Nash, R. An Introduction to Convolutional Neural Networks. arXiv:1511.08458.
45. Li, Z.; Liu, F.; Yang, W.; Peng, S.; Zhou, J. A Survey of Convolutional Neural Networks: Analysis, Applications, and Prospects. *IEEE Trans. Neural Networks Learn. Syst.* **2021**, *33* (12), 6999–7019.
46. Hochreiter, S.; Schmidhuber, J. Long Short-Term Memory. *Neural Comput.* **1997**, *9* (8), 1735–1780.
47. Cho, K.; Van Merriënboer, B.; Gulcehre, C.; Bahdanau, D.; Bougares, F.; Schwenk, H.; Bengio, Y. Learning Phrase Representations using RNN Encoder-Decoder for Statistical Machine Translation. arXiv:1406.1078.
48. Schuster, M.; Paliwal, K. K. Bidirectional Recurrent Neural Networks. *IEEE Trans. Signal Process.* **1997**, *45* (11), 2673–2681.

49. Vaswani, A.; Shazeer, N.; Parmar, N.; Uszkoreit, J.; Jones, L.; Gomez, A. N.; Kaiser, Ł.; Polosukhin, I. Attention is All You Need. *Adv. Neural Inf. Process. Syst.* **2017**, *30*, 1–11.
50. Devlin, J.; Chang, M.-W.; Lee, K.; Toutanova. K. BERT: Pre-Training of Deep Bidirectional Transformers for Language Understanding. arXiv:1810.04805.
51. Radford, A.; Wu, J.; Child, R.; Luan, D.; Amodei, D.; Sutskever, I. Language Models are Unsupervised Multitask Learners. *OpenAI Blog* **2019**, *1* (8), 9.
52. Zhang, S.; Fan, R.; Liu, Y.; Chen, S.; Liu, Q.; Zeng, W. Applications of Transformer-based Language Models in Bioinformatics: A Survey. *Bioinf. Adv.* **2023**, *3* (1), vbad001.
53. Jaume-Santero, F.; Bornet, A.; Valery, A.; Naderi, N.; Vicente Alvarez, D.; Proios, D.; Yazdani, A.; Bournez, C.; Fessard, T.; Teodoro, D. Transformer Performance for Chemical Reactions: Analysis of Different Predictive and Evaluation Scenarios. *J. Chem. Inf. Model.* **2023**, *63* (7), 1914–1924.
54. Shamshad, F.; Khan, S.; Zamir, S. W.; Khan, M. H.; Hayat, M.; Khan, F. S.; Fu, H. Transformers in Medical Imaging: A Survey. arXiv:2201.09873.
55. Yi, H.-C.; You, Z.-H.; Huang, D.-S.; Kwoh, C. K. Graph Representation Learning in Bioinformatics: Trends, Methods and Applications. *Briefings Bioinf.* **2022**, *23* (1), bbab340.
56. Masters, M. R.; Mahmoud, A. H.; Wei, Y.; Lill, M. A. Deep Learning Model for Efficient Protein–Ligand Docking with Implicit Side-Chain Flexibility. *J. Chem. Inf. Model.* **2023**, *63* (6), 1695–1707.
57. Shen, C.; Zhang, X.; Deng, Y.; Gao, J.; Wang, D.; Xu, L.; Pan, P.; Hou, T.; Kang, Y. Boosting Protein–Ligand Binding Pose Prediction and Virtual Screening based on Residue–Atom Distance Likelihood Potential and Graph Transformer. *J. Med. Chem.* **2022**, *65* (15), 10691–10706.
58. Fung, V.; Zhang, J.; Juarez, E.; Sumpter, B. G. Benchmarking Graph Neural Networks for Materials Chemistry. *Npj Comput. Mater.* **2021**, *7* (1), 84.
59. Schütt, K. T.; Sauceda, H. E.; Kindermans, P.-J.; Tkatchenko, A.; Müller, K.-R. SchNet — A Deep Learning Architecture for Molecules and Materials. *J. Chem. Phys.* **2018**, *148* (24), 241722.
60. Schütt, K.; Kessel, P.; Gastegger, M.; Nicoli, K.; Tkatchenko, A.; Müller, K.-R. SchNetPack: A Deep Learning Toolbox for Atomistic Systems. *J. Chem. Theory Comput.* **2018**, *15* (1), 448–455.
61. Cinelli, L. P.; Marins, M. A.; Da Silva, E. A. B.; Netto, S. L. *Variational Methods for Machine Learning with Applications to Deep Networks*, Springer, 2021.
62. Goodfellow, I.; Pouget-Abadie, J.; Mirza, M.; Xu, B.; Warde-Farley, D.; Ozair, S.; Courville, A.; Bengio, Y. Generative Adversarial Networks. *Commun. ACM* **2020**, *63* (11), 139–144.
63. Creswell, A.; White, T.; Dumoulin, V.; Arulkumaran, K.; Sengupta, B.; Bharath, A. A. Generative Adversarial Networks: An Overview. *IEEE Signal Process. Mag.* **2018**, *35* (1), 53–65.
64. Aggarwal, A.; Mittal, M.; Battineni, G. Generative Adversarial Network: An Overview of Theory and Applications. *Int. J. Inf. Manage. Data Insights* **2021**, *1* (1), 100004.

65. Jiang, Y.; Chang, S.; Wang, Z. TransGAN: Two Pure Transformers Can Make One Strong GAN, and That Can Scale Up. *Adv. Neural Inf. Process. Syst.* **2021**, *34*, 14745–14758.
66. Zhang, H.; Goodfellow, I.; Metaxas, D.; Odena, A. Self-Attention Generative Adversarial Networks. In *Proc. 36th Int. Conf. Machine Learning, PMLR,* **2019**, pp. 7354–7363.
67. Wang, H.; Wang, J.; Wang, J.; Zhao, M.; Zhang, W.; Zhang, F.; Xie, X.; Guo, M. GraphGAN: Graph Representation Learning with Generative Adversarial Nets. In *Proc. AAAI Conf. Artificial Intelligence,* AIAA, **2018**, pp. 2508–2515.
68. Jani, V.; Patel, W.; Shah, S. A Theoretical Evaluation of Machine Learning to Predict Enzyme Family. In *2022 Int. Conf. Electronics and Renewable Systems (ICEARS),* IEEE, 2022, pp. 1387–1392.
69. Bonetta, R.; Valentino, G. Machine Learning Techniques for Protein Function Prediction. *Proteins Struct. Funct. Bioinf.* **2020**, *88* (3), 397–413.
70. Tan, J.-X.; Lv, H.; Wang, F.; Dao, F.-Y.; Chen, W.; Ding, H. A Survey for Predicting Enzyme Family Classes using Machine Learning Methods. *Curr. Drug Targets* **2019**, *20* (5), 540–550.
71. Tao, Z.; Dong, B.; Teng, Z.; Zhao, Y. The Classification of Enzymes by Deep Learning. *IEEE Access* **2020**, *8*, 89802–89811.
72. Dou, Z.; Sun, Y.; Jiang, X.; Wu, X.; Li, Y.; Gong, B.; Wang, L. Data-Driven Strategies for the Computational Design of Enzyme Thermal Stability: Trends, Perspectives, and Prospects: Data-Driven Strategies for Enzyme Thermostability Design. *Acta Biochim. Biophys. Sin.* **2023**, *55* (3), 343.
73. Cui, Y.; Sun, J.; Wu, B. Computational Enzyme Redesign: Large Jumps in Function. *Trends Chem.* **2022**, *4* (5), 409–419.
74. Ming, Y.; Wang, W.; Yin, R.; Zeng, M.; Tang, L.; Tang, S.; Li, M. A Review of Enzyme Design in Catalytic Stability by Artificial Intelligence. *Briefings Bioinf.* **2023**, *24* (3), bbad065.
75. Friedberg, I. Automated Protein Function Prediction — The Genomic Challenge. *Briefings Bioinf.* **2006**, *7* (3), 225–242.
76. Finley, S. D.; Broadbelt, L. J.; Hatzimanikatis, V. Computational Framework for Predictive Biodegradation. *Biotechnol. Bioeng.* **2009**, *104* (6), 1086–1097.
77. Kodama, Y.; Shumway, M.; Leinonen, R. The Sequence Read Archive: Explosive Growth of Sequencing Data. *Nucleic Acids Res.* **2012**, *40* (D1), D54–D56.
78. McDonald, A. G.; Tipton, K. F. Enzyme Nomenclature and Classification: The State of the Art. *FEBS J.* **2023**, *290* (9), 2214–2231.
79. Cornish-Bowden, A. Current IUBMB Recommendations on Enzyme Nomenclature and Kinetics. *Perspect. Sci.* **2014**, *1* (1–6), 74–87.
80. Dobson, P. D.; Doig, A. J. Predicting Enzyme Class from Protein Structure Without Alignments. *J. Mol. Biol.* **2005**, *345* (1), 187–199.
81. Quester, S.; Schomburg, D. Enzymedetector: An Integrated Enzyme Function Prediction Tool and Database. *BMC Bioinf.* **2011**, *12* (1), 1–13.

82. Altschul, S. F.; Madden, T. L.; Schäffer, A. A.; Zhang, J.; Zhang, Z.; Miller, W.; Lipman, D. J. Gapped BLAST and PSI-BLAST: A New Generation of Protein Database Search Programs. *Nucleic Acids Res.* **1997**, *25* (17), 3389–3402.
83. Altschul, S. F.; Gish, W.; Miller, W.; Myers, E. W.; Lipman, D. J. Basic Local Alignment Search Tool. *J. Mol. Biol.* **1990**, *215* (3), 403–410.
84. Shah, I.; Hunter, L. Predicting Enzyme Function from Sequence: A Systematic Appraisal. In *Proc. Int. Conf. Intelligent Systems for Molecular Biology*, NIH Public Access, **1997**, pp. 276–283.
85. Radivojac, P.; Clark, W. T.; Oron, T. R.; Schnoes, A. M.; Wittkop, T.; Sokolov, A.; Graim, K.; Funk, C.; Verspoor, K.; Ben-Hur, A. A Large-Scale Evaluation of Computational Protein Function Prediction. *Nat. Methods* **2013**, *10* (3), 221–227.
86. Yu, T.; Cui, H.; Li, J. C.; Luo, Y.; Jiang, G.; Zhao, H. Enzyme Function Prediction using Contrastive Learning. *Science* **2023**, *379* (6639), 1358–1363.
87. Ryu, J. Y.; Kim, H. U.; Lee, S. Y. Deep Learning Enables High-Quality and High-Throughput Prediction of Enzyme Commission Numbers. *Proc. Natl. Acad. Sci. USA* **2019**, *116* (28), 13996–14001.
88. Li, Y.; Wang, S.; Umarov, R.; Xie, B.; Fan, M.; Li, L.; Gao, X. DEEPre: Sequence-based Enzyme EC Number Prediction by Deep Learning. *Bioinformatics* **2018**, *34* (5), 760–769.
89. Zou, Z.; Tian, S.; Gao, X.; Li, Y. mlDEEPre: Multi-Functional Enzyme Function Prediction with Hierarchical Multi-Label Deep Learning. *Front. Genet.* **2019**, *9*, 714.
90. Placzek, S.; Schomburg, I.; Chang, A.; Jeske, L.; Ulbrich, M.; Tillack, J.; Schomburg, D. BRENDA in 2017: New Perspectives and New Tools in BRENDA. *Nucleic Acids Res.* **2017**, *45* (D1), D380–D388.
91. Visani, G. M.; Hughes, M. C.; Hassoun, S. Enzyme Promiscuity Prediction using Hierarchy-Informed Multi-Label Classification. *Bioinformatics* **2021**, *37* (14), 2017–2024.
92. Huang, Y.; Niu, B.; Gao, Y.; Fu, L.; Li, W. CD-HIT Suite: A Web Server for Clustering and Comparing Biological Sequences. *Bioinformatics* **2010**, *26* (5), 680–682.
93. Nallapareddy, M. V.; Dwivedula, R. ABLE: Attention Based Learning for Enzyme Classification. *Comput. Biol. Chem.* **2021**, *94*, 107558.
94. Heinzinger, M.; Littmann, M.; Sillitoe, I.; Bordin, N.; Orengo, C.; Rost, B. Contrastive Learning on Protein Embeddings Enlightens Midnight Zone. *NAR Genomics Bioinf.* **2022**, *4* (2), lqac043.
95. UniProt: The Universal Protein Knowledgebase in 2021. *Nucleic Acids Res.* **2021**, *49* (D1), D480–D489.
96. Rives, A.; Meier, J.; Sercu, T.; Goyal, S.; Lin, Z.; Liu, J.; Guo, D.; Ott, M.; Zitnick, C. L.; Ma, J. Biological Structure and Function Emerge from Scaling Unsupervised Learning to 250 Million Protein Sequences. *Proc. Natl. Acad. Sci. USA* **2021**, *118* (15), e2016239118.
97. Semwal, R.; Aier, I.; Tyagi, P.; Varadwaj, P. K. DeEPn: A Deep Neural Network based Tool for Enzyme Functional Annotation. *J. Biomol. Struct. Dyn.* **2021**, *39* (8), 2733–2743.

98. Gao, R.; Wang, M.; Zhou, J.; Fu, Y.; Liang, M.; Guo, D.; Nie, J. Prediction of Enzyme Function based on Three Parallel Deep CNN and Amino Acid Mutation. *Int. J. Mol. Sci.* **2019**, *20* (11), 2845.
99. Zhang, T.; Tian, Y.; Yuan, L.; Chen, F.; Ren, A.; Hu, Q.-N. Bio2Rxn: Sequence-based Enzymatic Reaction Predictions by a Consensus Strategy. *Bioinformatics* **2020**, *36* (11), 3600–3601.
100. Memon, S. A.; Khan, K. A.; Naveed, H. HECNet: A Hierarchical Approach to Enzyme Function Classification using a Siamese Triplet Network. *Bioinformatics* **2020**, *36* (17), 4583–4589.
101. Khan, K. A.; Memon, S. A.; Naveed, H. A Hierarchical Deep Learning Based Approach for Multi-Functional Enzyme Classification. *Protein Sci.* **2021**, *30* (9), 1935–1945.
102. Strodthoff, N.; Wagner, P.; Wenzel, M.; Samek, W. UDSMProt: Universal Deep Sequence Models for Protein Classification. *Bioinformatics* **2020**, *36* (8), 2401–2409.
103. Kroll, A.; Engqvist, M. K.; Heckmann, D.; Lercher, M. J. Deep Learning Allows Genome-Scale Prediction of Michaelis Constants from Structural Features. *PLoS Biol.* **2021**, *19* (10), e3001402.
104. Davidi, D.; Noor, E.; Liebermeister, W.; Bar-Even, A.; Flamholz, A.; Tummler, K.; Barenholz, U.; Goldenfeld, M.; Shlomi, T.; Milo, R. Global Characterization of in Vivo Enzyme Catalytic Rates and their Correspondence to in Vitro k_{Cat} Measurements. *Proc. Natl. Acad. Sci. USA* **2016**, *113* (12), 3401–3406.
105. Alley, E. C.; Khimulya, G.; Biswas, S.; AlQuraishi, M.; Church, G. M. Unified Rational Protein Engineering with Sequence-based Deep Representation Learning. *Nat. Methods* **2019**, *16* (12), 1315–1322.
106. Friedman, J. H. Greedy Function Approximation: A Gradient Boosting Machine. *Ann. Stat.* **2001**, *29*, 1189–1232.
107. Kroll, A.; Rousset, Y.; Hu, X.-P.; Liebrand, N. A.; Lercher, M. J. Turnover Number Predictions for Kinetically Uncharacterized Enzymes using Machine and Deep Learning. *Nat. Commun.* **2023**, *14* (1), 4139.
108. Wittig, U.; Kania, R.; Golebiewski, M.; Rey, M.; Shi, L.; Jong, L.; Algaa, E.; Weidemann, A.; Sauer-Danzwith, H.; Mir, S. SABIO-RK — Database for Biochemical Reaction Kinetics. *Nucleic Acids Res.* **2012**, *40* (D1), D790–D796.
109. Li, F.; Yuan, L.; Lu, H.; Li, G.; Chen, Y.; Engqvist, M. K.; Kerkhoven, E. J.; Nielsen, J. Deep Learning-based k_{cat} Prediction Enables Improved Enzyme-Constrained Model Reconstruction. *Nat. Catal.* **2022**, *5* (8), 662–672.
110. Li, F.; Chen, Y.; Anton, M.; Nielsen, J. GotEnzymes: An Extensive Database of Enzyme Parameter Predictions. *Nucleic Acids Res.* **2023**, *51* (D1), D583–D586.
111. Kroll, A.; Lercher, M. J. Machine Learning Models for the Prediction of Enzyme Properties should be Tested on Proteins not used for Model Training. bioRxiv:10.1101/2023.02.06.526991.
112. Yu, H.; Deng, H.; He, J.; Keasling, J.; Luo, X. Highly Accurate Enzyme Turnover Number Prediction and Enzyme Engineering with PreKcat. *Research Square*, **2023**.

113. Elnaggar, A.; Heinzinger, M.; Dallago, C.; Rehawi, G.; Wang, Y.; Jones, L.; Gibbs, T.; Feher, T.; Angerer, C.; Steinegger, M. ProtTrans: Toward Understanding the Language of Life Through Self-Supervised Learning. *IEEE Trans. Pattern Anal. Mach. Intell.* **2021**, *44* (10), 7112–7127.
114. Honda, S.; Shi, S.; Ueda, H. R. SMILES Transformer: Pre-Trained Molecular Fingerprint for Low Data Drug Discovery. arXiv:1911.04738.
115. Geurts, P.; Ernst, D.; Wehenkel, L. Extremely Randomized Trees. *Mach. Learn.* **2006**, *63*, 3–42.
116. Ferrer, M.; Golyshina, O.; Beloqui, A.; Golyshin, P. N. Mining Enzymes from Extreme Environments. *Curr. Opin. Microbiol.* **2007**, *10* (3), 207–214.
117. Gado, J. E.; Knotts, M.; Shaw, A. Y.; Marks, D.; Gauthier, N. P.; Sander, C.; Beckham, G. T. Deep Learning Prediction of Enzyme Optimum pH. bioRxiv:10.1101/2023.06.22.544776.
118. Reimer, L. C.; Sardà Carbasse, J.; Koblitz, J.; Ebeling, C.; Podstawka, A.; Overmann, J. Bac Dive in 2022: The Knowledge Base for Standardized Bacterial and Archaeal Data. *Nucleic Acids Res.* **2022**, *50* (D1), D741–D746.
119. Sayers, E. W.; Beck, J.; Bolton, E. E.; Bourexis, D.; Brister, J. R.; Canese, K.; Comeau, D. C.; Funk, K.; Kim, S.; Klimke, W. Database Resources of the National Center for Biotechnology Information. *Nucleic Acids Res.* **2021**, *49* (D1), D10–D17.
120. Meier, J.; Rao, R.; Verkuil, R.; Liu, J.; Sercu, T.; Rives, A. Language Models Enable Zero-Shot Prediction of the Effects of Mutations on Protein Function. *Adv. Neural Inf. Process. Syst.* **2021**, *34*, 29287–29303.
121. Li, G.; Buric, F.; Zrimec, J.; Viknander, S.; Nielsen, J.; Zelezniak, A.; Engqvist, M. K. Learning Deep Representations of Enzyme Thermal Adaptation. *Protein Sci.* **2022**, *31* (12), e4480.
122. Zhao, J.; Yan, W.; Yang, Y. DeepTP: A Deep Learning Model for Thermophilic Protein Prediction. *Int. J. Mol. Sci.* **2023**, *24* (3), 2217.
123. Ahmed, Z.; Zulfiqar, H.; Khan, A. A.; Gul, I.; Dao, F.-Y.; Zhang, Z.-Y.; Yu, X.-L.; Tang, L. iThermo: A Sequence-based Model for Identifying Thermophilic Proteins using a Multi-Feature Fusion Strategy. *Front. Microbiol.* **2022**, *13*, 790063.
124. Chen, Z.; Zhao, P.; Li, F.; Leier, A.; Marquez-Lago, T. T.; Wang, Y.; Webb, G. I.; Smith, A. I.; Daly, R. J.; Chou, K.-C. iFeature: A Python Package and Web Server for Features Extraction and Selection from Protein and Peptide Sequences. *Bioinformatics* **2018**, *34* (14), 2499–2502.
125. Pei, H.; Li, J.; Ma, S.; Jiang, J.; Li, M.; Zou, Q.; Lv, Z. Identification of Thermophilic Proteins Based on Sequence-based Bidirectional Representations from Transformer-Embedding Features. *Appl. Sci.* **2023**, *13* (5), 2858.
126. Li, G.; Rabe, K. S.; Nielsen, J.; Engqvist, M. K. Machine Learning Applied to Predicting Microorganism Growth Temperatures and Enzyme Catalytic Optima. *ACS Synth. Biol.* **2019**, *8* (6), 1411–1420.
127. Song, J.; Li, F.; Takemoto, K.; Haffari, G.; Akutsu, T.; Chou, K.-C.; Webb, G. I. PREvaIL, an Integrative Approach for Inferring Catalytic Residues using Sequence, Structural, and Network Features in a Machine-Learning Framework. *J. Theor. Biol.* **2018**, *443*, 125–137.

128. Torng, W.; Altman, R. B. High Precision Protein Functional Site Detection using 3D Convolutional Neural Networks. *Bioinformatics* **2019**, *35* (9), 1503–1512.
129. Das, S.; Scholes, H. M.; Sen, N.; Orengo, C. CATH Functional Families Predict Functional Sites in Proteins. *Bioinformatics* **2021**, *37* (8), 1099–1106.
130. Feehan, R.; Franklin, M. W.; Slusky, J. S. Machine Learning Differentiates Enzymatic and Non-Enzymatic Metals in Proteins. *Nat. Commun.* **2021**, *12* (1), 3712.
131. Shen, X.; Zhang, S.; Long, J.; Chen, C.; Wang, M.; Cui, Z.; Chen, B.; Tan, T. A Highly Sensitive Model based on Graph Neural Networks for Enzyme Key Catalytic Residue Prediction. *J. Chem. Inf. Model.* **2023**, *63* (14), 4277–4290.
132. Ribeiro, A. J. M.; Holliday, G. L.; Furnham, N.; Tyzack, J. D.; Ferris, K.; Thornton, J. M. Mechanism and Catalytic Site Atlas (M-CSA): A Database of Enzyme Reaction Mechanisms and Active Sites. *Nucleic Acids Res.* **2018**, *46* (D1), D618–D623.
133. Kreutter, D.; Schwaller, P.; Reymond, J.-L. Predicting Enzymatic Reactions with a Molecular Transformer. *Chem. Sci.* **2021**, *12* (25), 8648–8659.
134. Watanabe, N.; Yamamoto, M.; Murata, M.; Vavricka, C. J.; Ogino, C.; Kondo, A.; Araki, M. Comprehensive Machine Learning Prediction of Extensive Enzymatic Reactions. *J. Phys. Chem. B* **2022**, *126* (36), 6762–6770.
135. Kanehisa, M.; Goto, S. KEGG: Kyoto Encyclopedia of Genes and Genomes. *Nucleic Acids Res.* **2000**, *28* (1), 27–30.
136. Asgari, E.; Mofrad, M. R. Continuous Distributed Representation of Biological Sequences for Deep Proteomics and Genomics. *PloS One* **2015**, *10* (11), e0141287.
137. Öztürk, H.; Ozkirimli, E.; Özgür, A. A Novel Methodology on Distributed Representations of Proteins using their Interacting Ligands. *Bioinformatics* **2018**, *34* (13), i295–i303.
138. Jiang, J.; Liu, L.-P.; Hassoun, S. Learning Graph Representations of Biochemical Networks and its Application to Enzymatic Link Prediction. *Bioinformatics* **2021**, *37* (6), 793–799.
139. Kroll, A.; Ranjan, S.; Engqvist, M. K.; Lercher, M. J. A General Model to Predict Small Molecule Substrates of Enzymes based on Machine and Deep Learning. *Nat. Commun.* **2023**, *14* (1), 2787.
140. Mirzaei, M.; Rastegar Shariat Panahi, M.; Ghahremani, E.; Rezae, E.; Seid Moradi, R.; Farnoosh, G.; Latifi, A. M. Improving the Activity and Stability of OPH Enzyme using DNA Shuffling Method. *New Cell. Mol. Biotechnol. J.* **2020**, *10* (39), 91–102.
141. Farnoosh, G.; Latifi, A. M. A Review on Engineering of Organophosphorus Hydrolase (OPH) Enzyme. *J. Appl. Biotechnol. Rep.* **2014**, *1* (1), 1–10.
142. Farnoosh, G.; Khajeh, K.; Latifi, A. M.; Aghamollaei, H. Engineering and Introduction of De Novo Disulphide Bridges in Organophosphorus Hydrolase Enzyme for Thermostability Improvement. *J. Biosci.* **2016**, *41*, 577–588.
143. Repecka, D.; Jauniskis, V.; Karpus, L.; Rembeza, E.; Rokaitis, I.; Zrimec, J.; Poviloniene, S.; Laurynenas, A.; Viknander, S.; Abuajwa, W. Expanding

Functional Protein Sequence Spaces using Generative Adversarial Networks. *Nat. Mach. Intell.* **2021**, *3* (4), 324–333.

144. Luo, Y.; Jiang, G.; Yu, T.; Liu, Y.; Vo, L.; Ding, H.; Su, Y.; Qian, W. W.; Zhao, H.; Peng, J. ECNet is an Evolutionary Context-Integrated Deep Learning Framework for Protein Engineering. *Nat. Commun.* **2021**, *12* (1), 5743.
145. Ramírez-Palacios, C.; Marrink, S. J. Super High-Throughput Screening of Enzyme Variants by Spectral Graph Convolutional Neural Networks. *J. Chem. Theory Comput.* **2023**, *19*, 4668–4677.
146. Harbola, A.; Negi, D.; Manchanda, M.; Kesharwani, R. K. Bioinformatics and Biological Data Mining, in *Bioinformatics: Methods and Applications*, Elsevier, 2022, pp. 457–471.
147. Han, X.; Zhang, Z.; Ding, N.; Gu, Y.; Liu, X.; Huo, Y.; Qiu, J.; Yao, Y.; Zhang, A.; Zhang, L. Pre-Trained Models: Past, Present and Future. *AI Open* **2021**, *2*, 225–250.
148. Stahlschmidt, S. R.; Ulfenborg, B.; Synnergren, J. Multimodal Deep Learning for Biomedical Data Fusion: A Review. *Briefings Bioinf.* **2022**, *23* (2), bbab569.

Part V
Computational Biophysics and Chemistry and Diseases

https://doi.org/10.1142/9789819808199_0013

Chapter 13

Computational Analysis of Hereditary Spastic Paraplegia Mutations in the Kinesin Motor Domains of KIF1A and KIF5A*

Vidhyanand Mahase[†], Adebiyi Sobitan[†], Christina Johnson[†], Farion Cooper II[†], Yixin Xie[‡], Lin Li[‡,§] and Shaolei Teng[†,¶]

[†]*Department of Biology, Howard University, Washington, D.C., 20059 USA*
[‡]*Computational Science Program, University of Texas at El Paso, El Paso, Texas 79902, USA*
[§]*Department of Physics, University of Texas at El Paso, El Paso, Texas 79902, USA*
[¶]*shaolei.teng@howard.edu*

Hereditary spastic paraplegias (HSPs) are a genetically heterogeneous collection of neurodegenerative disorders categorized by progressive lower-limb spasticity and frailty. The complex HSP forms are characterized by various neurological features including progressive spastic weakness, urinary sphincter dysfunction, extra pyramidal signs and intellectual disability (ID). The kinesin superfamily proteins (KIFs) are microtubule-dependent molecular motors involved in intracellular transport. Kinesins directionally transport membrane vesicles, protein complexes, and mRNAs along neurites, thus playing important roles in neuronal development and function. Recent genetic studies have identified kinesin mutations in patients with HSPs. In this study, we used the computational approaches to

*To cite this article, please refer to its earlier version published in *J. Theor. Comput. Chem.* **2020**, *19* (6), 2041003, DOI: 10.1142/S0219633620410035
[¶]Corresponding author.
Vidhyanand Mahase and Adebiyi Sobitan contributed equally to this work.

investigate the 40 missense mutations associated with HSP and ID in *KIF1A* and *KIF5A*. We performed homology modeling to construct the structures of kinesin–microtubule binding domain and kinesin–tubulin complex. We applied structure-based energy calculation methods to determine the effects of missense mutations on protein stability and protein–protein interaction. The results revealed that the most of disease-causing mutations could change the folding free energy of kinesin motor domain and the binding free energy of kinesin–tubulin complex. We found that E253K associated with ID in KIF1A decrease the protein stability of kinesin motor domains. We showed that the HSP mutations located in kinesin–tubulin complex interface, such as K253N and R280C in KIF5A, can destabilize the kinesin–tubulin complex. The computational analysis provides useful information for understanding the roles of kinesin mutations in the development of ID and HSPs.

Keywords: Kinesin mutation; kinesin–tubulin complex; intellectual disability; protein stability; protein–protein interaction.

1. Introduction

Hereditary spastic paraplegia (HSP) describes an assorted group of inherited neurodegenerative or neurodevelopmental disorders in which the prominent neurological symptoms and signs are spasticity and weakness in the lower extremities. HSP is a multiplex with more than 70 genetic subtypes.[1] It is involved in all patterns of hereditary inheritance and non-Mendelian parental conduction. The most shared pathological feature is a retrograde distal axonopathy of the elongated descending motor fibers of the corticospinal tract and posterior columns caused by molecular mechanisms that affect membrane vesicular trafficking, organelle morphogenesis and distribution, axonal transport, lipid metabolism, mitochondrial functions, and different steps of the myelination process.[1,2] Complex forms of HSP comprises of intellectual disability (ID), epilepsy, and extrapyramidal features such as parkinsonism, chorea, dystonia.[2] ID is the most frequent cause of severe disability and a leading socioeconomic healthcare problem in western countries.[3] Natural forms of the disease are usually autosomal dominant (AD-HSP).[4] Autosomal recessive HSPs (AR-HSPs) have a unique clinical paragon, including peripheral neuropathy, movement disorders, cognitive dysfunction and an assortment of other features.[3]

Recent studies revealed that nonsynonymous Single Nucleotide Polymorphisms (nsSNPs) in kinesins are associated with the neurological disorders including HSP and ID.[5] Kinesins are molecular motor proteins that carry cargo along microtubule tracks, thus playing important roles in intracellular transport and cell division.[6] Kinesins use the chemical energy of ATP to generate motile force to directional move various cargos, including mRNAs, protein complexes and membrane organelles, along microtubules.[7] Kinesins are responsible for the transport in the neuronal axons and dendrites. They control different sorts of transports in synaptic proteins, Golgi–plasma membrane, Golgi–ER and lysosome, and play important functions for the elongation of neurites and polarization of neurons. For example, KIF1A is known to transfer cargo toward the positive ends of the microtubules and translate axons in the transport pathway. Disruption of kinesin-related axonal transport may cause vesicle-trafficking problems, alter specific packages interactions and create flaws in retrograde survival signs, essentially leading to neuronal death and loss of human brain function.[8] Kinesins consist of a conserved motor domain (MD), a stalk domain and a cargo-binding region.[9] The motor domain hydrolyzes ATP and reversibly binds to microtubules to generate directed motion.[10] Each kinesin member of the family uses ATP hydrolysis to execute various functions, using both its own electrical motor domains to produce distinct properties. Based on the location of motor domain, kinesin family proteins are comprised of three major groups: N-terminal motor domain KIFs (N-KIFs), middle motor domain KIFs (M-KIFs), and C-terminal motor domain KIFs (C-KIFs).[9] The different kinesins are distinguished by the mechanisms that they use to recognize and bind to their specific cargos. KIF1A and KIF5A are N-KIF motors composed of an N-terminal motor domain followed by a neck coil–coil domain.[11] Different from KIF5A, KIF1A has a pleckstrin homology (PH) domain in its C-terminal. Kinesins can regulate many broader biological processes such as higher brain function, tumor suppression and developmental patterning.

Numerous experiments have demonstrated that KIF1A and KIF5A perform critical roles in controlling spindle association, finalizing cytokinesis chromosome alignment, and completing chromosome condensation.[12] Mutations in the KIF1A and KIF5A genes were described in two different clinically and genetically heterogeneous groups of neurodegenerative diseases.[13,14] The mutations in kinesin motor domain can cause abnormal accumulations of target proteins in the nervous system and disrupt the transport of cargoes through axons.[5] For example, the genetic screening of KIF5A in patients with neurodegenerative disorders identified that the missense mutations in the motor and stalk domains motors are implicated in axonal Charcot-Marie-Tooth type 2 and spastic paraplegia type 10.[15] Understanding the roles of kinesin mutations can provide important information for identifying the targets causes neurological disorders.

Crystallographic investigation showed that kinesin motor domain has a single α/β arrowhead-shaped structure, which shares high similarity with the catalytic core of the motor protein myosin.[16] The alanine-scanning mutagenesis revealed that the kinesin residues that alter microtubule binding are located in three loops that cluster together in a patch of the motor domain.[17] The improvement of protein structure modeling makes it possible to predict the effects of kinesin mutations by mapping them on the corresponding kinesin models. The structure-based approaches can be used to quantitatively assess the effects of disease-causing mutations on protein stability and protein–protein interaction.[18,19] In a recent work, we investigated the electrostatic force changes upon kinesin mutations.[20] We showed that the mutations in motor domains can significantly affect the electrostatic component of the binding force and the force changes can be used to discriminate the damaging mutations and neutral variations. In this study, we used the computational approaches to investigate the kinesin mutations associated with neurological disorders including HSP and ID. We applied structure-based energy calculation methods to predict the effects of missense mutations on stability of kinesin motors and kinesin–microtubule binding affinity. The analytical results provide useful information for characterizing the kinesin missense mutations and elucidating the genetic mechanisms by which the coding mutations cause neurological disorders.

2. Methods and computational details

2.1. *Structure preparation*

In this study, we investigated the protein structures of human KIF1A and KIF5A. Both proteins have no available structures in Protein Data Bank (PDB).[21] However, the experimentally determined structures of mouse kinesin motor are available. The sequences of KIF1A (Entry: Q12756) and KIF5A (Entry: Q12840) were downloaded from UniProt.[22] The pairwise sequence alignment of N-terminal motor domains (1-360 aa) of KIF1A and KIF5A was performed using EMBOSS Water.[23] SWISS-MODEL[24] was used to construct the human kinesin structures based on their homologues structures in mouse. Only the top model from multiple models provided by SWISS-MODEL was selected as the corresponding structure. The structure model of human KIF1A was built based on the mouse KIF1A head–microtubule complex structure (PDB ID: 2HXH; sequence similarity = 95.9%). Crystal Structure of the mouse KIF5C motor domain (PDB ID: 3WRD; sequence similarity = 91.2%) was used as the template for KIF5A. The high sequence similarities between the target proteins and the corresponding templates suggest that the models obtained by SWISS-MODEL are very reliable. Profix[25] was used to fix the experimentally determined structures missing heavy atoms. NAMD[26] was applied for a 10,000-step minimization to each kinesin motor structure, and the CHARMM[27] force field and the Generalized Born (GB) implicit solvent model were used in the minimizations. The structure alignment of KIF1A and KIF5A motor domains was carried out using FATCAT.[28]

The structure of the human kinesin–tubulin complex is not available experimentally. However, it was shown that despite their differences, all kinesins bind to microtubule using the same binding interface.[29] Thus, we built the kinesin–tubulin complex structures by homology. The cryo-EM map of microtubule-bound kinesin-5 motor domain (PDB ID: 4AQW)[30] was used as a complex template for all kinesin–tubulin complexes. Chimera[31] was utilized to build complex structures by aligning the N-termini of each kinesin to the structure of recombinant neuronal human tubulin (PDB ID 5JCO).[32] The protein structural pictures were generated using the PyMOL molecular visualization system (http://www.pymol.org/).

2.2. *Mutation collection*

The missense mutations in the kinesin motor domains were extracted from dbNSFP v3.4.[33] To determine the effects of the mutations on protein structure, we collected all mutations that can be mapped onto the corresponding motor domain structures. Human Gene Mutation Database (HGMD)[34] and ClinVar[35] 2017 versions were used for searching for disease-causing mutations. A total of 40 damaging mutations, including 22 KIF1A mutations and 18 KIF5A mutations, were investigated in this study. Of the 22 KIF1A mutations, 14 are associated with ID,[34] 3 contribute to HSP,[34] 5 are marked as pathogenic in ClinVar.[35] All 18 KIF5A mutations are related to HSP.[34]

2.3. *Free energy changes calculations*

Two types of energy calculations, the change of the folding free energy and the change of the binding free energy, were carried out to predict the effects of kinesin mutations on protein stability of kinesin motor domain and binding affinity of kinesin–tubulin complex.

The effects of missense mutations on protein stability were assessed by calculating the change of the folding free energy. The Single Amino Acid Folding free Energy Changes (SAAFEC) method[36] was used for folding energy calculations based on unbound kinesin structures. SAAFEC uses the Molecular Mechanics Poisson–Boltzmann Surface Accessibility (MM/PBSA) approach combined with knowledge-based (KB) terms of protein biophysical characteristics. The folding free energy (ΔG) is computed as

$$\Delta G_{(\text{folding})} = G_{(\text{folded})} - G_{(\text{unfolded})}. \tag{1}$$

Thus, the folding energy difference ($\Delta\Delta G$) between the structure with the mutation (MUT) and wild-type (WT) structure is calculated using

$$\Delta\Delta G_{(\text{stability})} = \Delta G_{(\text{folding})}\text{MUT} - \Delta G_{(\text{folding})}\text{WT}, \tag{2}$$

$$\Delta\Delta G_{(\text{stability})} = [G_{(\text{folded})}\text{MUT} - G_{(\text{unfolded})}\text{MUT}] - [G_{(\text{folded})}\text{WT} - G_{(\text{unfolded})}\text{WT}]. \tag{3}$$

For free folding energy calculation, the unfolded protein is considered as two parts including a three-residues segment with the mutation in the center and all other residues.[37] Thus, the folding energy change can be computed as

$$\Delta\Delta G_{(\text{stability})} = [G_{(\text{folded})}\text{MUT} - G_3\text{MUT}] - [G_{(\text{folded})}\text{WT} - G_3\text{WT}]. \tag{4}$$

A positive value of $\Delta\Delta G$ suggests that the mutation can destabilize the monomer protein and a positive value of $\Delta\Delta G$ indicates that it makes the protein more stable. The change of the folding free energy caused by amino acids mutations is an indicator of how mutations affect the stability of the corresponding kinesin. Any deviation of the protein stability from wild type may be affecting the function of kinesins.[20]

The binding energy difference between the mutant and wild-type structures was used to estimate the effect of the kinesin mutations on protein–microtubule interactions. The Single Amino Acid Mutation change of Binding Energy (SAAMBE) method[38] was used to compute the change of the binding free energy of kinesins and tubulins. The binding free energy was the difference between the energy of the complex (AB) and the individual monomers (A and B)

$$\Delta\Delta G_{(\text{binding})} = \Delta G_{(\text{folding: AB})} - \Delta G_{(\text{folding: A})} - \Delta G_{(\text{folding: B})}. \tag{5}$$

Changes in the binding energy ($\Delta\Delta\Delta$G) between the mutant and wild-type kinesins are calculated by

$$\Delta\Delta\Delta G_{(\text{binding})} = \Delta\Delta G_{(\text{binding})}\text{MUT} - \Delta\Delta G_{(\text{binding})}\text{WT} \tag{6}$$

A positive value of $\Delta\Delta\Delta G$ indicates that the mutation weakens the binding affinity, whereas a negative value suggests that the mutation strengthens the binding affinity. Any alteration of the binding affinity is expected to affect kinesin–tubulin interaction.

We utilized mCSM-PPI2[39] Predictions to identify the mutations located in interface of kinesin–tubulin complex. The Distance to

Interface was used to classify the mutations as two groups: interface (Distance < 12 Å) and noninterface (Distance > 12 Å).

2.4. *Mutation pathogenicity and conservation*

We applied Rare Exome Variant Ensemble Learner (REVEL)[40] to predict the damaging effects of missense mutation on kinesin function. REVEL applies random forest on the data sets of rare variants and incorporates 18 damaging prediction scores to construct the prediction models. REVEL shows better performance for analysis of mutation pathogenicity compared with other ensemble tools such as CADD[41] and DANN.[42] We utilized the PhyloP[43] to analyze the evolutionary conservation of kinesins residues. PhyloP[43] computes evolutionary conservation at distinct alignment sites. The scores are compared to the evolution expected under neutral drift. Positive scores reflect slower evolution than expected, and such residues that are predicted to be conserved. Negative scores suggest faster evolution than expected, and such sites that are predicted to be nonconserved. PhyloP scores are useful to estimate signs of selection at particular nucleotides or classes of nucleotides.

3. Results and discussion

3.1. *KIF1A and KIF5A motor domains*

In this work, we applied homology modeling methods to create motor domain structures of KIF1A and KIF5A. Both kinesin proteins are N-terminal motor domain KIFs (Fig. 1(a)). The sequence alignment in Fig. 1(b) suggests that their motor domains have a 56.9% similarity. All of 40 damaging mutations are in the conserved region and 22 mutations (55%) have identical residues in the sequence alignment (Fig. 1(b)). Since the kinesins have similar microtubule binding sites,[29] the human kinesin–tubulin complex structures were built by aligning each kinesin to the structures available in other species. The mutant structures of kinesin motor domains and kinesin–tubulin were built by introducing the corresponding mutations to wild-type structures. The kinesin–tubulin complex structures of KIF1A and KIF5A are shown as examples in Fig. 1(c). The kinesin motor domain

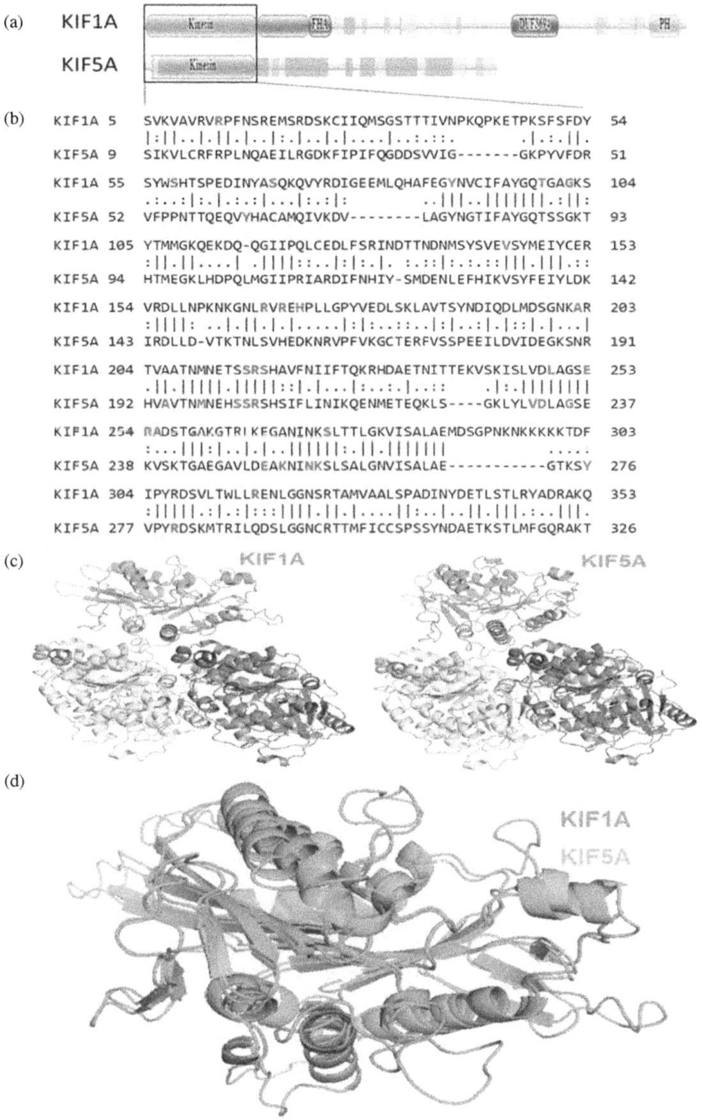

Fig. 1. (Color online) KIF1A and KIF5A motor domains. (a) Schematic functional domains of KI1A and KIF5A from Pfam.[44] (b) The pairwise sequence alignment of N-terminal motor domains (1–360 aa) of KIF1A and KIF5A from EMBOSS Water.[23] The positions of disease-causing mutations are colored with red. (c) The overall structure of KIF1A and KIF5A motor domains (green) with a tubulin dimer (A chain: orange; I chain: yellow). (d) The structure alignment of KIF1A and KIF5A motor domains from FATCAT[28] (KIF1A: green; KIF5A: cyan).

has a positively charged binding interface, so it can form a binding funnel with the tubulin dimer chains which have a negatively charged binding interface. The structure alignment shows that the structures of KIF1A and KIF5A motor domains are significantly similar ($P = 0.00e+00$; raw score $= 691.73$) and have 315 equivalent positions with an RMSD of 2.74 (Fig. 1(d)).

3.2. *Effects of mutations on protein stability*

We computed the changes of the folding free energy ($\Delta\Delta G$) and changes of binding free energy ($\Delta\Delta\Delta G$) introduced by 40 pathogenic mutations in the data set. The energy changes of 22 KIF1A and 18 KIF5A disease-causing mutations were shown in Table 1. The $\Delta\Delta G$ was calculated to access the effects of kinesin mutations on motor stability. Of 40 KIF1A and KIF5A mutations, 30 (75%) have positive $\Delta\Delta G$ values and the remaining 10 (25%) have negative values, suggesting the most disease-causing mutations can decrease protein stability. 34 of 40 (85%) pathogenic mutations have large effects (—$\Delta\Delta G| > 0.5\,\text{kcal/mol}$) on protein stability. The largest folding energy change ($\Delta\Delta G = 12.948\,\text{kcal/mol}$) is introduced by E253K associated with ID in *KIF1A*. The HSP mutation E251K related to HSP in *KIF5A* has the second largest $\Delta\Delta G$ value at 7.924 kcal/mol. The results implicate that both mutations can significantly destabilize the kinesin motor domains. In contrast, G102D implicated in ID in *KIF1A* mutation has the largest negative folding free energy change at −4.498 kcal/mol. The HSP mutations located in R280 of *KIF5A* cause the significant folding energy changes (R280H: $\Delta\Delta G = -3.136\,\text{kcal/mol}$; R280C: $\Delta\Delta G = -2.732\,\text{kcal/mol}$; R280L: $\Delta\Delta G = -2.133\,\text{kcal/mol}$). The negative $\Delta\Delta G$ values suggest that these mutations may stabilize the kinesin monomer structure.

3.3. *Effects of mutations on protein–protein interaction*

The $\Delta\Delta\Delta G$ was computed to estimate the effects of kinesin mutations on kinesin–tubulin Interaction. Of 40 pathogenic mutations, 36 (90%) have $\Delta\Delta\Delta G > 0$ and remaining 4 (10%) have

Table 1. Effects of kinesin pathogenic mutations in KIF1A and KIF5A.

Mutation†	Energy Change (kcal/mol)		Distance to interface (Å)	REVEL	phyloP	Disease*
	$\Delta\Delta G$	$\Delta\Delta\Delta G$				
KIF1A						
R13H	0.603	1.268	19.140	0.953	7.648	HSP
S58L	0.794	0.403	25.030	0.938	9.750	ID
S69L	1.365	0.743	30.250	0.576	6.219	HSP
Y89D	2.519	0.778	20.880	0.954	9.083	ID
T99M	0.370	0.113	13.710	0.933	9.741	ID
G102D	−4.498	1.574	19.530	0.980	7.706	ID
V144F	−0.225	0.455	20.080	0.946	7.706	ID
R167C	1.837	0.721	7.850	0.917	5.514	ID
R169K	0.948	0.257	3.270	0.690	5.989	Pathogenic
H171P	−0.022	1.089	5.220	0.950	7.815	Pathogenic
A202P	1.730	0.777	22.000	0.860	7.706	ID
S215R	0.579	0.296	15.050	0.980	7.728	ID
R216H	1.961	0.857	10.460	0.950	7.620	ID
R216P	0.656	1.932	10.460	0.960	7.620	ID
R216C	0.517	0.884	10.460	0.960	9.628	ID
S217C	1.370	0.789	14.440	0.930	9.628	Pathogenic
L249Q	0.924	0.008	12.790	0.990	9.039	ID
E253K	12.948	−0.789	4.660	0.950	7.670	ID
R254W	2.959	0.730	2.570	0.810	3.040	Pathogenic
A255V	1.543	2.194	3.200	0.810	9.693	HSP
S274L	1.397	0.610	7.850	0.840	9.746	Pathogenic
R316W	3.954	0.577	20.370	0.910	2.704	ID
KIF5A						
Y63C	0.773	0.098	29.440	0.900	9.325	HSP
A194P	4.261	0.681	16.650	0.903	9.823	HSP
M198T	1.028	0.080	15.580	0.925	7.938	HSP
S202N	−1.104	0.366	17.390	0.846	9.900	HSP
S203C	1.096	−0.012	15.950	0.882	7.827	HSP
R204W	2.654	−0.149	12.710	0.852	0.568	HSP
R204Q	−0.075	0.996	12.710	0.924	9.900	HSP
V231L	0.015	0.446	18.650	0.900	9.814	HSP
D232N	5.392	0.053	17.830	0.902	9.814	HSP
G235E	−1.798	1.748	10.610	0.906	9.768	HSP
E251K	7.924	−0.681	10.500	0.890	10.003	HSP
K253N	−1.826	1.401	4.010	0.701	1.092	HSP
N256S	0.067	2.110	4.150	0.916	9.325	HSP
K257N	0.886	1.711	5.850	0.825	3.160	HSP

(*Continued*)

Table 1. (*Continued*)

	Energy Change (kcal/mol)					
Mutation†	$\Delta\Delta G$	$\Delta\Delta\Delta G$	Distance to interface (Å)	REVEL	phyloP	Disease*
Y276C	1.257	0.287	1.390	0.882	9.197	HSP
R280C	−2.732	2.140	1.880	0.902	3.122	HSP
R280H	−3.136	1.416	1.880	0.883	9.864	HSP
R280L	−2.133	1.987	1.880	0.883	9.864	HSP

Notes: †Bold: Mutations with $|\Delta\Delta G| > 0.5$ kcal/mol and $|\Delta\Delta\Delta G| > 0.5$ kcal/mol; *Intellectual disability (ID), Hereditary spastic paraplegia (HSP), Pathogenic: ClinVar clinical significance value.

negative $\Delta\Delta\Delta G$ values, indicating that most of disease-causing mutations can decrease kinesin–tubulin binding affinity. 26 of 40 (65%) pathogenic mutations have large effects (—$\Delta\Delta\Delta G| > 0.5$ kcal/mol) on protein–protein interaction stability. The largest binding energy change ($\Delta\Delta\Delta G = 2.194$ kcal/mol) is introduced by A255V in KIF1A. The R280C in KIF5A has the second largest $\Delta\Delta\Delta G$ value at 2.14 kcal/mol, and $\Delta\Delta\Delta G$ value caused by R280H located in the same position is 1.416 kcal/mol. These three mutations are related to HSP. The large positive $\Delta\Delta\Delta G$ values indicate that these mutations can significantly reduce the binding affinity of kinesin–tubulin complex. In contrast, E253K associated with ID in KIF1A has the largest negative $\Delta\Delta\Delta G$ at -0.789 kcal/mol, suggesting that the mutation makes kinesin–tubulin complex more stable.

We identified 19 interface mutations (Distance to Interface < 12 Å) and 21 noninterface mutations (Distance to Interface > 12 Å). We observed the $|\Delta\Delta G|$ and $|\Delta\Delta\Delta G|$ medians of interface mutations are higher compared to those values of noninterface mutations (Fig. 2). Not surprisingly, the difference of $|\Delta\Delta\Delta G|$ between the interface mutations and noninterface mutations is statistically significant (P-value = 0.0003), which indicates that the interface mutations have large effects on binding affinity of kinesin–tubulin complex. However, the difference of effects on protein stability ($|\Delta\Delta G|$) between interface mutations and noninterface mutations is statistically indistinguishable (P-value = 0.3363). In addition, 23 mutations

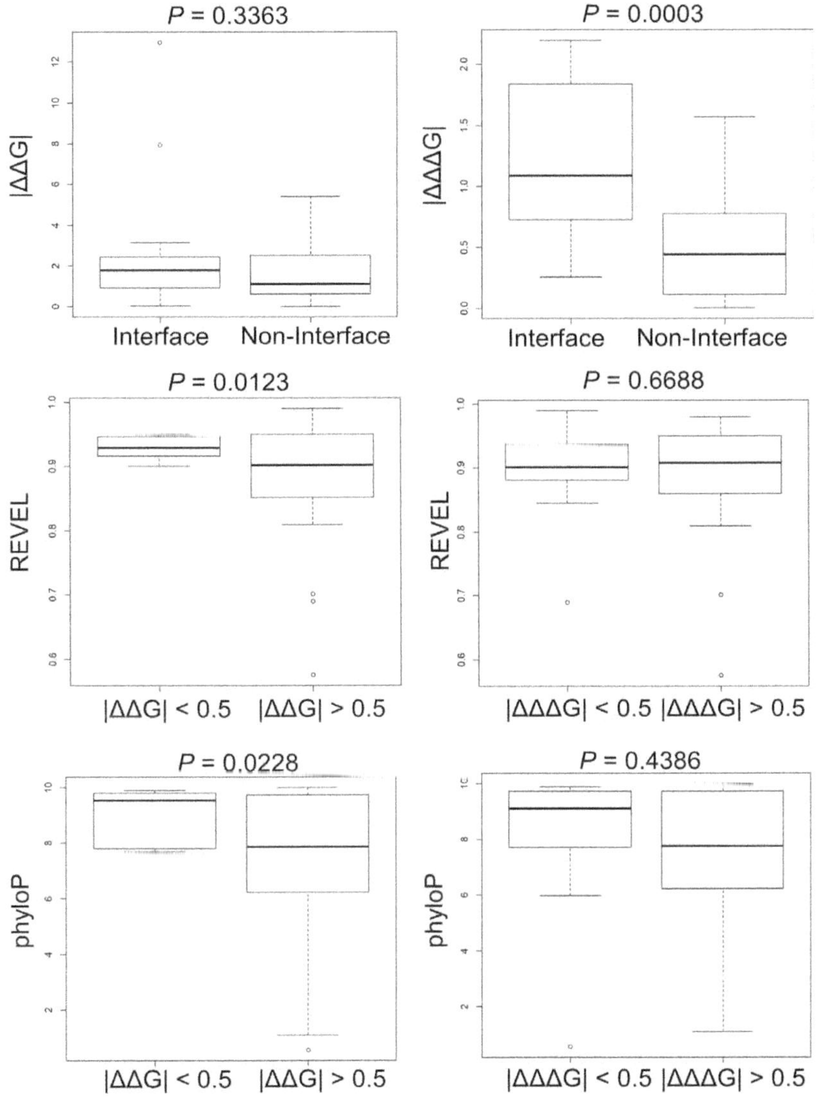

Fig. 2. Boxplots for the energy changes (kcal/mol) of kinesin pathogenic mutations.

cause large changes on both folding energy ($|\Delta\Delta G| > 0.5$) and binding energy ($|\Delta\Delta\Delta G| > 0.5$). We computed Pearson correlation coefficient r between $\Delta\Delta G$ and $\Delta\Delta\Delta G$. We observed a moderate negative association ($r = -0.6307$) between $\Delta\Delta G$ and $\Delta\Delta\Delta G$ for all

40 mutations. The association is stronger in 19 interface mutations ($r = -0.7849$), which implies that the effects of kinesin mutations on motor domain stability and kinesin–tubulin binding affinity are negatively corrected in interface region.

3.4. *Mutation pathogenicity and conservation*

REVEL was applied to predict the harmful effects of kinesin missense mutations. The REVEL score can range from 0 to 1, with 1 reflecting the most likelihood of pathogenicity. 38 (95%) of 40 disease-causing kinesin mutations with REVEL score > 0.75 are predicted to have damaging effects on protein function. The REVEL scores of mutations with large folding energy changes ($|\Delta\Delta G| > 0.5|$,kcal/mol) and binding energy changes ($|\Delta\Delta\Delta G| > 0.5\,\text{kcal/mol}$) were compared with those mutations with small effects, respectively. Figure 2 shows the difference between the mutations with large effects on protein stability (—$\Delta\Delta G| > 0.5\,\text{kcal/mol}$) and the mutations with $|\Delta\Delta G| < 0.5\,\text{kcal/mol}$ is statistically significant (P-value $= 0.0123$). However, we did not find a statistically significant difference for the effects of two groups ($|\Delta\Delta\Delta G| > 0.5$ and $|\Delta\Delta\Delta G| < 0.5$) on the binding affinity of kinesin–tubulin complex (P-value $= 0.6688$).

The PhyloP scores were used to access the conservation of kinesin mutations (Table 1). As shown in Fig. 2, the PhyloP medians of the mutations with small folding energy changes ($|\Delta\Delta G| < 0.5\,\text{kcal/mol}$) and binding energy changes ($|\Delta\Delta\Delta G| < 0.5\,\text{kcal/mol}$) are lower than those mutations with large energy changes. There is a statistically significant difference for the effects of variants on protein stability (P-value $= 0.0228$), which suggests that the KIF1A and KIF5A mutations with small effects on kinesin monomer domains are more conserved in mammalians. The findings indicate that the residues with low mutability have limited effect on protein stability, which is consistent with the previous study showing that the large shifts of mutational effects on stability are rare during protein evolution.[45] We did not observe a statistically significant result on the protein–protein interaction of kinesin–tubulin complex (P-value $= 0.4386$).

3.5. *Intellectual disability (ID) and KIF1A mutations*

Intellectual disability (ID) is the most common neurodevelopmental disorder characterized by cognitive delays in intellectual functioning and adaptive behavior. The mutant genes are important risk factors for the ID. In this study, we investigated 14 kinesin mutations associated with ID in *KIF1A*. KIF1A is a microtubule plus end-directed motor protein that belongs to the kinesin-3 family and is involved in the anterograde transport of synaptic vesicle precursors supported by the GTP-Rab3 adaptor.[46] Various heterozygous de novo missense mutations in *KIF1A* have been identified in patients who present intellectual disability along with atrophy and spastic paraplegia. Lee *et al.* applied exome sequencing to sequence 14 patients with ID and other complex neurologic phenotypes such as epilepsy, peripheral neuropathy and spastic paraparesis.[47] They identified 11 de novo missense mutations in the motor domain of KIF1A. They found two recurrent mutations, T99M and E253K, in unrelated cases. E253K was carried by the most severely affected patients. In this study, E253K caused the largest folding energy change ($\Delta\Delta G$) at 12.948 kcal/mol, suggesting that the mutation can significantly destabilize the KIF1A motor domain. As shown in Fig. 3(a), E253 is located in a loop and can form a back-door salt bridge. The mutation of negatively charged glutamic acid to positively charged lysine may disturb the salt bridge. The folding energy changes caused by the nearby mutations R254W and A255V are $\Delta\Delta G = 2.959$ kcal/mol and $\Delta\Delta G = 1.543$ kcal/mol, respectively. These mutations may break the back-door salt bridge and alter the local structure of KIF1A thereby destabilizing the motor domain. The A255V mutation is located at the interaction interface of KIF1A and tubulin (Fig. 3(a)). It introduced the largest binding energy change at $\Delta\Delta\Delta G = 2.194$ kcal/mol, indicating that the mutation could significantly destabilize kinesin–tubulin complex. Interestingly, the binding energy change caused by E253K is $\Delta\Delta\Delta G = -0.789$ kcal/mol. The results suggest that even nearby mutations can have different effects on kinesin–tubulin interaction. We also noted that the ID mutations in the same region have different effects on the stability of the KIF1A motor domain (Fig. 3(b)). T99M has a small effect on protein stability

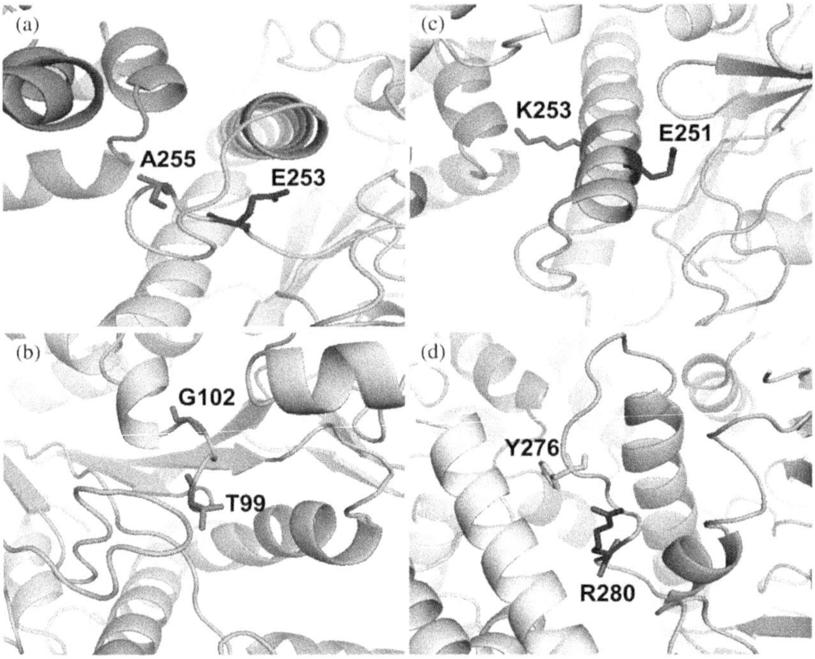

Fig. 3. (Color online) Structural representation of pathogenic mutations in KIF1A (a & b) and KIF5A (c & d) motor domains (green) with a tubulin dimer (A chain: orange; I chain: yellow). (a) Locations of E253 (blue) and A255 (red) in KIF1A. (b) Locations of T99 (pink) and G102 (marine) in KIF1A. (c) Locations of E251 (blue) and K253 (hot pink) in KIF5A. (d) Locations of Y276 (cyan) and R280 (deep blue) in KIF5A.

($\Delta\Delta G = 0.37\,\text{kcal/mol}$). In contrast, G102D has the largest negative folding free energy change at $\Delta\Delta G = -4.498\,\text{kcal/mol}$. G102 is in a nucleotide-binding pocket which is important for ATP binding (Fig. 3(b)). The mutation of nonpolar glycine to polar aspartic acid may induce a significant conformational change for this binding pocket and disrupt the ATP binding of the motor domain.

3.6. *Hereditary spastic paraplegias and KIF5A mutations*

Hereditary spastic paraplegias (HSPs) are a genetically heterogeneous collection of neurodegenerative disorders categorized by progressive lower-limb spasticity and frailty.[48–50] HSP is characterized by the axonal degeneration of motor and sensory neurons that is at the peak at distal ends of the longest axons of the central nervous system. In this research, we studied the effects of 21 mutations associated with HSP on kinesin structure and function. Of these 21 mutations, 3 are located in KIF1A and 18 are located in KIF5A. Recent reports suggest that the most common form of HSP, pure *SPG4,* is caused by the damaged regulation of the microtubule cytoskeleton, of which the kinesin family is involved.[50,51] Particularly, *SPG10* is an autosomal dominant form of HSP that is caused by mutations in *KIF5A*.[52] KIF5A is a neuronally-enriched isoform of the Kinesin-1 family of cellular transport motors. KIF5A is believed to be in control of intracellular vesicular movements for the processive nature of their movement.[52] The kinesin heavy chain (KHC) proteins, of which *KIF5A* is a part of, constitute a large portion of the multi-subunit complex that perform as a plus-end–directed microtubule motor implicated in anterograde transport of membranous organelles in nerve axons.[33,50,53] Currently, there are many loci at which the *KIF5A* gene can be examined for phenotypic effects in multiple forms of HSP. 23 separate mutations in the motor domain of KIF5A have been identified in patients with a complex form of HSP.[41] E251K in KIF5A has been identified in HSP patients with axonal neuropathy but no pyramidal signs, which is related to the phenotype of Charcot–Marie–Tooth disease type 2.[15] Here, E251K introduced the second largest $\Delta\Delta G$ value at 7.924 kcal/mol. As shown in Fig. 3(c), E251 resides in a helix and participates in the salt bridge confirmation. When the negatively charged glutamic acid is replaced by the positively charged lysine, the mutation may disturb the salt bridge and destabilize the motor domain. Interestingly, K253N, which lies two residues away from E251K, can stabilize the KIF5A monomer structure ($\Delta\Delta G = -1.826$ kcal/mol). The mutation is located at the interaction interface of KIF5A (Fig. 3(c)). The binding energy change caused by K253N is 1.401 kcal/mol suggesting that the mutation of positively charged lysine to uncharged asparagine acid could weaken the binding infinity of KIF5A kinesin–tubulin complex. K253N is close to the nucleotide binding domain, suggesting it may impact nucleotide binding and hydrolysis release. Three mutations, R280C, R280H, and R280L, located in R280 of KIF5A (Fig. 3(d)) have similar effects with K253N. They can stabilize the motor domain but destabilize the protein complex. In addition to the reduced kinesin–tubulin binding affinity, K253 and R280 sites have been associated with the microtubule gliding velocity and ATPase activity.[14] However, Y276C can reduce the stability of the KIF5A motor domain ($\Delta\Delta G = 1.257\,\text{kcal/mol}$) and has an opposite effect compared with these three mutations in R280. The results suggest the different effects of HSP mutations in KIF5A on kinesin mechanochemistry.

4. Conclusion

In this work, we employed the homology modeling approach to construct the kinesin–microtubule binding domain and kinesin–tubulin complex structures of KIF1A and KIF5A. The structure-based energy calculation methods were applied to determine the effects of kinesin missense mutations on protein stability and protein–protein interaction. We observed most of disease-causing mutations (75%) can decrease the stability of kinesin motor domain and 85% pathogenic mutations have large effects ($|\Delta\Delta G| > 0.5\,\text{kcal/mol}$) on protein stability. The binding energy calculation shows that 90% mutations have $\Delta\Delta\Delta G > 0$ and 65% mutations have large effects ($|\Delta\Delta\Delta G| > 0.5\,\text{kcal/mol}$) which indicate that the majority of deleterious mutations can decrease kinesin–tubulin binding affinity. We compared the damaging and conservation scores of kinesin mutations with large energy changes with those mutations with small effects. The results suggest that the mutations with large effects on the binding affinity of kinesin–tubulin complex could cause significant damaging effects on kinesin function (P-value $= 0.0486$). We found the kinesin mutations with small effects on the stability of kinesin motor domain are more conserved in mammalians (P-value $= 0.0228$). The results revealed the mutations associated with neurological disorders could alter folding free energy. For example, ID-related mutation E253K in KIF1A caused the $\Delta\Delta G$ at $12.948\,\text{kcal/mol}$, suggesting that the mutation can significantly destabilize the protein stability of kinesin–microtubule binding domains. The harmful mutations associated with HSP, such as K253N and R280C in

KIF5A, located in complex interface can reduce the binding affinity of kinesin–tubulin complex. The results provide useful evidence for characterizing the kinesin mutations implicated in ID and HSP and elucidating the genetic mechanisms by which the point mutations cause these neurological disorders.

Acknowledgments

We thank Professor Emil Alexov at Clemson University for supports and critical comments. We acknowledge the Howard University Junior Faculty Writing & Creative Works Summer Academy. This work was supported by the Howard University startup funds (U100193) and National Science Foundation DBI 2000296 and IIS 1924092. This project was supported (in part) by the National Institute on Minority Health and Health Disparities of the National Institutes of Health under Award Number 2U54MD007597. This research was also supported by Grant SC1GM132043-01 from National Institutes of Health; Grant 5U54MD007592 from the National Institutes on Minority Health and Health Disparities (NIMHD), a component of the NIH. The content is solely the responsibility of the authors and does not necessarily represent the official views of the National Institutes of Health. First two authors contributed equally to this work.

References

1. Guidubaldi A *et al.*, Novel mutations in SPG11 cause hereditary spastic paraplegia associated with early-onset levodopa-responsive Parkinsonism, *Mov Disord* **26**:553–556, 2011.
2. de Souza PVS, de Rezende Pinto WBV, de Rezende Batistella GN, Bortholin T, Oliveira ASB, Hereditary spastic paraplegia: Clinical and genetic hallmarks, *Cerebellum* **16**:525–551, 2017.
3. Abou Jamra R *et al.*, Adaptor protein complex 4 deficiency causes severe autosomal-recessive intellectual disability, progressive spastic paraplegia, shy character, and short stature, *Am J Hum Genet* **88**:788–795, 2011.
4. Faber I, Branco LMT, França Júnior MC, Cognitive dysfunction in hereditary spastic paraplegias and other motor neuron disorders, *Dement Neuropsychol* **10**:276–279, 2016.

5. De Vos KJ, Grierson AJ, Ackerley S, Miller CCJ, Role of axonal transport in neurodegenerative diseases, *Annu Rev Neurosci* **31**:151–173, 2008.
6. Hirokawa N, Takemura R, Molecular motors and mechanisms of directional transport in neurons, *Nat Rev Neurosci* **6**:201–214, 2005.
7. Hirokawa N, Noda Y, Intracellular transport and kinesin superfamily proteins, KIFs: Structure, function, and dynamics, *Physiol Rev* **88**:1089–1118, 2008.
8. Goldstein AY, Wang X, Schwarz TL, Axonal transport and the delivery of presynaptic components, *Curr Opin Neurobiol* **18**:495–503, 2008.
9. Miki H, Okada Y, Hirokawa N, Analysis of the kinesin superfamily: Insights into structure and function, *Trends Cell Biol* **15**:467–476, 2005.
10. Goshima G, Cell cycle-dependent dynamics and regulation of mitotic kinesins in drosophila S2 cells, *Mol Biol Cell* **16**:3896–3907, 2005.
11. Hirokawa N, Noda Y, Tanaka Y, Niwa S, Kinesin superfamily motor proteins and intracellular transport, *Nat Rev Mol Cell Biol* **10**:682–696, 2009.
12. Seeger MA, Zhang Y, Rice S E, Kinesin tail domains are intrinsically disordered, *Proteins Struct Funct Bioinforma* **80**: 2437–2446, 2012.
13. Klebe S *et al.*, KIF1A missense mutations in SPG30, an autosomal recessive spastic paraplegia: Distinct phenotypes according to the nature of the mutations, *Eur J Hum Genet* **20**:645–649, 2012.
14. Dutta M, Diehl MR, Onuchic JN, Jana B, Structural consequences of hereditary spastic paraplegia disease-related mutations in kinesin, *Proc Natl Acad Sci USA* **115**: E10822–E10829, 2018.
15. Crimella C *et al.*, Mutations in the motor and stalk domains of KIF5A in spastic paraplegia type 10 and in axonal Charcot-Marie-Tooth type 2, *Clin Genet* **82**:157–164, 2012.
16. Kull FJ, Sablin EP, Lau R, Fletterick RJ, Vale RD, Crystal structure of the kinesin motor domain reveals a structural similarity to myosin, *Nature* **380**:550–555, 1996.
17. Woehlke G *et al.*, Microtubule interaction site of the kinesin motor, *Cell* **90**: P207–P216, 1997.
18. Teng S, Srivastava AK, Schwartz CE, Alexov E, Wang L, Structural assessment of the effects of amino acid substitutions on protein stability and protein protein interaction, *Int J Comput Biol Drug Des* **3**: 334–349, 2010.
19. Teng S, Madej T, Panchenko A, Alexov E, Modeling effects of human single nucleotide polymorphisms on protein-protein interactions, *Biophys J* **96**: 2178–2188, 2009.
20. Li L *et al.*, Forces and disease: Electrostatic force differences caused by mutations in kinesin motor domains can distinguish between disease-causing and non-disease-causing mutations, *Sci Rep* **7**:8237, 2017.
21. Berman H, Henrick K, Nakamura H, Markley JL, The worldwide Protein Data Bank (wwPDB): Ensuring a single, uniform archive of PDB data, *Nucleic Acids Res* **35**:D301–D303, 2007.
22. Wu CH *et al.*, The Universal Protein Resource (UniProt): An expanding universe of protein information, *Nucleic Acids Res* **34**: D187–191, 2006.
23. Madeira F *et al.*, The EMBL-EBI search and sequence analysis tools APIs in 2019, *Nucleic Acids Res* **47**:W636–W641, 2019.

24. Biasini M *et al.*, SWISS-MODEL: Modelling protein tertiary and quaternary structure using evolutionary information, *Nucleic Acids Res* **42**:W252–W258, 2014.
25. Xiang Z, Advances in homology protein structure modeling, *Curr Protein Pept Sci* **7**:217–227, 2006.
26. Nelson MT *et al.*, NAMD: A parallel, object-oriented molecular dynamics program, *Int J High Perform Comput Appl* **10**:251–268, 1996.
27. Vanommeslaeghe K *et al.*, CHARMM general force field: A force field for drug-like molecules compatible with the CHARMM all-atom additive biological force fields, *J Comput Chem* **31**:671–690, 2010.
28. Ye Y, Godzik A, FATCAT: A web server for flexible structure comparison and structure similarity searching, *Nucleic Acids Res* **32**:W582–W585, 2004.
29. Mizuno N *et al.*, Dynein and kinesin share an overlapping microtubule-binding site, *EMBO J* **23**:2459–2467, 2004.
30. Goulet A *et al.*, The structural basis of force generation by the mitotic motor kinesin-5, *J Biol Chem* **287**:44654–44666, 2012.
31. Pettersen EF *et al.*, UCSF Chimera — A visualization system for exploratory research and analysis, *J Comput Chem* **25**:1605–1612, 2004.
32. Vemu A *et al.*, Structure and dynamics of single-isoform recombinant neuronal human tubulin, *J Biol Chem* **291**:12907–12915, 2016.
33. Liu X, Wu C, Li C, Boerwinkle E.dbNSFP v3.0: A one-stop database of functional predictions and annotations for human nonsynonymous and splice-site SNVs, *Hum Mutat* **37**: 235–241, 2016.
34. Stenson PD *et al.*, The human gene mutation database: 2008 update, *Genome Med* **1**: 13, 2009.
35. Landrum MJ *et al.*, ClinVar: Public archive of interpretations of clinically relevant variants, *Nucleic Acids Res* **44**:D862–D868, 2016.
36. Getov I, Petukh M, Alexov E, SAAFEC: Predicting the effect of single point mutations on protein folding free energy using a knowledge-modified MM/PBSA approach, *Int J Mol Sci* **17**:512, 2016.
37. Capriotti E, Fariselli P, Casadio R., A neural-network-based method for predicting protein stability changes upon single point mutations, *Bioinformatics* **20**(Suppl 1):i63–68, 2004.
38. Petukh M, Dai L, Alexov E, SAAMBE: Webserver to predict the charge of binding free energy caused by amino acids mutations, *Int J Mol Sci* **17**:547, 2016.
39. Rodrigues CHM, Myung Y, Pires DEV, Ascher DB, mCSM-PPI2: predicting the effects of mutations on protein-protein interactions, *Nucleic Acids Res* **47**: W338–W344, 2019.
40. Ioannidis NM *et al.*, REVEL: An ensemble method for predicting the pathogenicity of rare missense variants, *Am J Hum Genet* **99**:877–885, 2016.
41. Chun S, Fay JC, Identification of deleterious mutations within three human genomes, *Genome Res* **19**:1553–1561, 2009.
42. Quang D, Chen Y, Xie X, DANN: A deep learning approach for annotating the pathogenicity of genetic variants, *Bioinformatics* **31**:761–763, 2015.

43. Wu J, Li Y, Jiang R, Integrating multiple genomic data to predict disease-causing nonsynonymous single nucleotide variants in exome sequencing studies, *PLoS Genet* **10**:e1004237, 2014.
44. El-Gebali S *et al.*, The Pfam protein families database in 2019, *Nucleic Acids Res* **47**:D427–D432, 2019.
45. Ashenberg O, Gong LI, Bloom JD, Mutational effects on stability are largely conserved during protein evolution, *Proc Natl Acad Sci USA* **110**:21071–21073, 2013.
46. Blom N, Gammeltoft S, Brunak S, Sequence and structure-based prediction of eukaryotic protein phosphorylation sites, *J Mol Biol* **294**:1351–1362, 1999.
47. Lee JR *et al.*, De novo mutations in the motor domain of KIF1A cause cognitive impairment, spastic paraparesis, axonal neuropathy, and cerebellar atrophy, *Hum Mutat* **36**:69–78, 2015.
48. Xue Y *et al.*, GPS 2.0, a tool to predict kinase-specific phosphorylation sites in hierarchy, *Mol Cell Proteomics*, **7**:1598–1608, 2008.
49. Li A, Xue Y, Jin C, Wang M, Yao X, Prediction of nepsilon-acetylation on internal lysines implemented in Bayesian discriminant method, *Biochem Biophys Res Commun* **350**:818–824, 2006.
50. Deng W *et al.*, GPS-PAIL: Prediction of lysine acetyltransferase-specific modification sites from protein sequences, *Sci Rep* **6**:39787, 2016.
51. Blom N, Sicheritz-Pontén T, Gupta R, Gammeltoft S, Brunak S, Prediction of post-translational glycosylation and phosphorylation of proteins from the amino acid sequence, *Proteomics* **4**:1633–1649, 2004.
52. Cooper DN, Ball EV, Krawczak M, The human gene mutation database, *Nucleic Acids Res* **26**:285–287, 1998.
53. Liu Z *et al.*, GPS-YNO2: computational prediction of tyrosine nitration sites in proteins, *Mol Biosyst* **7**:1197–1204, 2011.

https://doi.org/10.1142/9789819808199_0014

Chapter 14

In-Silico Analysis to Identify the Role of MEN1 Missense Mutations in Breast Cancer*

Satishkumar Ranganathan Ganakammal[†,§,¶], Mahesh Koirala[‡,¶], Bohua Wu[‡] and Emil Alexov[†,‡]

[†]*Department of Healthcare Genetics*
School of Nursing, Clemson University
Clemson SC, USA
[‡]*Department of Physics*
Clemson University, Clemson SC, USA
[§]*satishk@clemson.edu*

Background: The multiple endocrine neoplasia type 1 (*MEN1*) gene located on chromosome 11q13 encodes menin protein. Previously reported mutations were thought to result in loss of function of menin protein and that they are associated with multiple endocrine neoplasia 1 disorder. However, recently menin has also been characterized as an oncosuppressor protein and it was suggested that mutations in it are associated with various other tumors. Studies indicate that the menin protein stimulates the estrogen receptor (ER) that in turn increases the predisposition for inherited breast cancer.

Methods: Here, we used our supervised in-house combinatory in-silico predictor method to investigate the impact of unclassified missense mutations in MEN1 gene found in breast cancer tissue. We also examined the biophysical and biochemical properties to predict the effects of these missense variants on the menin protein stability and interactions. The results are compared with the effects of known pathogenic mutations in menin causing neoplasia.

*To cite this article, please refer to its earlier version published in *J. Theor. Comput. Chem.* **2020**, *19* (6), 2041002, DOI: 10.1142/S0219633620410023

¶Satishkumar Ranganathan Ganakammal and Mahesh Koirala are contributed equally to this work.

Results: Our analysis indicates that some of the variants found in breast cancer tissue show similar pattern of destabilizing the menin protein and its interactions as the pathogenic variants associated with neoplasia. Taking together with the results of our in-silico consensus predictor, we classify missense mutations in menin protein found in breast cancer tissue into pathogenic and benign, and thus, suggesting as an indicator for early detection of elevated breast cancer risk.

Keywords: Menin protein; missense mutation; breast cancer; folding free energy change; binding free energy change; disease-causing mutation; machine learning.

1. Introduction

Variants in *MEN1* gene have been found to have a strong association with multiple endocrine neoplasia type 1 (*MEN1*) that follows an autosomal dominant (AD) inheritance pattern in various tumor syndromes.[1] MEN1 triggered neoplasia type1 is involved in the development of various tumors such as parathyroid adenomas, duodenopancreatic neuroendocrine tumors, and pituitary (anterior) adenomas with a 94% penetrance by the age of 40.[2] MEN1 encodes for menin, a 610 amino acid oncosuppressor protein, found in the nucleus. Menin protein interacts with other proteins such as JunD/AP1, SMAD3, NFκB, estrogen receptor (ER) and others to participate in various transcription and cell signaling processes.[3]

The inheritance of germline MEN1 mutation such as a point mutation or deletion (leading to loss of heterozygosity (LOH)), disrupts the signaling pathway of menin protein, and thus predisposes an individual to develop tumor. The germline mutations occurring in MEN1 that truncate the menin protein result in dysfunctional product and thus, affecting its role in tumor suppressor activity.[3] Studies based on mutagenesis performed previously indicate that the LOH of menin protein disrupts its interaction with histone methyltransferase (HMT) mixed lineage leukemia protein (MLL). This mechanism in turn impacts histone associated trimethylation (H3K4me3) transcription activation process thus fails to suppress the development of endocrine tumors.[4]

The oncosuppressor menin protein plays a vital role in the suppression of ER-positive breast cancer as well.[5] Previous studies outline that the upregulation of menin protein which binds to the ER and thus enhances the ER activity which implicates that it acts as a growth stimulator in ER-positive breast cancer.[6] Menin also regulates ERα-mediated transcription by enhancing H3K4 (histone 3 lysine 4) methylation process leading to breast carcinogenesis and various other tumor progression mechanisms[7] LOH and disruption in menin protein expression in a Dutch cohort strongly shows involvement of MEN1 in breast cancer carcinogenesis.[8]

The goal of this study is to classify missense MEN1 variants reported to be identified in breast cancer tissues from Catalogue of Somatic Mutations in Cancer (COSMIC) database[9] with unknown pathogenicity. This is done by applying our combinatory in-silico predictor[10] approach to annotate the pathogenicity of these variants. We also analyzed the impact of these variants on biochemical and biophysical properties of the menin protein along with the in-house built menin-ER and menin-MLL complexes. In parallel, the same is done for curated pathogenic variants causing neoplasia. Comparing the observations made for mutations associated with neoplasia and variants in COSMIC database along with predictions made by our in-silico predictor, we provide classification of pathogenic menin mutations associated with breast tumorigenesis.

2. Materials and methods

2.1. *Selection of the MEN1 variants*

We collected 19 unique missense variants in MEN1 gene (GRCh37/hg19 version) from COSMIC database that are reported to be identified in breast cancer samples. Among these 19 variants, only two are listed in ClinVar[11] and they are annotated as variants of unknown significance (VUS).

In parallel, we obtained 8 MEN1 missense variants associated with MEN1 from ClinVar database categorized as pathogenic. We will use these pathogenic mutations to compare their effect on menin stability and interactions with the effect caused by variants taken from

Table 1. List of all the variants used in our study. The MEN1 missense variants associated with breast cancer are obtained from COSMIC database, whereas the MEN1 pathogenic missense variants associated with multiple endocrine neoplasia are obtained from ClinVar (that is used as a benchmarking set).

Protein change	Clinical significance	Variant type	rs# (dbSNP)	Phenotype list	Source
		Breast Cancer Associated Variants (COSMIC)			
S606C	Unknown	Probably somatic		Breast cancer	COSMIC
S606Y	Unknown	Probably somatic		Breast cancer	COSMIC
L605V	Unknown	Probably somatic		Breast cancer	COSMIC
M563I	Unknown	Probably somatic		Breast cancer	COSMIC
K562M	Unknown	Probably somatic		Breast cancer	COSMIC
P534A	Unknown	Probably somatic		Breast cancer	COSMIC
T530I	Unknown	Probably somatic	rs750591216	Breast cancer	COSMIC
R490Q	Unknown	Probably somatic		Breast cancer	COSMIC
Q398E	Unknown	Probably somatic		Breast cancer	COSMIC
A289E	Unknown	Probably somatic		Breast cancer	COSMIC
L272R	Unknown	Probably somatic		Breast cancer	COSMIC
S258L	Unknown	Probably somatic	rs386134259	Breast cancer	COSMIC
V220M	Unknown	Probably somatic	rs794728621	Breast cancer	COSMIC
H204Y	Unknown	Probably somatic		Breast cancer	COSMIC
E200K	Unknown	Probably somatic		Breast cancer	COSMIC
C170Y	Unknown	Probably somatic		Breast cancer	COSMIC
R108P	Unknown	Probably somatic		Breast cancer	COSMIC
L89R	Unknown	Probably somatic		Breast cancer	COSMIC
L36F	Unknown	Probably somatic		Breast cancer	COSMIC
		Neoplasia Associated Variants (ClinVar)			
W441R	Pathogenic	Probably germline	rs104894259	Multiple endocrine neoplasia, type 1	ClinVar
A373D	Pathogenic	Probably germline	rs1555164707	Multiple endocrine neoplasia, type 1	ClinVar
E260K	Pathogenic	Probably germline	rs104894268	Multiple endocrine neoplasia, type 1	ClinVar
V189E	Pathogenic	Probably germline	rs104894262	Multiple endocrine neoplasia, type 1	ClinVar
A165P	Pathogenic	Probably germline	—	Multiple endocrine neoplasia, type 1	ClinVar
R280K	Pathogenic	Probably germline	—	Multiple endocrine neoplasia, type 1	ClinVar
D423N	Pathogenic	Probably germline	rs104894264	Hereditary cancer-predisposing syndrome; Multiple endocrine neoplasia, type 1	ClinVar

Table 1. (*Continued*)

Protein change	Clinical significance	Variant type	rs# (dbSNP)	Phenotype list	Source
H139D	Pathogenic	Probably germline	rs104894263	Hereditary cancer-predisposing syndrome; Multiple endocrine neoplasia, type 1	ClinVar
MEN1 Benign Variants (VariSNP)					
P519S	Benign	Probably germline	rs150202288	NA	VariSNP
V555L	Benign	Probably germline	rs562257963	NA	VariSNP
I377T	Benign	Probably germline	rs115859693	NA	VariSNP
E371D	Benign	Probably germline	rs149383809	NA	VariSNP
G508D	Benign	Probably germline	rs375804228	NA	VariSNP

COSMIC database and five benign variants from VariSNP database,[12] since there were no benign variants reported in ClinVar. Table 1 summarizes the list of all variants used in this study.

2.2. *In-house algorithm for pathogenicity prediction*

Recently we reported a combinatory in-silico predictor of pathogenicity[10] which combines the eight best pathogenicity prediction algorithms such as Polyphen2 (Polymorphism Phenotyping V-2),[13] LRT (Likelihood ratio test),[14] MetaSVM,[15] VEST3,[16] PROVEAN (Protein Variation Effect Analyzer),[17] REVEL,[18] Eigen,[19] CADD (Combined Annotation Dependent Depletion).[20] We used this method to characterize the pathogenicity of the missense variant.

2.3. *Annotating MEN1 missense variants*

We also annotated these variants with allele frequency information from 1000 genome[21] and Genome Aggregation Database (gnomAD)[22] to see the distribution of these variants among various population groups. We also annotated the variants with GERP + + scores[23] to evaluate the evolutionary conservation on a functional sequence.

We used this score along with another biochemical feature, the folding energy change, to build an in-house classification method (using KNN classification method) to characterize the pathogenicity of missense variants.

2.4. *3D model of menin protein and generation of Menin-ER and Menin-MLL complexes*

The crystal structure of menin protein was obtained from Protein Data Bank (PDB).[24] The PDB file (ID: 3U84) contains two chains, chain A and chain B, and has missing residues. The structure was re-modeled using SWISS-model[25] resulting in a full-length protein. (Fig. 1(a)). The structure of ER was modeled with SWISS-modeling using its amino acid sequence taken from UniProt (ID: Q99527)[26] and a template (PDB ID: 4ZNH) with high sequence similarity (sequence similarity of 0.59) to build the model (Fig. 1(b)). The structure of MLL was taken from PDB file (ID: 4GQ6). Then the structures of menin, ER and MLL were used to model the complexes menin-ER and menin-MLL. For this purpose, we used the docking methods ZDOCK[27] and HawkDock.[28] Among the different models given by both methods, one best model is chosen from each method, each with similar position and near to the binding site of the protein. The binding site is predicted using web server meta -PPISP[29] which gives

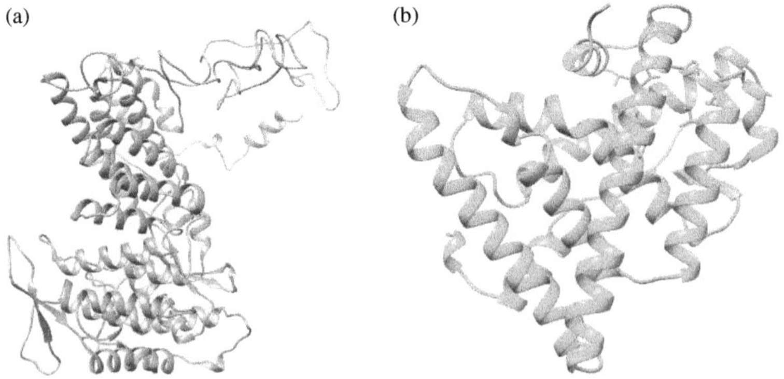

Fig. 1. 3D protein structure. (a) 3D structure of menin; (b) 3D structure of ER.

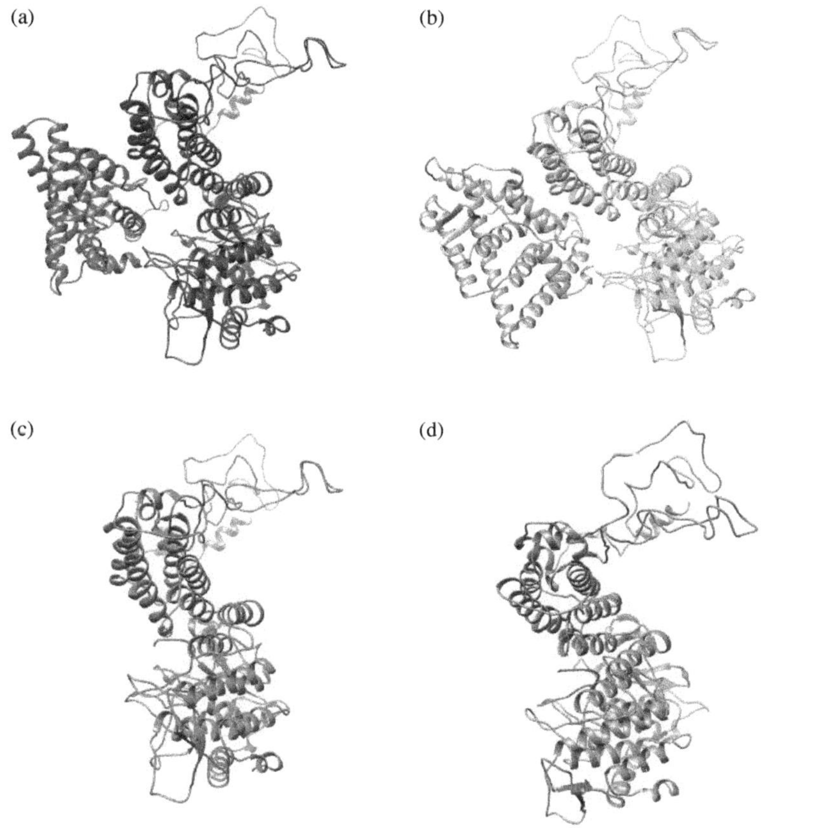

Fig. 2. 3D structure of menin–ER and menin–MLL complex. (a) 3D structure of menin-ER complex from ZDOCK (blue colored chain menin and purple colored chain ER; (b) 3D structure of menin-ER complex from HawkDock (cyan colored chain menin and pink colored chain ER); (c) 3D structure of menin-MLL complex from ZDOCK (orange colored chain menin and purple colored chain ER); (d) 3D structure of menin-MLL complex from HawkDock (green colored chain menin and blue colored chain ER).

consensus results based on three individual web servers: cons-PPISP,[30] PINUP[31] and Promate[32] (Figs. 2(a)–2(d)).

2.5. *Computing folding free energy change due to mutation*

Folding free energy change due to mutation was computed via in-house tool, the SAAFEC[33] algorithm, along with third party webservers as DUET,[34] CUPSAT,[35] mCSM,[36] SDM,[37] and I-Mutant 2.0.[38]

2.6. *Computing binding free energy change due to mutation*

The change of the binding free energy of menin-ER and menin-MLL complexes was computed with in-house algorithm, the SAAMBE[39] method, along with third party tools as BeAtMuSiC,[40] mCSM-PPI2[41] and MutaBind.[42]

2.7. *K-Nearest neighbors (KNN) classification*

K-nearest neighbors algorithm was used to classify the missense mutations using biochemical property along with rate of evolutionary conservation. The dataset includes total of 32 missense mutations. We split this dataset set into two subset (1) train dataset that comprises of 13 (5 Benign and 8 pathogenic) mutations; (2) test dataset consists of all the 19 mutations from COSMIC database (nonclassified/unknown effect mutations). The KNN classification was performed using R program and various numbers of K values were tested to obtain the best performance on the training dataset.

3. Results and Discussion

3.1. *Predicting pathogenicity of MEN1 variants*

To evaluate the pathogenicity of the 19 MEN1 missense variants taken from COSMIC database we utilized the in-house combinatory approach described in method section.[10] Table 2 summarizes the score and the assessment from the approach. According to this, 15 out of 19 variants are classified as "***pathogenic***" and 4 variants are categorized as "***benign***". In order to add more evidence to our categorization, we mined UniProt database to see if any of 15 predicted pathogenic variants has reported to have damaging effect on the menin protein, including associations with other disorders. We noticed that some of the variants from COSMIC database classified as pathogenic (by our in-house combinatory approach) such as V220M and A289E are also listed in UniProt database as causing damage to menin protein and resulting in neoplasia. This

Table 2. Summary of annotation of the 19 variants from COSMIC database. This consists of in-silico prediction scores that is used by combinatory approach to make our pathogenicity classification. We also have GERP + + score that shows higher rates of evolutionary conservation for pathogenic variants compared to the benign.

AA mutation	Polyphen2	LRT	Meta SVM	VEST3	PROVEAN	REVEL	Eigen	CADD	Our prediction	GREP + +	gnomAD	Clnsig
S606C	0.965	0.024	1.062	0.437	−2.320	0.565	4.351	25.9	Pathogenic	4.57		
S606Y	0.948	0.024	1.064	0.451	−1.770	0.578	4.324	26.8	Pathogenic	4.57		
L605V	0.821	0.000	1.042	0.178	0.070	0.365	2.095	22.8	Benign	4.57		
M563I	0.932	0.000	1.069	0.326	−3.060	0.883	5.017	29.5	Pathogenic	4.48		
K562M	0.998	0.000	1.042	0.788	−4.930	0.858	5.063	29	Pathogenic	4.48		
P534A	0.849	0.723	0.772	0.144	−0.720	0.318	2.038	6.224	Benign	2.34		
T530I	0.011	0.148	0.551	0.048	−1.800	0.296	0.519	5.558	Benign	2.63	4.035E-06	
R490Q	0.988	0.000	1.045	0.521	−0.340	0.622	4.880	27.1	Pathogenic	4.54		
Q398E	0.001	0.052	0.165	0.157	0.520	0.351	1.187	3.064	Benign	3.89		
A289E	0.994	0.000	1.079	0.967	−2.700	0.967	6.548	29.7	Pathogenic	4.24		
L272R	1.000	0.000	1.044	0.979	−5.100	0.954	6.031	30	Pathogenic	4.39		
S258L	0.979	0.000	1.063	0.974	−3.100	0.913	6.209	27.1	Pathogenic	4.80		VUS
V220M	0.896	0.002	1.088	0.785	−1.820	0.799	3.983	24.4	Pathogenic	4.80		VUS
H204Y	0.975	0.000	1.054	0.911	−5.020	0.969	8.195	27.4	Pathogenic	4.76		
E200K	0.985	0.000	1.053	0.954	−3.550	0.948	7.484	34	Pathogenic	4.76		
C170Y	0.995	0.000	1.029	0.979	−9.330	0.953	7.550	29.3	Pathogenic	4.8		
R108P	0.700	0.000	1.081	0.668	−1.280	0.655	4.143	25.4	Pathogenic	5.02		
L89R	0.998	0.000	1.036	0.996	−4.640	0.923	6.063	24.5	Pathogenic	5.02		
L36F	0.994	0.000	1.048	0.826	−2.540	0.901	5.751	26.8	Pathogenic	4.89		

Notes: Clnsig: Clinical Significance from ClinVar Database.
VUS: Variant of Unknown Significance.

provides additional confidence that our predictions are correct and indeed V220M and A289E are pathogenic mutations.

We did not find any allele frequency information associated with 18 variants except for one variant (T530I) categorized as "***benign***" that shows an allele frequency of $< 5\%$ from both 1000 genome[21] and gnomAD[22] databases. The results from our conservation analysis from GERP++[23] on our dataset, show that all the pathogenic variants have evolutionary conservation rate of larger than 4, while the benign variants have evolutionary conservation rates ranging between 2 and 4.6. Studies have shown that high rates of evolutionary conservation along with other factors such as biochemical changes could contribute to the disruption of protein function.[43]

3.2. *Mapping the missense mutations onto the Menin-MLL and Menin-ER complex*

All the 19 variants taken from the COSMIC database, out of which 15 were predicted to be pathogenic and 4 to be benign, are mapped in the complex structures of menin-ER and menin-MLL complex, along with neoplasia-causing mutations and 5 benign mutations from VariSNP database[12] (Fig. 3). It can be seen that all the variants are scattered within the structure and no variants are found near the interface of the protein complexes, both menin-ER and menin-MLL complexes. Neoplasia mutations are found in helixes, beta sheet and loops whereas, COSMIC pathogenic mutations are found in helixes and loops and benign mutations are found in loops except mutation L605Y which is found in helix.

3.3. *Effect of missense mutation on protein stability*

We find that most of the mutations destabilize the menin protein (Table 3a). The changes in stability are greater for the pathogenic mutations compared with the benign mutations. Out of fifteen pathogenic mutations, eight are predicted to cause folding free energy change greater than or equal to 1.50 kcal/mol, four to result in folding free energy change less than 1.5 kcal/mol and three to less

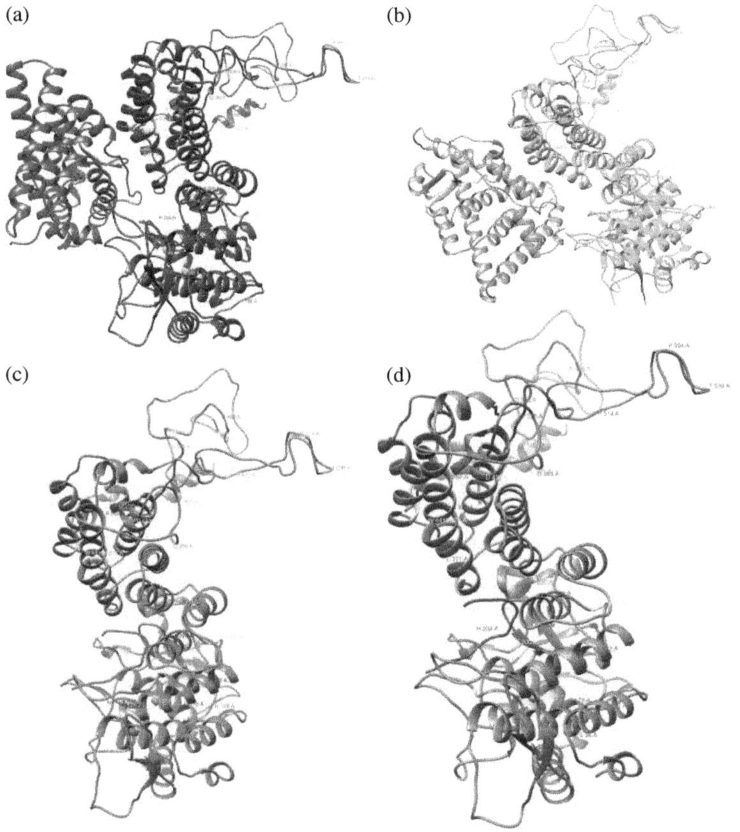

Fig. 3. Mapping of pathogenic and benign mutations to menin-ER and menin-MLL complexes. (a) mapping of mutations to menin-ER complex from ZDOCK (red colored pathogenic mutations from COSMIC, yellow colored pathogenic mutations associated with neoplasia from ClinVar, green colored benign mutations from COSMIC and VariSNP); (b) mapping of mutations to menin-ER complex from HawkDock (red colored pathogenic mutations from COSMIC, yellow colored pathogenic mutations associated with neoplasia from ClinVar, green colored benign mutations from COSMIC and VariSNP); (c) mapping of mutations to menin-MLL complex from ZDOCK (blue colored pathogenic mutations from COSMIC, yellow colored pathogenic mutations associated with neoplasia from ClinVar, green benign mutations from COSMIC and VariSNP), (d) mapping of mutations to menin-MLL complex from HawkDock (red colored pathogenic mutations from COSMIC, yellow colored pathogenic mutations associated with neoplasia from ClinVar, magenta colored benign mutations from COSMIC and VariSNP).

Table 3a. Calculation of folding free energy change (kcal/mol) of variants from COSMIC & VariSNP database. The calculated folding free energy changes in kcal/mol of menin protein due to mutations from COSMIC and VariSNP database. The positive value indicates destabilization and negative value indicates stabilization.

AA mutation	Our prediction	SAAFEC	mCSM	SDM	DUET	CUPSAT	I-Mutant	Average
		Missense variants from COSMIC DB						
S606C	Pathogenic	2.23	0.27	−0.75	−0.20	−0.73	2.18	1.56
S606Y	Pathogenic	0.29	0.43	−0.54	0.16	0.95	0.51	0.47
L605Y	Benign	−2.53	0.48	0.6	0.52	0.53	−0.03	0.53
M563I	Pathogenic	−1.08	1.10	0.32	0.68	−1.18	−0.48	−0.91
K562M	Pathogenic	5.90	−0.48	0.06	0.56	−0.01	−0.23	2.13
P534A	Benign	2.57	0.27	−0.69	−0.11	1.62	0.98	1.36
T530I	Benign	0.23	0.16	−1.31	−0.41	2.97	0.11	0.87
R490Q	Pathogenic	2.50	0.67	0.32	0.59	−0.61	0.56	0.93
Q398E	Benign	2.07	0.03	−0.97	−0.46	1.25	−1.44	1.12
A289E	Pathogenic	11.39	2.71	2.97	3.04	2.94	0.75	3.96
L272R	Pathogenic	−4.26	2.30	3.02	2.36	5.55	0.51	2.75
S258L	Pathogenic	1.80	0.03	−0.22	−0.18	1.7	0.95	1.12
V220M	Pathogenic	−0.65	0.75	0.14	0.57	2.2	2.15	1.16
H204Y	Pathogenic	−1.42	−1.32	−0.02	−1.35	−1.52	−0.25	−1.14
E200K	Pathogenic	−4.06	2.14	1.15	2.20	−4.5	0.5	1.50
C170Y	Pathogenic	−1.00	1.01	1.9	1.32	4.5	0.78	1.90
R108P	Pathogenic	6.49	0.31	2.23	0.87	0.71	0.81	1.90
L89R	Pathogenic	−2.33	1.54	1.76	1.47	5.16	2.1	2.41
L36F	Pathogenic	1.17	1.51	1.39	1.77	2.02	0.84	1.45

Table 3a. (*Continued*)

AA mutation	Our prediction	SAAFEC	mCSM	SDM	DUET	CUPSAT	I-Mutant	Average
VariSNP benign variants								
P519S	Benign	1.00	0.95	−0.72	0.47	−2.93	1.43	0.96
V555L	Benign	−2.58	0.63	0.86	0.45	0.41	0.35	0.54
I377T	Benign	−0.99	2.46	3.46	2.80	4.46	4.18	3.47
E371D	Benign	−4.08	1.04	1.39	1.04	0.31	0.81	0.92
G508D	Benign	−0.88	0.40	−0.26	−0.06	0.53	1.34	0.76

Table 3b. Calculation of folding free energy change (kcal/mol) of variants from ClinVar database. The calculated folding free energy change in kcal/mol of menin protein taken from ClinVar associated with neoplasia. The positive value indicates destabilization and negative value indicates stabilization.

AA mutation	Conditions	Clinical significance	SAAFEC	mCSM	SDM	DUET	CUPSAT	Average
A373D	Neoplasia	Pathogenic	6.39	1.93	3.23	2.82	2.25	3.22
E260K	Neoplasia	Pathogenic	−3.32	−0.26	0.77	−0.46	2.13	−1.35
V189E	Neoplasia	Pathogenic	5.73	2.83	3.34	3.21	4.24	3.87
A165P	Neoplasia	Pathogenic	0.42	−0.99	4.45	−0.25	0.17	1.67
R280K	Neoplasia	Pathogenic	1.95	0.66	0.1	0.46	−0.17	0.85
W441R	Neoplasia	Pathogenic	−1.92	2.56	1.11	2.21	5.34	2.81
D423N	Neoplasia	Pathogenic	−0.98	1.66	0.83	1.59	3.7	1.94
H139D	Neoplasia	Pathogenic	5.81	0.32	1.65	0.49	1.25	1.90

than 1 kcal/mol. Out of nine benign mutations, only one is predicted to cause change in folding free energy greater than 1.5 kcal/mol, six are predicted to cause change in folding free energy less than 1 kcal/mol and two greater than 1 kcal/mol and but less than 1.5 kcal/mol. The results indicate that pathogenic mutations tend to cause larger changes of folding free energy compared with predicted for benign mutations. Some of the pathogenic mutations as A289E, R108P and L36F are predicted by all predictors to result in similar change of stability which provides confidence that they are correct. These results can be compared with the change in folding free energy due to pathogenic mutations associated with neoplasia reported to ClinVar (Table 3b). It is shown out of eight mutations associated with neoplasia, six have folding free energy change greater than 1.5 kcal/mol and two have less than 1.5 kcal/mol. This shows that pathogenic mutations associated with neoplasia and mutations from COSMIC database predicted to be pathogenic have similar effect on protein stability.

3.4. *Effect of missense mutation on binding affinity*

Tables 4a and 4b show the change in binding free energy due to mutations in menin-ER and menin-MLL complexes, respectively. For both complexes, all the mutations destabilize the protein binding but the changes in binding energy are greater for pathogenic mutation compared with benign mutations. Out of 15 pathogenic mutations, 9 from menin-ER complex are predicted to cause binding free energy change greater than 0.5 kcal/mol and 6 to less than 0.5 kcal/mol. In menin-MLL complex, 11 mutations are predicted to cause change in binding free energy greater than 0.5 kcal/mol and only four to less than 0.5 k cal/mol. Mutations as A289E, C170Y and L272R are predicted to have largest effect on binding free energy change for all methods. Most of the benign mutations of both complexes are predicted to cause change in binding energy less than 0.5 kcal/mol which is lower in comparison to pathogenic mutations. These can be compared with the change in binding free energy due to mutations associated with neoplasia reported in ClinVar (Tables 4c and 4d).

Table 4a. Calculation of binding free energy change of menin-ER complex dues to COSMIC and VariSNP mutations. The calculated binding free energy changes in kcal/mol of menin-ER complex due to mutations from COSMIC and VariSNP database. The positive value indicates destabilization and negative value indicates stabilization.

AA mutation	Our prediction	SAAMBE	BeAtMuSiC	mCSM-PPI2	MutaBind	Average
Unclassified mutations from COSMIC database						
S606C	Pathogenic	0.30	−0.09	0.19	0.60	0.37
S606Y	Pathogenic	0.29	0.05	−0.04	0.55	0.30
L605Y	Benign	0.29	0.01	−0.24	0.62	0.31
M563I	Pathogenic	0.39	0.25	0.49	1.75	0.72
K562M	Pathogenic	0.52	0.15	0.25	1.05	0.50
P534A	Benign	−0.27	0.13	0.59	−0.24	0.36
T530I	Benign	0.20	0.09	−0.01	0.51	0.27
R490Q	Pathogenic	0.37	0.25	0.11	0.75	0.45
Q398E	Benign	0.42	0.07	0.14	0.47	0.28
A289E	Pathogenic	0.57	0.68	−0.55	2.63	1.30
L272R	Pathogenic	0.34	0.92	0.38	2.00	0.91
S258L	Pathogenic	0.40	0.53	0.21	1.53	0.67
V220M	Pathogenic	0.26	0.19	−0.22	0.71	0.39
H204Y	Pathogenic	0.34	0.38	0.32	0.44	0.37
E200K	Pathogenic	−0.24	0.39	0.77	0.90	0.69
C170Y	Pathogenic	0.57	0.23	−0.34	2.75	1.18
R108P	Pathogenic	0.70	0.16	1.08	1.27	0.80
L89R	Pathogenic	0.54	0.48	0.20	1.16	0.60
L36F	Pathogenic	0.29	0.34	0.43	1.55	0.73
Benign mutations from VariSNP Database						
P519S	Benign	−0.22	−0.37	−0.16	−0.43	−0.29
V555L	Benign	0.40	0.22	0.15	0.49	0.32
I377T	Benign	0.27	1.27	0.10	1.53	0.79
E371D	Benign	0.22	0.19	−0.13	1.68	0.70
G508D	Benign	0.70	0.33	−0.27	0.75	0.59

It is shown that out of eight pathogenic mutations associated with neoplasia in menin-ER, four are predicted to cause binding free energy to change greater than or equal to 1.0 kcal/mol, three to cause change in binding free energy greater than 0.5 kcal/mol and less than 1.0 kcal/mol and one to less than 0.5 kcal/mol. For menin-MLL complex, five are predicted to cause binding energy change to greater

Table 4b. Calculation of binding free energy change of menin-MLL complex dues to COSMIC and VariSNP mutations. The calculated binding free energy changes in kcal/mol of menin-MLL complex due to mutations from COSMIC and VariSNP database. The positive value indicates destabilization and negative value indicates stabilization.

AA mutation	Our prediction	SAAMBE	BeAtMuSiC	mCSM-PPI2	MutaBind	Average
		COMSIC mutations				
S606C	Pathogenic	0.24	−0.09	0.19	0.76	0.40
S606Y	Pathogenic	0.44	0.05	−0.01	0.70	0.41
L605Y	Benign	0.34	0.01	−0.17	0.55	0.30
M563I	Pathogenic	0.39	0.25	0.45	1.59	0.67
K562M	Pathogenic	0.43	0.05	0.24	0.99	0.43
P534A	Benign	0.09	0.2	0.47	−0.19	0.25
T530I	Benign	0.30	0.09	−0.03	0.56	0.32
R490Q	Pathogenic	0.44	0.25	−0.06	0.91	0.53
Q398E	Benign	0.41	0.07	−0.18	0.36	0.28
A289E	Pathogenic	1.00	1.02	−0.70	2.20	1.40
L272R	Pathogenic	0.86	0.88	0.32	2.35	1.10
S258L	Pathogenic	0.18	0.64	0.23	1.32	0.59
V220M	Pathogenic	0.25	0.19	0.23	0.57	0.31
H204Y	Pathogenic	0.73	0.29	−0.05	1.12	0.71
E200K	Pathogenic	0.49	0.39	0.84	1.42	0.78
C170Y	Pathogenic					

than 1.0 kcal/mol, two to greater than 0.5 kcal mol and less than 1.0 kcal/mol and one to less than 0.5 kcal/mol. This shows the similar binding effects of pathogenic mutations associated with neoplasia and mutations predicted to be pathogenic and associated with breast cancer from COSMIC database.

Table 4c. Calculation of binding free energy change of menin-ER complex due to ClinVar mutations. The calculated binding free energy changes in kcal/mol of menin-ER complex due to mutations from ClinVar. The positive value indicates destabilization and negative value indicates stabilization.

AA mutation	Conditions	Clinical significance	SAAMBE	BeAtMuSiC	mCSM-PPI2	MutaBind	Average
A373D	Neoplasia	Pathogenic	0.58	1.29	−0.61	2.26	1.38
E260K	Neoplasia	Pathogenic	0.08	0.30	0.64	0.90	0.56
V189E	Neoplasia	Pathogenic	0.53	1.92	0.21	2.06	1.18
A165P	Neoplasia	Pathogenic	0.68	1.24	0.18	2.90	1.25
R280K	Neoplasia	Pathogenic	0.08	0.15	0.16	0.31	0.18
W441R	Neoplasia	Pathogenic	−0.74	1.23	1.83	2.33	1.80
D423N	Neoplasia	Pathogenic	0.12	0.16	0.98	1.20	0.62
H139D	Neoplasia	Pathogenic	0.51	0.51	0.34	1.63	0.75

Table 4d. Calculation of binding free energy change of menin-MLL complex due to ClinVar mutations. The calculated binding free energy changes in kcal/mol of menin-MLL complex due to mutations from ClinVar. The positive value indicates destabilization and negative value indicates stabilization.

AA mutation	Conditions	Clinical significance	SAAMBE	BeAtMuSiC	mCSM-PPI2	MutaBind	Average
A373D	Neoplasia	Pathogenic	0.21	0.78	−0.65	2.11	1.03
E260K	Neoplasia	Pathogenic	0.40	0.39	0.68	1.20	0.67
V189E	Neoplasia	Pathogenic	0.30	1.92	0.40	2.3	1.23
A165P	Neoplasia	Pathogenic	0.58	1.20	0.17	2.5	1.11
R280K	Neoplasia	Pathogenic	0.25	0.06	0.30	0.30	0.23
W441R	Neoplasia	Pathogenic	0.53	1.01	0.55	2.79	1.22
D423N	Neoplasia	Pathogenic	0.63	0.23	0.74	1.43	0.76
H139D	Neoplasia	Pathogenic	1.70	0.79	0.75	1.84	1.27

3.5. *Biophysical properties based pathogenicity classification using KNN model*

Here we use KNN method along with biophysical features such as folding free energy ($\Delta\Delta G$) and conservation score (GERP + +) to characterize the 19 MEN1 missense variants from COSMIC database. The dataset has a total of 32 mutation that encompasses 23 pathogenic mutations and 9 benign mutations. These 32 mutations were partitioned into 13 mutations (composed from ClinVar and VariSNP data sources) and 19 mutations with unknown significance (obtained from COSMIC database) and then subjected to the KNN classifications. The 13 well curated mutations were randomly split into 80% as train dataset and 20% as test data and we preformed 10-fold cross validation. At 100% accuracy we identified the best K value is 5. Using the validated KNN classification model with a $K = 5$ we performed the categorization of the 19 mutations with unknown significance with both features (GERP + + and $\Delta\Delta$G) and just folding free energy ($\Delta\Delta$G). Table 5 summarizes the predictions results from our KNN classification and compares with the results obtained from our combinatory approach. When we use the both the GERP + + scores and $\Delta\Delta$G as features, two mutations that were predicted benign, Q398E and L605V, in our combinatory in-silico

Table 5. KNN classification results of the 19 mutations from COMSIC database. The training dataset was constructed using well curated mutations from ClinVar and VariSNP. The features used for this supervised learning method is evolutionary conservation score (GERP + +) and folding energy change ($\Delta\Delta$G). And with a $K = 5$ we were able to classify at a higher accuracy.

				KNN classification features	
AA change	GERP + +	$\Delta\Delta$G	Combinatory in-silico prediction	GERP + + and $\Delta\Delta$G	$\Delta\Delta$G
S606C	4.57	1.56	P	P	P
S606Y	4.57	0.47	P	P	***B***
A289E	4.24	3.96	P	P	P
L272R	4.39	2.75	P	P	P
S258L	4.8	1.12	P	P	P
V220M	4.8	1.16	P	P	P
H204Y	4.76	−1.14	P	P	B
Q398E	3.89	1.12	B	***P***	B
L605V	4.57	0.53	B	***P***	B
M563I	4.48	−0.91	P	P	B
K562M	4.48	2.13	P	P	P
P534A	2.34	1.36	B	B	B
T530I	2.63	0.87	B	B	B
R490Q	4.54	0.93	P	P	P
E200K	4.76	1.5	P	P	P

Table 5. (*Continued*)

AA change	GERP ++	ΔΔG	Combinatory in-silico prediction	KNN classification features: GERP ++ and ΔΔG	KNN classification features: ΔΔG
C170Y	4.8	1.9	P	P	P
R108P	5.02	1.9	P	P	P
L89R	5.02	2.41	P	P	P
L36F	4.89	1.45	P	P	P

Notes: ***P*** denotes Pathogenic and ***B*** denotes Benign.
Highlighted are the difference in prediction between KNN classification and combinatory in-silco prediction approach.

approach were predicted to be pathogenic. But when only ΔΔG was used as a feature, three mutations, S606Y, H204Y and M563I that were predicted as benign using KNN classification differ from the results of our combinatory in-silico method prediction. Previous studies on contribution of biophysical property such as folding free energy change (ΔΔG) has been used characterization of the pathogenicity of the missense variants in MEN1.[44] Our results from KNN classification shows that the biophysical property such as folding free energy combined with evolutionary conservation rates provide more evidence in characterizing the mutation deleteriousness that comparable to the results from our in-silico combinatory method (Table 5). However, given the limited size of dataset, it is difficult to identify the full potential of our KNN classification method with folding free energy change (ΔΔG) and GERP ++ as features to categorize between variants classes.

4. Conclusions

The study investigated the effects of MEN1 missense mutations and their role in breast cancer. Our utilization of combinatory in-silico predictor approach characterized the pathogenicity of the 19 missense variants with unknown clinical significance taken from ClinVar. The results from the biophysical analysis using the menin protein along with its complexes (menin-MLL and menin-ER) aided in providing more evidence to the pathogenic impact caused by the variants in breast cancer tissues. The results from our in-house KNN classification method show strong correlation with the results from our in-silico combinatory approach. Furthermore, the benchmarking of the results using the neoplasia exclusive variants indicates that mutations clearly alter the protein stability of menin, and the changes are comparable with predicted pathogenic mutations taken from COSMIC database. Taken together, we provide classification of variants found in breast cancer tissues as pathogenic and benign based on several features.

Acknowledgment

The work on E.A was supported by grants from NIH, Grant Nos. R01GM092927 and R01GM125639.

References

1. Imachi H, Murao K, Dobashi H, Bhuyan MM, Cao X, Kontani K, Niki S, Murazawa C, Nakajima H, Kohno N, Menin, a product of the MENI gene, binds to estrogen receptor to enhance its activity in breast cancer cells: Possibility of a novel predictive factor for tamoxifen resistance, *Breast Cancer Res Treat* **122**:395–407, 2010.
2. Chandrasekharappa SC, Guru SC, Manickam P, Olufemi S-E, Collins FS, Emmert-Buck MR, Debelenko LV, Zhuang Z, Lubensky IA, Liotta LA, Positional cloning of the gene for multiple endocrine neoplasia-type 1, *Science* **276**:404–407, 1997.
3. Lemos MC, Thakker RV, Multiple endocrine neoplasia type 1 (MEN1): Analysis of 1336 mutations reported in the first decade following identification of the gene, *Hum Mutat* **29**:22–32, 2008.
4. Matkar S, Thiel A, Hua X, Menin: A scaffold protein that controls gene expression and cell signaling, *Trends Biochem Sci* **38**:394–402, 2013.
5. Imachi H, Yu X, Nishiuchi T, Miyai Y, Masugata H, Murao K, Raloxifene inhibits menin-dependent estrogen receptor activation in breast cancer cells, *J Endocrinol Invest* **34**:813–815, 2011.
6. Dreijerink KMA, Groner AC, Vos ESM, Font-Tello A, Gu L, Chi D, Reyes J, Cook J, Lim E, Lin CY, Enhancer-mediated oncogenic function of the menin tumor suppressor in breast cancer, *Cell Rep* **18**:2359–2372, 2017.
7. Dreijerink KMA, Mulder KW, Winkler GS, Höppener JWM, Lips CJM, Timmers HTM, Menin links estrogen receptor activation to histone H3K4 trimethylation, *Cancer Res* **66**:4929–4935, 2006.

8. Dreijerink KMA, Goudet P, Burgess JR, Valk GD, Breast-cancer predisposition in multiple endocrine neoplasia type 1, *N Engl J Med* **371**:583, 2014.
9. Tate JG, Bamford S, Jubb HC, Sondka Z, Beare DM, Bindal N, Boutselakis H, Cole CG, Creatore C, Dawson E, COSMIC: The catalogue of somatic mutations in cancer, *Nucleic Acids Res* **47**:D941–D947, 2018.
10. Ganakammal SR, Alexov E, Evaluation of performance of leading algorithms for variant pathogenicity predictions and designing a combinatory predictor method: Application to Rett syndrome variants, *PeerJ* **7**:e8106, 2019.
11. Landrum MJ, Chitipiralla S, Brown GR, Chen C, Gu B, Hart J, Hoffman D, Jang W, Kaur K, Liu C, ClinVar: Improvements to accessing data, *Nucleic Acids Res* **48**(D1):D835–D844, 2019.
12. Schaafsma GCP, Vihinen M, Vari SNP, A benchmark database for variations from db SNP, *Hum Mutat* **36**:161–166, 2015.
13. Adzhubei IA, Schmidt S, Peshkin L, Ramensky VE, Gerasimova A, Bork P, Kondrashov AS, Sunyaev SR, A method and server for predicting damaging missense mutations, *Nat Methods* **7**:248, 2010.
14. Chun S, Fay JC, Identification of deleterious mutations within three human genomes, *Genome Res* **19**(9):1553–1561, 2009, doi: 10.1101/gr.092619.109.
15. Dong C, Wei P, Jian X, Gibbs R, Boerwinkle E, Wang K, Liu X, Comparison and integration of deleteriousness prediction methods for nonsynonymous SNVs in whole exome sequencing studies, *Hum Mol Genet* **24**:2125–2137, 2014.
16. Carter H, Douville C, Stenson PD, Cooper DN, Karchin R, Identifying Mendelian disease genes with the variant effect scoring tool, *BMC Genomics* **14**:S3, 2013.
17. Choi Y, Sims GE, Murphy S, Miller JR, Chan AP, Predicting the functional effect of amino acid substitutions and indels, *PLoS one* **7**:e46688, 2012.
18. Ioannidis NM, Rothstein JH, Pejaver V, Middha S, McDonnell SK, Baheti S, Musolf A, Li Q, Holzinger E, Karyadi D, REVEL: An ensemble method for predicting the pathogenicity of rare missense variants, *Am J Hum Genet* **99**:877–885, 2016.
19. Ionita-Laza I, McCallum K, Xu B, Buxbaum JD, A spectral approach integrating functional genomic annotations for coding and noncoding variants, *Nat Genet* **48**:214, 2016.
20. Rentzsch P, Witten D, Cooper GM, Shendure J, Kircher M, CADD: Predicting the deleteriousness of variants throughout the human genome, *Nucleic Acids Res* **47**:D886–D894, 2018.
21. Clarke L, Zheng-Bradley X, Smith R, Kulesha E, Xiao C, Toneva I, Vaughan B, Preuss D, Leinonen R, Shumway M, The 1000 Genomes Project: Data management and community access, *Nat Methods* **9**:459, 2012.
22. Karczewski K, Francioli L, The genome Aggregation Database (gnomAD). *MacArthur Lab*, 2017.

23. Davydov EV, Goode DL, Sirota M, Cooper GM, Sidow A, Batzoglou S, Identifying a high fraction of the human genome to be under selective constraint using GERP + +, *PLoS Comput. Biol.* **6**:e1001025, 2010.
24. Berman HM, Bourne PE, Westbrook J, Zardecki C, The protein data bank, in *Protein Structure*, CRC Press, pp. 394–410, 2003.
25. Schwede T, Kopp J, Guex N, Peitsch MC, SWISS-MODEL: An automated protein homology-modeling server, *Nucleic Acids Res* **31**:3381–3385, 2003.
26. Consortium U. UniProt: A hub for protein information, *Nucleic Acids Res* **43**: D204–D212, 2014.
27. Pierce BG, Wiehe K, Hwang H, Kim B-H, Vreven T, Weng Z. ZDOCK server: Interactive docking prediction of protein–protein complexes and symmetric multimers, *Bioinformatics* **30**:1771–1773, 2014.
28. Weng G, Wang E, Wang Z, Liu H, Zhu F, Li D, Hou T, HawkDock: A web server to predict and analyze the protein–protein complex based on computational docking and MM/GBSA, *Nucleic Acids Res* **47**(W1):W322–W330, 2019.
29. Qin S, Zhou H-X, meta-PPISP: A meta web server for protein–protein interaction site prediction, *Bioinformatics* **23**:3386–3387, 2007.
30. Chen H, Zhou H, Prediction of interface residues in protein–protein complexes by a consensus neural network method: Test against NMR data, *Proteins: Struct, Funct, Bioinf* **61**:21–35, 2005.
31. Liang S, Zhang C, Liu S, Zhou Y, Protein binding site prediction using an empirical scoring function, *Nucleic Acids Res* **34**:3698–3707, 2006.
32. Neuvirth H, Raz R, Schreiber G, ProMate: A structure based prediction program to identify the location of protein–protein binding sites, *J Mol Biol* **338**:181–199, 2004.
33. Getov I, Petukh M, Alexov E, SAAFEC: Predicting the effect of single point mutations on protein folding free energy using a knowledge-modified MM/PBSA approach, *Int J Mol Sci* **17**:512, 2016.
34. Pires DEV, Ascher DB, Blundell TL, DUET: A server for predicting effects of mutations on protein stability using an integrated computational approach, *Nucleic Acids Res* **42**:W314–W319, 2014.
35. Parthiban V, Gromiha MM, Schomburg D, CUPSAT: Prediction of protein stability upon point mutations, *Nucleic Acids Res* **34**:W239–W242, 2006.
36. Pires DEV, Ascher DB, Blundell TL, mCSM: Predicting the effects of mutations in proteins using graph-based signatures, *Bioinformatics* **30**:335–342, 2013.
37. Pandurangan AP, Ochoa-Montaño B, Ascher DB, Blundell TL, SDM: A server for predicting effects of mutations on protein stability, *Nucleic Acids Res* **45**: W229–W235, 2017.
38. Capriotti E, Fariselli P, Casadio R, I-Mutant 2.0: Predicting stability changes upon mutation from the protein sequence or structure, *Nucleic Acids Res* **33**: W306–W310, 2005.
39. Petukh M, Dai L, Alexov E, SAAMBE: Webserver to predict the charge of binding free energy caused by amino acids mutations, *Int J Mol Sci* **17**:547, 2016.
40. Dehouck Y, Kwasigroch JM, Rooman M, Gilis D, BeAtMuSiC: Prediction of changes in protein–protein binding affinity on mutations, *Nucleic Acids Res* **41**: W333–W339, 2013.

41. Rodrigues CHM, Myung Y, Pires DEV, Ascher DB, mCSM-PPI2: Predicting the effects of mutations on protein–protein interactions, *Nucleic Acids Res* **47** (W1):W338–W344, 2019.
42. Li M, Simonetti FL, Goncearenco A, Panchenko AR. MutaBind estimates and interprets the effects of sequence variants on protein–protein interactions, *Nucleic Acids Res* **44**:W494–W501, 2016.
43. Shu Y, Leabman MK, Feng B, Mangravite LM, Huang CC, Stryke D, Kawamoto M, Johns SJ, DeYoung J, Carlson E, Evolutionary conservation predicts function of variants of the human organic cation transporter, OCT1, *Proc Natl Acad Sci* **100**:5902–5907, 2003.
44. Caswell RC, Owens MM, Gunning AC, Ellard S, Wright CF, Using structural analysis to assess the impact of missense variants in MEN1, arXiv: 661512.

https://doi.org/10.1142/9789819808199_0015

Chapter 15

Computational and Structural Studies of MeCP2 and Associated Mutants*

Tugba G. Kucukkal[†,‡,¶] and Rijul U. Amin[‡,§,||]

[†]*Department of Science, Technology and Mathematics*
Gallaudet University, 800 Florida Ave NE
Washington, DC 20002, USA
[‡]*Quest Student Research Institute*
14153 Robert Paris Ct Chantilly, VA 20151, USA
[§]*Department of Biological Sciences*
University of Pittsburgh
4200 Fifth Ave, Pittsburgh, PA 15260, USA
[¶]*tugba.kucukkal@gallaudet.edu*
[||]*rua6@pitt.edu*

Rett Syndrome is a rare genetic disorder exclusively seen in girls. Approximately 95% of RTT cases is caused by mutations in the *MeCP2* gene which codes for Methyl-CpG-binding protein 2 (MeCP2). In this review, first, a brief introductory review of Rett Syndrome, MeCP2 protein structure and function, mutation types and frequencies, and phenotype–genotype relationships were provided. After that, the current knowledge on the wild-type and mutant MeCP2 protein structure and dynamics as well as its binding to DNA is reviewed. The review particularly focuses on computational (such as molecular dynamics) and experimental (such as electrophoretic mobility shift assays) studies on the MeCP2 binding to different types of DNA as well as the computational and experimental (such as circular dichroism) studies on the stability changes upon mutations. In the end, a brief opinion on future outlook for further computational studies is provided.

---

*To cite this article, please refer to its earlier version published in *J. Theor. Comput. Chem.* **2020**, *19* (6), 2041001, DOI: 10.1142/S0219633620410011
§Corresponding author.

Keywords: MeCP2 protein; Rett syndrome; binding free energy; folding free energy change; molecular dynamics; computational; MeCP2 mutation.

1. Rett syndrome and MeCP2 protein

Rett syndrome (RTT) is a rare, severe, noninherited genetic neurological disorder affecting almost exclusively girls. The onset age varies but infants with RTT develop apparently normally for the first 6 months and most common symptoms start at 6–18 months with sudden or slow progress.[1–3] RTT causes multiple symptoms including slowed growth, lack of language abilities, problems with hands, muscles and coordination, irregular breathing patterns and seizures. These lifelong impairments affecting nearly every aspect of the individual's life may slightly worsen over time or stay stable. Although there is no cure for RTT, treatments including standard care and medication, physical, speech, behavioral and occupational therapy, and nutrition help improve symptoms[3–5] RTT affects all racial and ethnic groups and occurs in approximately 1 of every 10,000–15,000 female births. RTT can rarely affect males but prevalence in males is not known.[6]

RTT is caused by mutations in the *MeCP2* gene mapping to Xq28 region.[7] Approximately, 95% of girls diagnosed with RTT have a conformed MeCP2 mutation. The MeCP2 encodes Methyl-CpG-binding 2 protein (MeCP2) which helps regulate gene activity (expression) by modifying chromatin. MeCP2 has a histone-deacetylase-independent role, allowing it to compress chromatin even without TRD binding histone deacetylases.[8] It is present in cells throughout the body, although it is particularly abundant in brain cells playing a critical role in synaptic function. It may also be involved in processing mRNA through alternative splicing[9] and act as a splice site regulator[8] and as a possible transcriptional activator.[10] In RTT brains, the neurons have decreased dendritic arborization, fewer dendritic spines and increased cell packing densities.[11,12] Besides neurons, astrocytes and microglia also play a role in RTT pathology.[13]

MeCP2 belongs to a large family of DNA-binding proteins characterized by the presence of methyl binding domain (MBD) which selectively binds 5-methylcytosine residues. While it has a 2–3 times greater affinity for methylated DNA, MeCP2 is capable of binding to unmethylated DNA as well.[8,14] Through its transcriptional repression domain (TRD), it interacts with various co-repressor complexes.[15,16]

MeCP2 includes two main functional domains: the MBD and TRD. The MBD allows MeCP2 to have hydrophilic interactions with CpG rich-promoters in the genome in order to repress transcription.[17] The MBD of MeCP2, unlike many other proteins in the MBD family can bind to a single methylated-DNA dinucleotide.[18] The TRD interacts with histone deacetylases (HDAC1 and HDAC2) and a co-repressor Sin3A[19] in order to reorganize and compress chromatin to make it inaccessible for DNA transcription and prevent gene expression.

The structure of MeCP2 is highly intrinsically disordered, consisting of ~35% β-strand/turn, 5% α-helix, and almost 60% unstructured protein.[8] The minimal secondary structure, with shorter stretches of ordered composition alternated with longer stretches of disorder, allows the protein to have a wide variety of quick interactions with target DNA and protein molecules.[20]

2. Mutations in the MeCP2 and genotype–phenotype relationships

Mutations in the MeCP2 protein account for 95% of the Rett Syndrome cases. The International Database for Rett Mutations, RettBase lists 928 MeCP2 variations and estimates that 540 (55.4%) of these are pathogenic.[21] The most common RTT mutations with their percent frequency along with positions in the MeCP2 sequence are shown in Fig. 1. The largest number of mutations in MeCP2 results from frameshift insertion/deletions.[21] The eight most common mutations, R106W, R133C, T158M, R168X, R255X, R270X, R294X, R306C, affect around 67% of RTT females.[22] Generally, most missense mutations occur in the highly conserved MBD, most nonsense mutations occur in the TRD, and most deletions in C terminus of MeCP,[18,21] as shown in Fig. 1. Besides the MBD and

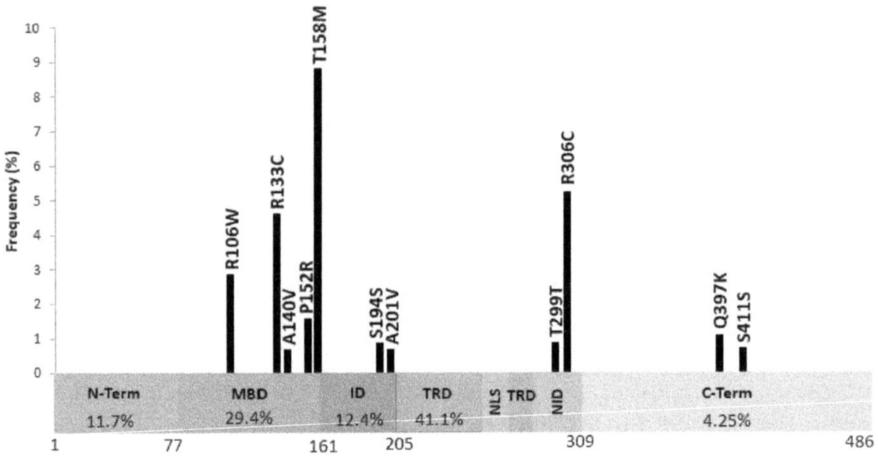

Fig. 1. MeCP2 domains with positions in the sequence and percent frequencies of all mutations in these regions are shown. Upper part shows the most common RTT mutations with their percent frequencies in all RTT mutations.

TRD, other locations for possible mutations include the corepressor interacting region (CRIR), located in between the MBD and TRD,[23] as well as the MeCP2 nuclear localization sequence, located in the TRD. The C-terminus of the polypeptide, from residues 309–486[21] is another common region for mutations causing RTT. However, in some RTT cases, there are no mutations in the MBD, TRD, or C-terminus, and it is likely that mutations in the 3′ or 5′ UTR cause the disease phenotype.[18]

The RTT disease phenotypes are classified as classical or atypical if the symptoms are milder or more severe than average. In females, it is generally difficult to find a correlation between a specific mutation and a phenotype due to random X chromosome inactivation. XCI leads to MeCP2 mosaic expression, with approximately half of the patient's cells expressing the WT MeCP2 and other half expressing mutated MeCP2.[24] However, if this expression is skewed, girls with the same mutation are liable to have different phenotype severity. In a similar vein, unaffected carriers of the mutation may show none or very mild RTT phenotype due to skewed or nonrandom XCI.[24] Cuddapah *et al.* found that severity of mutations T158M and R168X especially depend on XCI.[25] Cuddapah *et al.* also found that R133C,

R294X, R306C, 3′ truncations, and other point mutations led to less severe Rett phenotypes, while R106W, R168X, R255X, R270X, Splice Sites, Large Deletions, Insertions, and Deletions led to more severe phenotypes.[25] This coincides with the idea that frameshift mutations, mutations in the region between the MBD and TRD, as well as mutations that disrupt localization to the nucleus all can cause a large effect on RTT phenotype.

Although it has been difficult to find many correlations between specific mutations and phenotypes, it has been noted that these missense mutations and the late truncating mutations result in milder phenotypes than early truncating mutations.[19] Preserved Speech Variants (PSV) and high-functioning PSV patients only have late truncating or missense mutations.[26] Mutations in the C-terminus are less severe but can cause problems with protein stability or interactions.[18] Mutations within the NLS cause a more severe phenotype.[26] Zoghbi *et al.* found that respiratory dysfunction and HVA are more severe with nonsense mutations, while scoliosis was more severe with missense mutations.[19] Renieri *et al.* found that three or more of the following criteria increase likelihood of a milder phenotype: MECP2 late-truncating mutation (leaving intact the MBD and TRD), MECP2 missense mutation (especially R133, T158 and R306 residues) and visible characteristics such as normal head circumference, normal height, normal weight or overweight, scoliosis mild or absent, persistence of few words even in the regression period (stage 2).[26]

In males the mutations in MeCP2 are usually lethal due to males only carrying one copy of the X chromosome. However, few, rare cases of males with RTT have provided insight into specific genotype–phenotype relationships. Sheikh *et al.* divided the mutations depending on whether they cause complete mislocalization and failure of MeCP2 binding to DNA or whether MeCP2 partially or completely fail to compact chromatin.[27] The few males that survive often die young or show X-linked mental retardation.[27]

3. Computational and structural studies

The availability of the MBD and NID domain structures in the Protein Data Bank[28] as shown in Table 1, opens excellent avenues for

Table 1. Available PDB structures for MeCP2 protein in Protein Databank.[28]

PDB entry	Method	Resolution	Positions
1QK9	NMR		77–166
3C2I	X-ray	2.50 Å	77–167
5BT2	X-ray	2.20 Å	77–167
6C1Y	X-ray	2.30 Å	80–164
6OGJ	X-ray	1.80 Å	77–166
6OGK	X-ray	1.65 Å	77–167
5NAF	X-ray	2.49 Å	335–359

structure-based computational studies in understanding the molecular mechanisms of missense mutations occurring in these regions of the protein. Despite this, there are a limited number of computational studies available but together with biochemical experiments, considerable amount of structural information regarding stability and binding with the DNA was revealed as discussed below and summarized in Table 2.

3.1. *Missense mutations in the DNA-binding residues in MBD*

The X-ray crystallography structures of the MBD of MeCP2 indicate binding to methylated DNA. These cocrystal structures reveal that the core structure of the MBD is a wedge, from residues 103–145, consisting of 3 anti-parallel β-sheets and an α-helix on the side of the C-terminus, and framing these structures are disordered residues. The R133 and R111 are the residues appear to be making the most significant contacts with the DNA through hydrogen bonding with the two guanine bases in the CpG island in two strands. Also, the D121 interacts with one of the mC of the DNA. Within the protein, R133 forms a critical hydrogen bond with E137 and R111 makes a salt bridge with D121, which is conserved across the MDB-containing proteins and supporting the interaction with two arginine fingers.

Liu *et al.* investigated the DNA binding selectivities for MeCP2 MBD (residues 93 to 164) and several other MBD sequences through

Table 2. Summary of mutation effects on MBD stability and its binding to mDNA along with methods of study.

Mutation	Method	Binding to mDNA	MBD stability
R133A (Alanine scanning)	ITC[29]	Weakened	N/A
R111A (Alanine scanning)	ITC[29]	Abolished	N/A
R111G	EMSA[30] Distance-based Conformational Sampling (DBCS)[30] CD[35]	Nearly abolished Much weaker	Decreased
R133C	EMSA[17,30,34] DBCS[30] MD[32] CD[31,35]	Weakened or Weakened/ Abolished/Maintained depending on sequence [Ref. 34, see text]	Decreased
R133H	EMSA[30] DBCS[30]	Weakened	Decreased
R106Q	EMSA[30] CD[35] Distance-based CS[30]	Weakened	Not affected
R106W	EMSA[30] CD[35] DBCS[30]	Weakened	Not affected
T158M	EMSA[17,30] CD[31,35] DBCS[30] EMSA, Immunofluorescence (*in vivo* mice models)[34] Spectrometry based H/D Ex[36]	Abolished	Decreased
T158A	EMSA[17,30] CD[35] DBCS[30] EMSA, Immunofluorescence (*in vivo* mice models)[34]	Not affected	Decreased
D156E	EMSA[17,30] CD[35] DBCS[30]	Slightly weakened	Decreased
A140V	EMSA[17,30] CD[35] DBCS[30]	Slightly weakened	Decreased

(*Continued*)

Table 2. (*Continued*)

Mutation	Method	Binding to mDNA	MBD stability
S134C	EMSA[17,30] CD[35] DBCS[30]	Not affected	Decreased
L100V	EMSA[17,30] CD[35] DBCS[30]	Not affected	Decreased
P152R	EMSA[17,30] CD[35] DBCS[30]	Not affected	Decreased
F155S	EMSA[17,30] CD[35] DBCS[30] EMSA, Immunofluorescence (*in vivo* mice models)[34] Spectrometry-based H/D Ex[36]	Not affected	Decreased
Y120D	EMSA[37] MD[37]	Weakened	Not affected Decreased (apo protein)

isothermal titration calorimetry (ITC).[29] MeCP2 and other MBD-containing proteins were shown to bind to unmethylated CA or TG sites as was also shown by Yusufzai.[18] Recognition of mCG DNA was studied by mutating R133 and R111 to alanine. R111A was shown to disrupt the binding while R133A still retained binding with methylated DNA although weakened. Similarly, electrophoretic mobility shift assay (EMSA) experiments showed that a rare RTT mutation R111G profoundly disrupts the binding by reducing the binding affinity by 88-fold.[30] On the other hand, a common RTT mutation R133C was found to maintain a distinct but a weaker binding consistent with previous studies.[17,30,31] A similar but less pronounced effect was seen in R133H mutation. Molecular dynamics (MD) simulations of R133C based on the crystal structure 3C2I also showed retained but weakened binding as well as loss of helicity (positions 141–144) and loss of two salt bridges (E137-R133 and D121-K119) within the protein structure.[32] Interestingly, the R111G mutation was found to have both local and global effects as the simulations utilizing conformational sampling with CONCOORD[33] followed by short MD relaxation indicated that it substantially increased not only the mC-R111 distance but also affected the mC-R133 distance in nearly the same way as R133C.[30] Not surprisingly, simulations revealed reduction in hydrogen bonding with the methylated DNA for both the R133C and R111G with the latter being more pronounced.[30] Another EMSA-based study revealed that the reduced binding of R133C to methylated DNA is sequence dependent as R133C mutant of only the MBD (residues 77–167) abolished binding to methylated DNA while a longer sequence with the N-terminal (positions 1–205) retained wild-type binding.[34] Further investigating this, Brown *et al.* showed through cellular assays that full length mutant R133C (and T158M) completely abolished binding to DNA.[34]

3.2. *Other missense mutations in MBD*

Second most disruptive class of mutations on binding was found to be the R106Q and R106W.[30] Interestingly, these have a negligible effect on protein stability as shown through circular dichroism (CD) experiments despite the fact that the position 106 is involved in Asx-ST motif together with T158 and V159.[35] In addition, D156E and A140V were also found to have a weakening effect on the binding although less pronounced.[30] The modest effects on binding upon D156E and A140V were attributed to them impacting the protein stability as opposed to binding with the DNA.[30,35] This is curious as both the Asp and Glu as well as Ala and Val are structurally closely related. In fact, out of the 13 mutations studied, these two mutations along with the D156E, S134C, L100V and P152R were found to have a significant effect on the protein stability as revealed by measured folding free energy changes using CD.[35] All mutations were found to cause destabilization except the A140V which resulted in a slightly stabilized of the protein.

Moving on to rest of the MBD missense mutations, EMSA experiments showed that the S134C and P152R are not involved in binding. Supporting this, simulations indicated that upon mutation, position 134 becomes more flexible and loses two hydrogen bonds suggesting a decreased stability.[30] Similarly, P152R was also shown to have increased flexibility. This is consistent with the experimental folding free energy change measurements.[35]

The T158M and F155S mutations along with R106W were also studied through mass spectrometry-based Hydrogen/Deuterium exchange experiments for conformational flexibility.[36] Consistent with the structural considerations of disruption of the Asx-ST motif upon T158M mutation, it showed substantial increase in flexibility. Surprisingly, R106W was found to have identical flexibility as the wild-type. The F155 is not involved in any direct notable interactions within the protein but F155S mutation removes hydrophobicity from the position and was found to increase flexibility substantially compared to the wild-type.[36] In addition, Ghosh *et al.* concluded that unlike the wild-type, the F155S and T158M mutants along with R133C were found to have a strongly reduced thermal stability and also F155S and R106W were not further stabilized by binding to methylated DNA.[31]

Turning back to residues closely interacting with methylated DNA, positions D121 and an adjacent position Y120 were identified as of interest by D'Annessa *et al.*[37] A rare RTT mutation Y120D and phosphorylated Tyr at position 120 along with the wild-type were studied computationally and experimentally. Molecular dynamics simulations of MBD with and without DNA showed further stabilization of the MBD upon binding to methylated DNA as suggested previously.[31,36] However, since the solution NMR structure of apo protein shows very close structural similarity to the MBD bound to DNA, it is difficult to prove further stabilization by DNA through RMSD-based observations. Based on the solution NMR structure of the apo protein, it can be concluded that the structure is not induced by DNA binding but may well be further stabilized by binding. Simulations of MBD with the Y120D mutant bound to methylated DNA did not show any significant structural change as opposed to the wild-type or apparent disruption to its interactions with methylated DNA.[37] However, the simulations of the apo protein indicated destabilization upon Y120D mutation. This approach deciphers the structure of the mutant prior to binding with the methylated DNA. On the other hand, the assessment of the binding to methylated DNA through EMSA experiments showed that Y120D weakened the binding. In addition, the representative structures from the apo protein simulations were docked into methylated DNA to assess whether the D120Y could establish the original binding mode in the 3C2I crystallographic structure. The docked mutant D120Y structure was found to differ significantly from the original binding mode while the docked wild-type was able to establish sufficiently close binding to the original. This taken together with the weakened binding from the experimental assays suggests that Y120D affects the MBD structure and mobility which consequently affects its ability to properly bind to DNA.[37] Interestingly, in a subsequent study in mouse models, this mutant was found to alter chromatin binding.[38]

3.3. *Agreement and discrepancy between computational and experimental findings*

Apart from the biological findings, the computational methods for folding free energy including both fast methods and rigorous free energy perturbation method lack accuracy and poor agreement with the experiment were reported.[35] On the other hand, free energy perturbation results for the calculation of binding affinity between the DNA and the MBD of MeCP2 were found in reasonable agreement with the experiment.[30] However, not all EMSA experimental results for binding and the stability change results from CD are consistent with previous findings. For example, the EMSA and CD experiments indicated negligible change on binding and stability upon T158M, T158A and F155S mutations.[30] On the other hand, a previous study again utilizing EMSA found a significant disruption of binding upon these two common RTT mutations, the T158M and T158A.[17] This was attributed to the disruption of Asx-ST motif. In contrast, T158S was found to have retained substantial binding. In addition, this study draws attention to water mediation in binding to methylated DNA and suggests that recognition of methyl-CpG depends on water-mediation.[17]

3.4. *Mutations in the NCoR/SMRT domain*

Unlike the wealth of structural and computational studies for the MBD, little was known about the so-called NID domain (nuclear receptor corepressor/silencing mediator of retinoic acid and thyroid receptors (NCoR/SMRT) interaction domain) and the mutations thereof. The structure of the binary complex of MeCP2 and transducin beta-like (TBL) subunits of the silencing complex (TBL1 and TBLR1) was recently resolved. TBL1 and TBLR1 are two paralogs of the NCoR/SMRT.[39] Remarkably, four RTT missense mutations in the NID are the only residues that make the direct contacts with the TBLR1 out of the 25 residues of the peptide in the cocrystal structure. Position R306 makes hydrogen bonding and salt bridge interaction with TBLR1 and common RTT mutation R306C mutation would eliminate both contacts. Similarly, the other three RTT mutants P302R, K305R and K304E are expected to disrupt the binding. Surface plasmon resonance assays of disease associated TBLR1 mutations were found to disrupt binding with MeCP2NID.[39]

3.5. *Determination of pathogenicity*

Another avenue in which computational studies were utilized is the prediction of pathogenicity particularly in cases of newly identified, possibly RTT related mutations. Ghorbel *et al.* identified an unusual double mutation (new mutation G232V and a common mutation R294X) affecting the TRD domain of MeCP2 with a severe phenotype.[40] Pathogenicity of the G232V mutation was predicted by several different fast computational predictors.[40] Earlier, a similar approach was followed by Kharrat *et al.* to assess the pathogenicity of newly identified novel mutations and phenotype for each of the cases was discussed.[41]

4. Summary and outlook for computational studies

Perhaps due to Rett being a rare disease, despite the availability of structural information since 1999, MeCP2 has not been exploited in detail by both fast and rigorous computational studies. In addition to available MBD structures, the recently resolved cocrystal structure of

NID offers opportunities for further computational studies. Besides the structural data on MeCP2 domains in the Protein Data Bank, sufficient amount of detailed information is available on its binding to different types of DNA as well as on its thermal stability. Rigorous molecular dynamics simulations with advanced sampling have a great potential to shed light on wild-type and mutant binding profiles to different DNA models and simultaneously on the structural and dynamical changes on the protein structure. This is particularly so due to recent advances in force field accuracy and hardware efficiency. Recently, a rigorous free energy-based method (FEP+) has been applied to a large data set (700+ single point mutations in 10 proteins).[42] This indicates a great potential for rigorous free energy calculation methods,[43] which are already employed in drug design processes, in deciphering stability changes due to single point mutations.

This review does not focus on MeCP2 binding to different types of DNA, however, this together with water mediation in this context are areas where atomic simulations can be advantageous. Apart from rigorous time-dependent simulations, MeCP2 mutants prove to be an excellent but challenging problem for assessing structure-based fast methods in calculating protein stability changes upon single point mutations. In addition, computational pathogenicity prediction methods have a potential to provide insights on mutations that are marked as unknown in the RettBase or the identified polymorphisms and RTT mutations may be used to assess the current fast prediction methods.

Acknowledgment

Authors would like to thank Gallaudet University Center for Science and Technology Research.

Abbreviations

RTT	Rett Syndrome
EMSA	Electrophoretic Mobility Shift Assay
MD	Molecular Dynamics
CD	Circular dichroism

References

1. Percy AK, Lane JB, Rett syndrome: Clinical and molecular update, *Curr Opin Pediatr* **16**(6):670–677, 2004.
2. Neul JL, *et al.*, Developmental delay in Rett syndrome: Data from the natural history study, *J Neurodev Disord* **6**(1):20–20, 2014.
3. Hagberg B, Rett's syndrome: Prevalence and impact on progressive severe mental retardation in girls, *Acta Paediatr* **74**(3):405–408, 1985.
4. Percy A, MD. The American history of Rett syndrome, *Pediatr Neurol* **50**(1):1–3, 2014.
5. Vashi N, Justice MJ, Treating Rett syndrome: From mouse models to human therapies, *Mamm Genome off J Int Mamm Genome Soc* **30**(5–6):90–110, 2019.
6. Villard L, MECP2 mutations in males, *J Med Genet* **44**(7):417–423, 2007.
7. Zoghbi HY, Amir RE, Francke U, Van den Veyver IB, Wan M, Tran CQ, Rett syndrome is caused by mutations in X-linked MECP2, encoding methyl-CpG-binding protein 2, *Nat Genet* **23**(2):185–188, 1999.
8. Hite KC, Adams VH, Hansen JC, Recent advances in MeCP2 structure and function, *Biochem Cell Biol Biochim Biol Cell* **87**(1):219–227, 2009.
9. Chahrour M, Jung SY, Shaw C, Zhou X, Stephen TC, Wong Qin J, Zoghbi HY, MeCP2, key contributor to neurological disease, activates and represses transcription, *Science* **320**(5880):1224–1229, 2008.
10. Shin J, Ming G, Song H, By hook or by crook: Multifaceted DNA-binding properties of MeCP2, *Cell* **152**(5):940–942, 2013.
11. Armstrong DD, Neuropathology of Rett syndrome, *J Child Neurol* **20**(8):747–753, 2005.
12. Armstrong DD, Can we relate MeCP2 deficiency to the structural and chemical abnormalities in the Rett brain? *Brain Dev* **27**:S72–S76, 2005.
13. Ballas N, Lioy DT, Grunseich C, Mandel G, Non-cell autonomous influence of MeCP2-deficient glia on neuronal dendritic morphology, *Nat Neurosci* **12**:311, 2009.
14. Nikitina T, Shi X, Ghosh RP, Horowitz-Scherer RA, Hansen JC, Woodcock CL, Multiple modes of interaction between the methylated DNA binding protein MeCP2 and chromatin, *Mol Cell Biol* **27**(3):864–877, 2007.
15. Nan X, Campoy FJ, Bird A, MeCP2 is a transcriptional repressor with abundant binding sites in genomic chromatin, *Cell* **88**(4):471–481, 1997.
16. Laherty CD, Eisenman RN, Ng H-H, Bird A, Johnson CA, Nan X, Turner BM, Transcriptional repression by the methyl-CpG-binding protein MeCP2 involves a histone deacetylase complex, *Nature* **393**(6683):386–389, 1998.
17. Ho KL, McNae IW, Schmiedeberg L, Klose RJ, Bird AP, Walkinshaw MD, MeCP2 binding to DNA depends upon hydration at methyl-CpG, *Mol Cell* **29**(4):525–531, 2008.
18. Yusufzai TM, Wolffe AP, Functional consequences of Rett syndrome mutations on human MeCP2, *Nucleic Acids Res* **28**(21):4172–4179, 2000.
19. Van den Veyver IB, Zoghbi HY, Mutations in the gene encoding methyl-CpG-binding protein 2 cause Rett syndrome, *World Congr Rett Syndr 2000 Clin-Mol Biol Correl* **23**:S147–S151, 2001.

20. Ausió J, Paz de AM, Esteller M, MeCP2: The long trip from a chromatin protein to neurological disorders, *Trends Mol Med* **20**(9):487–498, 2014.
21. Krishnaraj R, Ho G, Christodoulou J, RettBASE: Rett syndrome database update, *Hum Mutat* **38**(8):922–931, 2017.
22. Fabio R, Colombo B, Russo S, Cogliati F, Masciadri M, Foglia S, Antonietti A, Tavian D, Recent insights into genotype–phenotype relationships in patients with Rett syndrome using a fine grain scale, *Res Dev Disabil* **35**:2976–2986, 2014.
23. Fendri-Kriaa N, Hsairi I, Kifagi C, Ellouze E, Mkaouar-Rebai E, Triki C, Fakhfakh F, A case of a Tunisian Rett patient with a novel double-mutation of the MECP2 gene, *Biochem Biophys Res Commun* **409**(2):270–274, 2011.
24. Hoffebuhr KC, Moses LM, Jerdonek MA, Naidu S, Hoffman EP, Associations between MeCP2 mutations, x-chromosome inactivation, and phenotype, *Ment Retard Dev Disabil Res Rev* **8**(2):99–105, 2002.
25. Cuddapah VA *et al.*, Methyl-CpG-binding protein 2 MECP2 mutation type is associated with disease severity in Rett syndrome, *J Med Genet* **51**(3):152, 2014.
26. Renieri A, Meloni I, Longo I, Ariani F, Mari F, Pescucci C, Cambi F, Rett syndrome: The complex nature of a monogenic disease, *J Mol Med* **81**(6):346–354, 2003.
27. Sheikh TI, Ausió J, Faghfoury H, Silver J, Lane JB, Eubanks JH, MacLeod P, Percy AK, Vincent JB, From function to phenotype: Impaired DNA binding and clustering correlates with clinical severity in males with missense mutations in MECP2, *Sci Rep* **6**:38590, 2016.
28. Berman HM, Westbrook J, Feng Z, Gilliland G, Bhat TN, Weissig H, Shindyalov IN, Bourne PE, The protein data bank, *Nucleic Acids Res* **28**(1):235–242, 2000.
29. Liu K, Xu C, Lei M, Yang A, Loppnau P, Hughes TR, Min J, Argonne National Lab. (ANL), A., IL (United States)., Advanced photon source (APS). Structural basis for the ability of MBD domains to bind methyl-CG and TG sites in DNA, *J Biol Chem* **293**(19):7344–7354, 2018.
30. Yang Y, Kucukkal TG, Li J, Alexov E, Cao W, Binding analysis of methyl-CpG binding domain of MeCP2 and Rett syndrome mutations, *ACS Chem Biol* **11**(10):2706–2715, 2016.
31. Ghosh RP, Horowitz-Scherer RA, Nikitina T, Gierasch LM, Woodcock CL, Rett syndrome-causing mutations in human MeCP2 result in diverse structural changes that impact folding and DNA interactions, *J Biol Chem* **283**(29):20523–20534, 2008.
32. Kucukkal TG, Alexov E, Structural, dynamical, and energetical consequences of Rett syndrome mutation R133C in MeCP2, *Comput Math Methods Med* **2015**:9, 2015.
33. de Groot BL, van Aalten DMF, Scheek RM, Amadei A, Vriend G, Berendsen HJC, Prediction of protein conformational freedom from distance constraints, *Proteins Struct Funct Genet* **29**(2):240–251, 1997.
34. Brown K *et al.*, The molecular basis of variable phenotypic severity among common missense mutations causing Rett syndrome, *Hum Mol Genet* **25**(3):558–570, 2015.

35. Kucukkal TG, Yang Y, Uvarov O, Cao W, Alexov E, Impact of Rett syndrome mutations on MeCP2 MBD stability, *Biochemistry* **54**(41):6357–6368, 2015.
36. Hansen JC, Wexler BB, Rogers DJ, Hite KC, Panchenko T, Ajith S, Black BE, DNA binding restricts the intrinsic conformational flexibility of methyl CpG binding protein 2 (MeCP2), *J Biol Chem* **286**(21):18938–18948, 2011.
37. D'Annessa I, Gandaglia A, Brivio E, Stefanelli G, Frasca A, Landsberger N, Di Marino D, Tyr120Asp mutation alters domain flexibility and dynamics of MeCP2 DNA binding domain leading to impaired DNA interaction: Atomistic characterization of a Rett syndrome causing mutation, *BBA Gen Subj* **1862**(5):1180–1189, 2018.
38. Gandaglia A *et al.*, A novel Mecp2Y120D knock-in model displays similar behavioral traits but distinct molecular features compared to the Mecp2-null mouse implying precision medicine for the treatment of Rett syndrome, *Mol Neurobiol* **56**(7):4838–4854, 2019.
39. Kruusvee V, Lyst MJ, Taylor C, Tarnauskaitė Ž, Bird AP, Cook AG, Structure of the MeCP2-TBLR1 complex reveals a molecular basis for Rett syndrome and related disorders. *Proc Natl Acad Sci USA* **114**(16):E3243–E3250, 2017.
40. Ghorbel R, Ghorbel R, Rouissi A, Fendri-Kriaa N, Ben Salah G, Belguith N, Ammar-Keskes L, Gouider-Khouja N, Fakhfakh F, First report of an unusual novel double mutation affecting the transcription repression domain of MeCP2 and causing a severe phenotype of Rett syndrome: Molecular analyses and computational investigation, *Biochem Biophys Res Commun* **497**(1):93–101, 2018.
41. Kharrat M *et al.*, Clinical, molecular, and computational analysis in patients with a novel double mutation and a new synonymous variant in MeCP2: Report of the first missense mutation within the AT-Hook1 cluster in Rett syndrome, *J Child Neurol* **32**(8):694–703, 2017.
42. Steinbrecher T, Zhu C, Wang L, Abel R, Negron C, Pearlman D, Feyfant E, Duan J, Sherman W, Predicting the amino acid single point mutations on protein stability-large-scale validation of MD-based relative free energy calculations, *J Mol Biol* **429**(7):948–963, 2017.
43. Armacost KA, Riniker S, Cournia Z, Novel directions in free energy methods and applications, *J Chem Inf Model* **60**(1):1–5, 2020.

www.ingramcontent.com/pod-product-compliance
Ingram Content Group UK Ltd.
Pitfield, Milton Keynes, MK11 3LW, UK
UKHW020401150926
8624IPUK00004B/285

9 789819 808182